THE NORTH AMERICAN GRASSHOPPERS

THE NORTH AMERICAN

Grasshoppers

VOLUME II

ACRIDIDAE

Oedipodinae

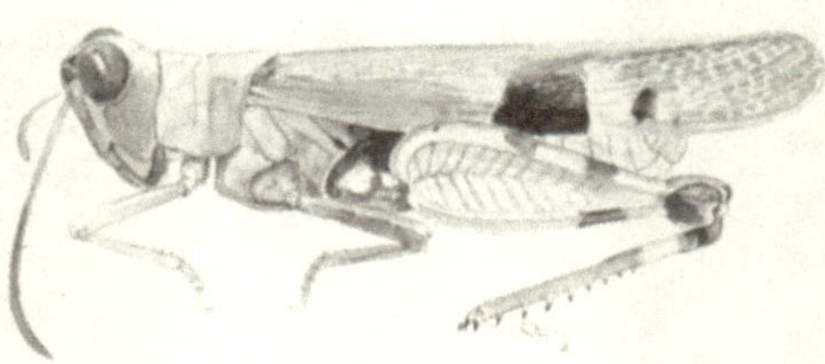

Daniel Otte

Harvard University Press

Cambridge, Massachusetts, and London, England · 1984

Drawings by Daniel Otte

10 9 8 7 6 5 4 3 2 1

Library of Congress Cataloging in Publication Data

Otte, Daniel.
The North American grasshoppers.

Bibliography: v. 1, p.
Includes index.
Contents: v. 1. Acrididae : Gomphocerinae and Acridinae—v. 2. Acrididae : Oedipodinae.
1. Locusts—North America—Identification—Collected works. 2. Insects—Identification—Collected works. 3. Insects—North America—Identification—Collected works. I. Title.
QL508.A2088 595.7'26 81-6806
ISBN 0-674-62660-5 (v. 1)
ISBN 0-674-62661-3 (v. 2)

ACKNOWLEDGMENTS

The completion of this second volume has been made possible with the assistance of many individuals and institutions. Again I must acknowledge my predecessors at the Academy of Natural Sciences who assembled the outstanding collection of grasshoppers upon which this work is based.

David Weissman and David Ferguson shared much valuable information about the taxonomy and biology of band-winged grasshoppers, for which I am grateful. The following institutions and individuals were helpful in lending me specimens and allowing me to examine specimens in their museums: the University of Michigan Museum of Zoology (I. J. Cantrall and T. J. Cohn); the California Academy of Sciences (D. Weissman and P. Arnaud); the United States National Museum (D. Nickle); the British Museum, Natural History (D. Ragge); Muséum d'Histoire Naturelle, Geneva (Bernd Hauser); Naturhistorisches Museum, Vienna (A. Kaltenbach).

I am especially thankful and indebted to Donald Azuma, who persevered for several years in the survey of the oedipodine taxonomic literature. His considerable help was responsible for hastening the completion of this volume. I am also indebted to Jamie Newlin, who plotted nearly all of the grasshopper distribution maps. His unfailing cheerfulness has made this onerous task a pleasant one. I also thank the typists who assisted me, especially Kim Kanevsky. The administration at the Academy has always been most supportive. I especially wish to mention Margaret Fisher and James Peterson, who have been helpful and encouraging in many small ways.

Finally I wish to acknowledge the constant support of my wife, Laurel, and daughters, Jennifer and Jessica, who have uncomplainingly tolerated my long absences in the field. The freedom to observe the various grasshopper species alive in the field has made my work enjoyable.

Contents

THE NORTH AMERICAN GRASSHOPPERS

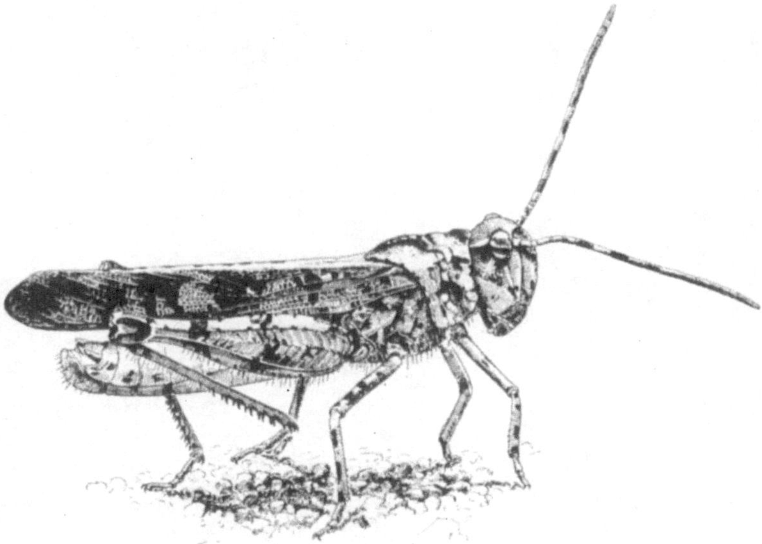

Spharagemon collare on hot sand

Introduction

The Oedipodinae, or band-winged grasshoppers, were first established as a family by Walker in 1870, with *Oedipoda* as the type genus. Since then they have been treated as a family, subfamily, or tribe. The oldest genus in the subfamily is *Locusta* Linnaeus, but because this name was so often misused, Dirsh (1975) decided to use *Oedipoda* Latreille as the basis of the subfamily name.

The systematic position of the Oedipodinae has for the most part been stable. For a time Dirsh and Uvarov disagreed on whether the group should stand as a separate subfamily or be subordinated under the Acridinae. Dirsh (1956, 1961) treated the subfamily as a group under the Acridinae, claiming that intermediate conditions between Acridinae and Oedipodinae could be found in any character used to separate them. But in his recent classification of acridomorphoid insects (1975) he restores the group to the rank of subfamily. The positions of some genera still remain problematical because they share characteristics with Acridinae or Gomphocerinae. In Volume I of this series (Otte 1981) I included one of these genera (*Stethophyma*) in the Gomphocerinae, and in this volume another (*Machaerocera*) is included in the Oedipodinae (see those genera for discussion).

Band-wings are among the most conspicuous grasshoppers on the continent. Many people in the East are familiar with the flight dances of the black-winged Carolina locusts along roadsides and other dusty places, or with the orange- and red-winged Arphias, which are so common in old fields, along roadsides, and on hill prairies. People in mountain areas of the western United States are familiar with the loud crackling and buzzing flights of *Circotettix* and *Trimerotropis* species.

The Oedipodinae range over all six zoogeographic zones, but they are most diverse in the Afrotropical, Palaearctic, and Nearctic regions. Presumably because of isolation from other regions during major periods of Oedipodine speciation and dispersal, the Neotropical and Australian regions are very poor in species.

North America is rivaled only by Africa in diversity; 169 valid North American species have been described. For Africa Johnston (1956) lists 265 valid species. Ritchie's recent work (1981, 1982) reduces by about half the number of African *Gastrimargus* species listed by Johnston but does not appreciably alter the number of *Oedaleus* species. We do not know whether the overall number of African species will increase as each genus is revised and as unexplored areas are visited, but N. Jago, a specialist of this fauna, believes it will.

Bei-Bienko and Mishchenko (1964) list 125 species for the Asian region, which includes the USSR, Europe, Near East, northern China, Korea, and Japan. Because this vast area has not been surveyed as thoroughly as North America, many more species will doubtless be discovered. South America and Australia have been well collected, but each continent has less than ten species of Oedipodinae.

Band-wings are most diverse in areas with warm, dry, sunny climates and are most abundant, in numbers of species and individuals, in habitats with a sparse covering of shrubs and grasses. In North America most species are found in prairies and desert grasslands of the southwestern United States and northern Mexico. Here it is common to find six to ten species coexisting.

Upon first inspection band-wings appear to be rather drab insects, but as the collector soon discovers, their red, orange, yellow, or blue wings, often banded with black, unfold like fans, and the inner surfaces of the hind femora are often beautifully banded with white, yellow, orange, red, or blue. As a boy I noticed the snapping and buzzing wing sounds they made during flight, but not until I visited the western United States as a graduate student did I come to know band-wings at their best. In a wheat field east of Boulder, Colorado, I once observed a spectacular event that demonstrated some of their unique qualities. Collecting along the edge of a stubble field on a hot July day, I was having difficulty finding grasshoppers in the dried-out grass; I had walked for nearly an hour when the quiet suddenly ended with a simultaneous display by hundreds of *Circotettix carlinianus.* I had encountered this species before in Wyoming and had watched males performing their hovering flights, but I had never seen such a mass display by insects. Apparently it was triggered by one male, and within seconds hundreds of males began their hovering and whirring flight from two to ten feet about the ground. The performance continued for twenty minutes or so and ended quite suddenly.

Catching band-wings can be challenging. Different species have different modes of escape, which demand varying skills of the collector. Some rely heavily on camouflage and remain hidden against their background until almost stepped upon. They then make short, usually looping flights that are difficult to follow, and during the last portion of the flight they often close their wings, making it difficult to see where they have landed. Other species are shy and difficult to approach and fly for great distances.

In the heat of the day some species are virtually impossible to catch; these are best collected in the morning or late afternoon when they are slower and when shadows make it easier to spot them. In species that are particularly shy one can get within striking distance by approaching them gradually along a spiraling path. This method apparently deceives the grasshopper into thinking you are going past him. You can also approach shy grasshoppers by moving slowly and ducking as you approach, which perhaps creates the illusion that you are not getting closer.

Taxonomic Characters and Variability

When one attempts to enumerate characters that can be used to separate species and genera, one becomes aware of how remarkably uniform the Oedipodinae are, compared to other grasshopper groups of similar size. Taxonomists, though often agreeing on which species should be placed together in a genus, are unable to tell what characters they used. Classification according to the look of a species resulted in the formation of groups of related species in some cases, but in others it produced unrelated assemblages.

Virtually all generic classification in the Oedipodinae has been based on the following characters: the nature of the median pronotoal crest; presence and nature of forewing spots or bands; hindwing banding pattern and venation; and fastigium shape. Of these characters the pronotum has been the most important in defining genera. Whether the pronotal crest is cut by one, two, or three transverse sulci or not cut at all was critical in early classifications of this group. When this character is analyzed in all North American species, it becomes apparent that it must be used with discretion; failure to relate the character to others created artificial separations of genera, species, and even populations that belong together and united taxa that were not closely related. Still the character is useful, and the reader will see that I have used it a number of times.

Some characters, not previously discovered or not used in Oedipodine classification, have been helpful in establishing relationships. One involves the shape of the male epiphallus. In Oedipodinae both the aedeagus and the epiphallus were for a long time considered relatively invariable features and therefore not useful in classification. They are indeed invariable when compared to many other groups, but some conspicuous features in the epiphallus give important clues to generic relationships. The trilobed configuration of the epiphallic lophi in *Xanthippus, Hippiscus, Pardalophora, Leprus, Hadrotettix,* and *Heliastus* first caused me to consider placing these genera within one group. Later, other characters were discovered that agree with such a clustering. Several of these genera were already thought to be closely related, but *Hadrotettix* and *Heliastus* were never suspected to be members. Reproductive behavior furnishes additional evidence of generic relationships: presence or absence of flight display, the nature of the flight display, signals used in courtship and male–male interactions, and habitat preference can all be used.

Males and females of Oedipodinae are very similar to one another, much more so than in the majority of Gomphocerinae. Even in size the sexes are more alike than in other grasshopper groups. The proportion of species with flightless females is especially low; flightless females are found in only two North American species.

Geographic variation within a species is perhaps most pronounced in wide-ranging species that have broad altitudinal tolerance. Examples include *Xanthippus corallipes, Conozoa texana,* and *Arphia conspersa.* Many species are variable in hindwing color. It is common in many genera to find yellow, orange, and reddish wings within a single species and sometimes within a population. Some species, on the other hand, show little such variation.

Body color is rather consistent in most species, but adults of some species tend to assume the prevailing color of the local soil. In most species the nymphs are much more variable in color than the adults.

Those species that are the most difficult to approach and that fly the farthest show the least amount of intrapopulational variation in body color. *Xanthippus, Pardalophora,* and some *Trimerotropis* are difficult to approach and are remarkably uniform in color; *Heliastus,* some *Lactista,* and some *Trimerotropis,* which rely more on protective coloration and allow one to approach within a few inches, are quite variable within a population.

Subspecies

I have used subspecies names sparingly, following the precepts of Hubbell (1954):

(1) Clinal variation is variation of characters in populations, not of populations.

(2) Clines themselves therefore cannot be taxonomic units.
(3) In a population showing only a stepped cline in a single character or in a group of correlated characters it is feasible and may be convenient to treat the segments between steps as subspecies.
(4) In populations in which clinal change is gradual or in which two or more non-coincident clines exist it is inadvisable to try to separate subspecies.
(5) In general, infraspecific variation is best treated by description, graphic presentation, and non-technical names, not by formal latinized names subject to rules of nomenclature.
(6) Nothing should ever be named for the sake of naming it, but only in order that something may be said about it.

In my opinion very few previously recognized grasshopper subspecies meet these criteria. In the Oedipodinae I have recognized subspecies only in *Trimerotropis verruculata* and *Spharagemon marmorata.*

Preparation of Specimens

In preparing band-wing grasshoppers for study it is essential to spread the hindwings. I relax dried specimens overnight in a water-saturated chamber, then spread the left wings on a pinning board, using paper strips and pins to hold the wings in place. The inner face of the hind femur often bears useful color characters, so it is advisable to set the femur in such a way that the inner bands can be seen. I usually lower one femur below the level of the abdomen. In already set specimens I remove the right femur and glue it, inner face up, on a small card mounted under the locality label. Since the hind tibia fits snugly against the femur and hides itself and the color of the femur, I set the tibia about 15–20 degrees from the femur.

Communication in Oedipodinae

Species distinctiveness in communication has been used to elucidate taxonomic problems among similar species of Oedipodinae. This distinctiveness exists because reproductive behavior and the communication patterns associated with it diverge relatively rapidly in a set of interacting species, while other kinds of behavior, such as grooming, feeding, and locomotion, evolve slowly and are likely to show fewer differences. Special efforts to examine closely related species have also furnished a basis upon which to consider the origins of signals and the details of their evolution (Otte 1970).

The principal visual signals given by Oedipodinae fall into two classes: flight signals and hind leg signals.

Flight Signals

In many species solitary males perform special flights to attract sexually receptive females. In their performances, males fly up from the ground, snapping their hindwings and flashing their contrasting wing colors, then settle back to the ground. The attraction of females has been seen in only a few species, presumably because the spectator must be present just at the moment when females become receptive. After mating, the females are nonreceptive for a relatively long period. The males remain sexually active more or less continuously and signal long before and long after females are receptive. Not all species snap their wings (crepitate); members of some groups do not make any specialized flight sounds. These species flash their wings mainly while approaching individuals of the opposite sex. Perhaps all Oedipodinae males at least occasionally approach females by flying; those that crepitate often do so, others merely display their colors more or less silently (Otte 1970).

Hind Leg Signals

This type of communication shows the most diversity in this subfamily. The major kinds of hind leg movements, illustrated in Fig. 1, are divided into two classes: those that involve the transmission of sound, or substrate vibrations, and silent movements, which function mainly as visual signals.

Sound signals. Rubbing one part of the body against another, usually the hind femur against the forewing, is a widely used signal and is frequently employed to attract females. Four distinct types of stridulation are common in the Oedipodinae. The most common, *O-stridulation* (ordinary stridulation), consists of relatively slow (countable or almost countable), high-amplitude, up-and-down movements of the hind femora (one or both, alternating or together) against the forewings. Also quite common is *V-stridulation* (vibratory stridulation), which consists of a low-amplitude rapid vibration of the femora (one or both together) against the forewing. A rare signal, here called *T-stridulation* (for "ticking") is made by raising the hind femora and then kicking out with the tibiae to produce a "ticking" sound as the tibia strikes the forewing. A fourth form is *W-stridulation* (for "wing-striking") in which the hind femora are vigorously shaken and are caused to strike the forewings. The movement is not nearly as fast as vibratory stridulation but is not countable. Sonograms of these forms of stridulation are shown for the genus *Arphia* in Fig. 2, along with the flight crepitation sounds.

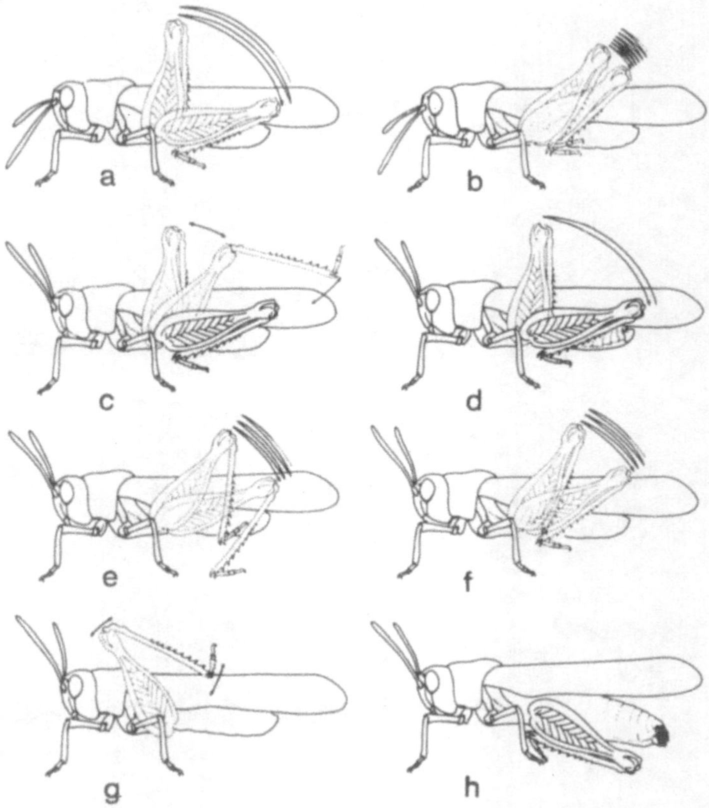

Fig. 1. Common forms of movement in the production of sound and visual signals in Oedipodinae. Different species use different combinations of signals. Signals *a–f* are used by males, *g* and *h* by females. *a,* O-stridulation (courtship); *b,* V-stridulation (courtship); *c,* T-stridulation (courtship); *d,* femur-tipping (silent movement used in both male–female and male–male interactions; *e,* W-stridulation, produced when the femora strike the forewings, or drumming when the tibiae strike the ground (usually a male–male signal, but it has become a courtship signal in a few species); *f,* W-stridulation, or silent shaking (same function as *e*); *g,* femur-raising (silent movement used mainly by females to repel male courtship advances); *h,* presenting (stance adopted by sexually receptive females when courted by males).

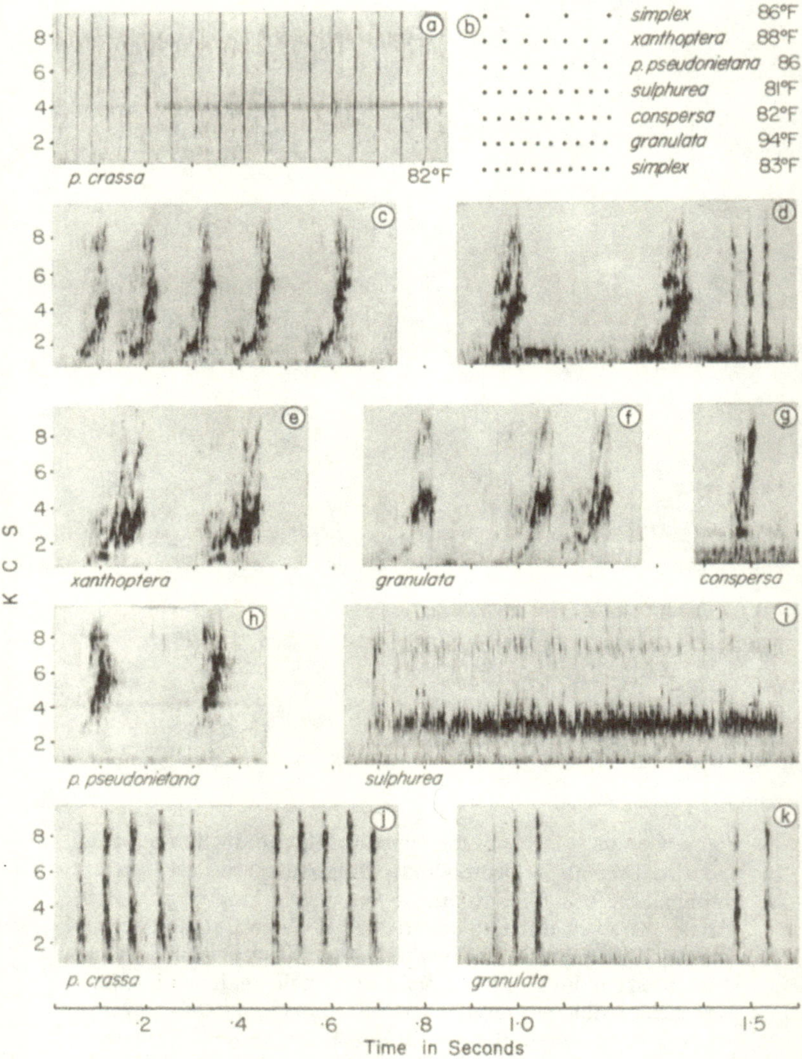

Fig. 2. Sonagrams of *Arphia* acoustical signals: *a,* crepitation of *pseudonietana; b,* comparison of crepitation patterns of six species; last line indicates the normal wingbeat rate of *simplex* during flight; *c,* O-stridulation in *sulphurea; d,* courtship O-stridulation in *simplex* followed by agonistic (male–male) W-stridulation in the male being courted; *e–h,* O-stridulation in four species; *i,* V-stridulation in *sulphurea; j, k,* W-stridulation by two agonistic males of *pseudonietana* and *granulata* (from Otte 1970).

Many species make drumming signals by shaking the hind legs vigorously causing the tibiae to drum against the ground.

Visual signals. There are three main forms of soundless (therefore presumably visual) signals. By far the most commonly observed form in this group is *femur-tipping,* which consists of raising and lowering the hind femora. It may occur singly or be repeated and is common in both male–female and male–male interactions. It seems to be the principal signal used by males to communicate their sex to either sex. In most species the contrasting light and dark bands on the inner face of the femur make the movement more conspicuous. A second, less stereotyped signal is *femur-raising,* in which males or females raise the femora, slowly or rapidly and to various heights, to repel the courtship advances of males. The height to which the femora are raised depends on the extent of the disturbance. If the intruder attempts to mount, it is kicked away with the tibiae. *Silent shaking* is very similar and probably homologous to W-stridulation and drumming. It differs only in that neither the wings nor the ground is struck. Sexually receptive females assume a posture called *presenting,* which may only make mounting by males easier. Whether or not it functions as a signal is not clear.

Contact signals. Two signals are commonly used by individuals attempting to repel others that have come into contact with them. In the first, *femur-jerking,* the hind femora are sharply jerked upward from a few degrees to 90 degrees or more. In the second, *kicking,* the hind tibiae kick out from their resting position against the femora.

Reproductive Isolation

Differences in crepitation pattern and wing color among species living together probably serve as reproductive isolating mechanisms when the species are closely related. But even among species that are incapable of mating, the differences in sound and visual flight signals may serve to reduce the number of needless interactions between species. Since male Oedipodinae often approach individuals of the wrong species, even distantly related species, one would expect that natural selection would make it easier to recognize their own species. In southern Colorado and northern New Mexico, where several Oedipodine species live together in various combinations, each species has a distinctive crepitation pattern (Fig. 3). The possible exception is the pair *Spharagemon equale* and *S. campestris,* which are normally separated by altitude. Even though the various species may have acquired their particular patterns of crepitation before establishing their present

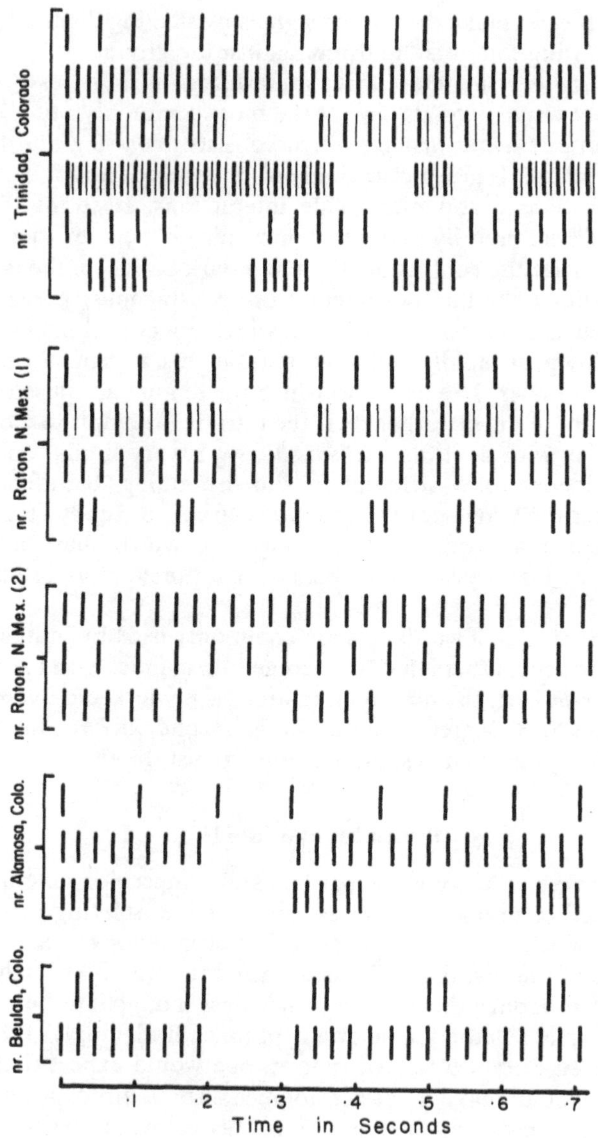

Fig. 3. Wing crepitation patterns in cohabiting groups of bandwings at five localities in New Mexico and Colorado. Each line indicates one loud wing-snapping sound (from Otte 1970).

sympatry and may be more or less distantly related to one another, the differences in crepitation patterns probably reduce the number of accidental pairs formed and may therefore be maintained or improved upon by natural selection.

No studies have shown that wing color or wing pattern results in reproductive isolation in grasshoppers. Since these characters are quite similar among close relatives, and are different only in a minority of cases, and since polymorphism in wing color exists within some species, it appears that color differences are usually not important in interspecific recognition.

In the Oedipodinae, courtship signals are more varied among related species than other nonsignaling kinds of behavior. Fig. 4 illustrates sequences of evolutionary changes postulated to have occurred in various lineages. The most common pattern during courtship is O-stridulation. But in various lineages V-stridulation, femur-shaking, and T-stridulation have also become incorporated into courtship, and in some species O-stridulation has been lost.

The considerable diversity in courtship signals is probably due to

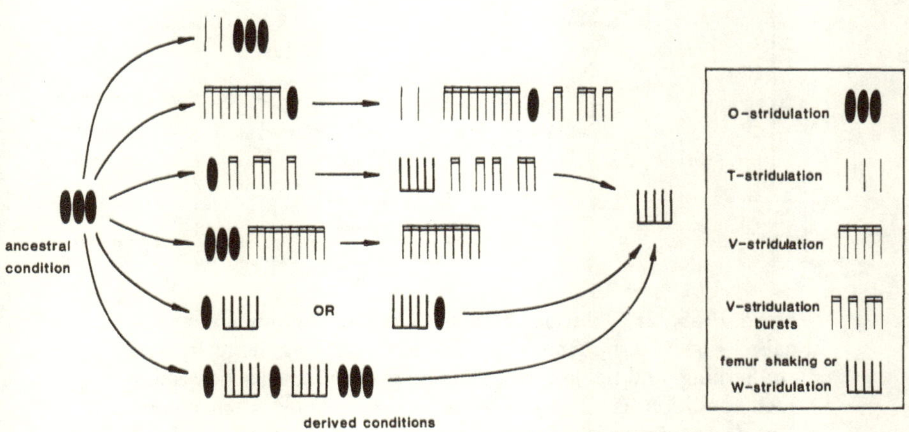

Fig. 4. Qualitative changes in courtship in Oedipodinae. Arrows indicate probable direction of evolution. The postulated ancestral condition, on the left, is possessed by a large majority of species, including congeners of species possessing conditions shown on the right. A number of different courtship patterns were evidently derived from this original condition. Thus, V-stridulation and femur-shaking were apparently incorporated into courtship independently a number of times. This was evidently followed, in several lineages, by the loss of O-stridulation (from Otte 1970).

selection for differences that will reduce wasteful interactions between species. There has evidently been little selection for distinctiveness in aggressive signals, perhaps because it is not important for males to discriminate between conspecific and heterospecific males, since separation from both kinds is advantageous. Possibly, if pairing mistakes

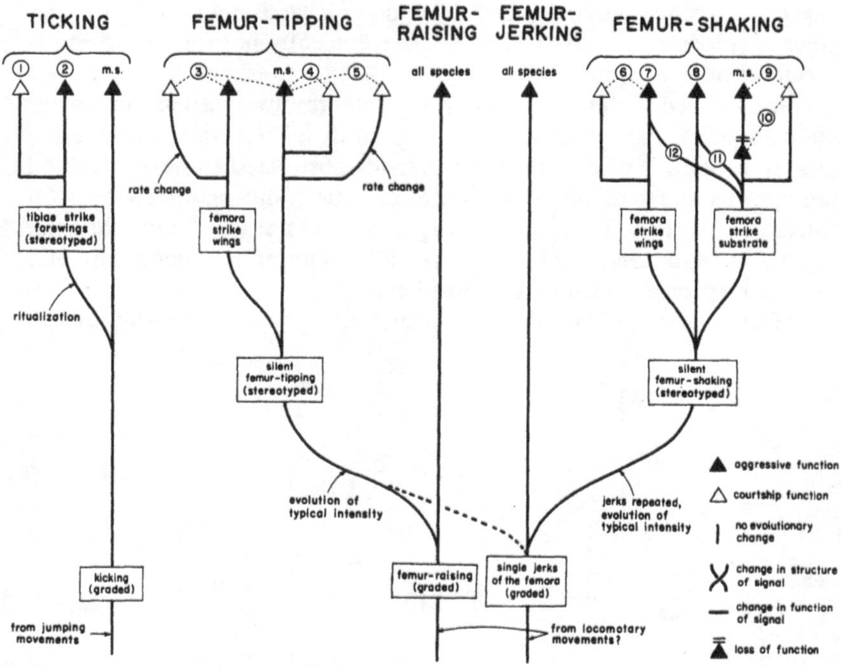

Fig. 5. Probable evolution of various kinds of communicative signals: T-stridulation, femur-tipping, femur-shaking, femur-raising and jerking, and ticking. Numbers refer to the species possessing each condition: (1) *Encoptolophus sordidus, Aiolopus thalassinus, Bryodema tuberculata, Dissosteira pictipennis,* and *Omocestus viridulus;* (2) *Psophus stridulus, Orphulella speciosa,* and *Omocestus rufipes;* (3) *Dissosteira carolina;* (4) *Chortophaga australior;* (5) *Trachyrhachys kiowa, Mestobregma plattei:* (6) *Lactista elota:* (7) *Lactista* species, *Arphia* species, *Circotettix undulatus:* (8) most Hippiscus group species, *Trimerotropis fontana;* (9) *Spharagemon bolli, S. cristatum, Trimerotropis maritima, T. californica* (Calif.); (10) *Spharagemon crepitans, Trimerotropis californica* (Ariz.); (11) *Trimerotropis fontana; (12) Circotettix undulatus, Trimerotropis verruculata suffusa* (Yosemite National Park).

between heterospecific males were costly enough, there would be selection for similarity of aggressive signals in different species.

Grasshoppers' signals evidently originate (1) by ritualization of various noncommunicative activities (orientation and locomotive movements have assumed a signal function in some species); (2) through structural changes of preexisting signals (one signal gives rise to two or more different signals); (3) through functional changes of preexisting signals (a signal changes from an agonistic to a courtship function, for example) or (4) as a more or less sudden appearance of a new motor pattern (Fig. 5; Otte 1970).

It seems likely that ritualized approaching flights evolved from unspecialized approaching flights and that calling flights by solitary males evolved from ritualized approaching flights. Listed below are conditions of flight and wing morphology in various species of Oedipodinae. These conditions are probably similar to the stages in the evolution of noisy and visually conspicuous female-attracting flights.

1. Flight not associated with pair formation; hindwings colorless; no flight sounds. Example: most grasshopper species.
2. No female-attracting flights, but males *occasionally* approach females by flying (unspecialized approaching flights); hindwings colorless; no flight sounds. Example: *Camnula pellucida.*
3. No female-attracting flights, but males *frequently* approach females by flying; hindwings brightly colored; no flight sounds. Examples: Psinidia genus group.
4. Short, female-attracting flights by solitary males occur, but only rarely; hindwings brightly colored; no flight sounds. Example: *Pardalophora* species.
5. Female-attracting flights performed by solitary males; hindwings brightly colored and contrastingly marked; crepitation loud. Examples: *Arphia* and *Circotettix* species.

THE SUBFAMILY

Oedipodinae

The Oedipodinae, Acridinae, and Gomphocerinae are closely related. Roberts (1941) could find no outstanding genitalic differences among the three groups. As a group they possess the following combination of features: ejaculatory sac is closed and relatively small; thin sigmoid flexure connects the endophallic plates and proximal end of ventral aedeagal valves; there are two pairs of aedeagal valves and an arch of the dorsal valves; the well-developed ventral lobe is partly sclerotized on outer side; rami of cingulum form a broad collar around base of aedeagus; apodemes of cingulum are relatively heavy, rodlike processes; ventral infold is relatively small and much reduced; ancorae of epiphallus are separate sclerites connected to lateral plates by membranes.

In separating the Oedipodinae from the Acridinae and Gomphocerinae, the following combination of characters is useful: forewings have an intercalary vein (see Fig. 8*p*) (exceptions: *Heliastus* and *Machaerocera*); in many species forewings have broad crossbands or large dark patches; hindwings are usually banded or brightly marked with yellows, oranges, reds, or blues and often have a black or smoky band bordering the colored area (clear or indistinctly colored wings exist in *Camnula, Encoptolophus, Trachyrhachys, Chortophaga,* and *Dissosteira*); inner face of hind femora lacks stridulatory pegs; forewings are rounded at apex; face is usually more vertical than backward-slanting; antennae are usually filiform; males and females are usually very similar morphologically; male flight displays are common; usual habitat is dry ground.

■ Key to Genera of Oedipodinae

1. Pronotal crest with numerous small teeth (Fig. 6*a*): **Tropidolophus**
   Pronotal crest not toothed (Fig. 6*b, c, d*): **2**
2. Pronotal crest cut by only one transverse sulcus. In some species this sulcus is barely visible, in others it is well marked (Fig. 6*b, c*): **3**
   Pronotal crest cut by at least two sulci. In some species of *Conozoa* anterior sulcus is quite indistinct or absent, but in lateral view crest remains at least slightly trilobate. No median ridge in some species, but the two sulci cross median line (Fig. 6*d*): **17**
3. Lower posterior margin of lateral lobes angulate or produced (Fig. 6*b;* Pl. 20): **Hippopedon**
   Lower posterior margin not angulate but sometimes having a tooth (Fig. 6*c*): **4**
4. Hindwings clear or with dusky or smoky area on fan (Fig. 6*e;* Pls. 5, 6): **5**
   Hindwings with outer dark band that may be distinct, indistinct, or interrupted, but with margins of band clearly defined (Fig. 6*f, g*): **6**
5. Forewings spotted and with pale streak along the Cu_2 and 1A veins (see Fig. 8*p* for names of veins). Hindwings without anterior dark streak and without expanded cells as in Fig. 29. Lateral lobe marked as in Fig. 6*h;* Pl. 11): **Camnula**
   Forewings usually with distinct crossbands over anterior two-thirds, and always without pale streak along Cu_2 and 1A. Hindwings usually with dark area along anterior margin and often with expanded cells in anterior fold (Fig. 6*i, j;* Pls. 5, 6) Chortophaga genus group: **35**
6. Hindwings either mostly black with outer pale band or mostly transparent with slight dark band only on fan (Fig. 6*g;* Pl. 13): **Dissosteira**
   Hindwings with broad outer dark band, or narrow band extending to anterior fold (Fig. 6*f*): **7**
7. Forewings reddish brown, body greenish. Hindwings brownish, becoming yellowish near base. Hind femora with black band covering narrowest part (Fig. 9*b;* Pls. 12, 13): **Eximacris**
   Not fitting above description: **8**
8. Forewings without crossbands or light and dark spots, but sometimes finely spotted; posterior (anal) margin sometimes pale (Fig. 7*a;* Pls. 1, 3): **9**
   Forewings spotted or banded (Fig. 7*b, c*): **15**

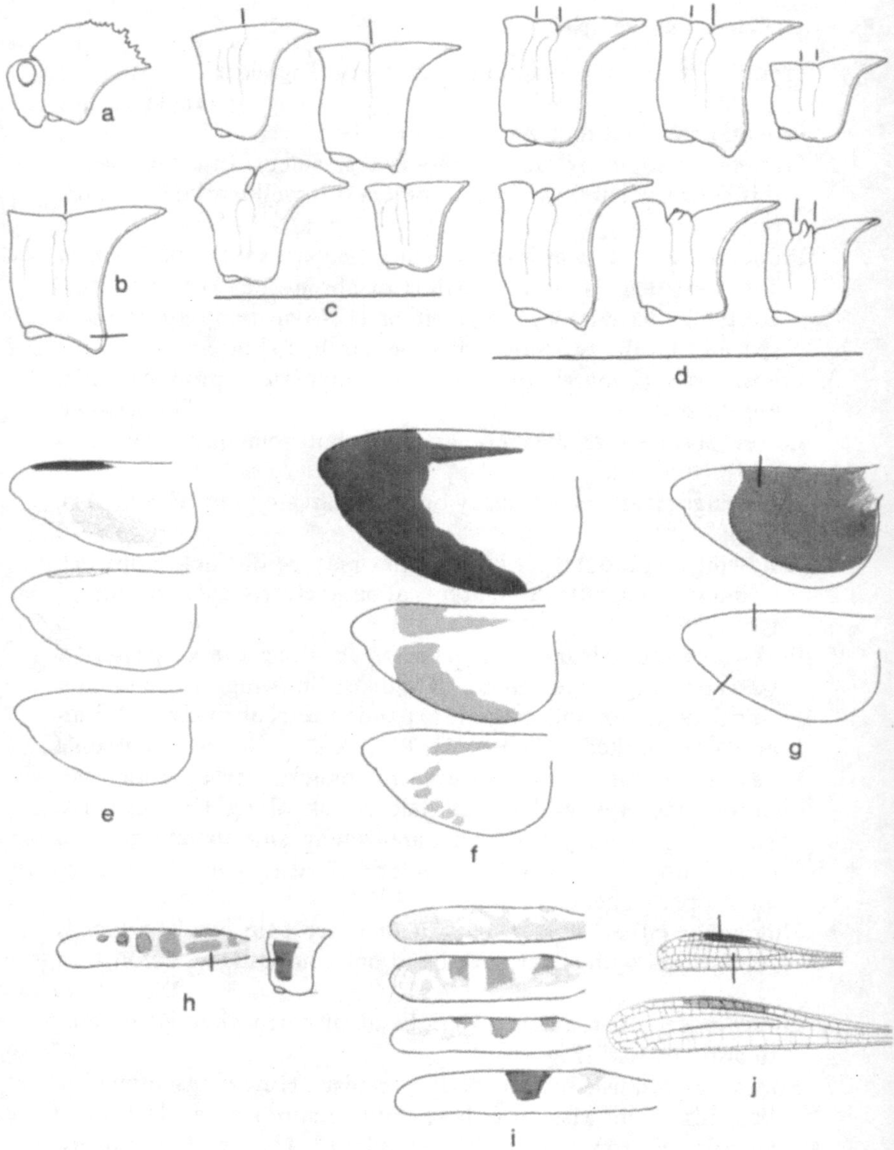

Fig. 6

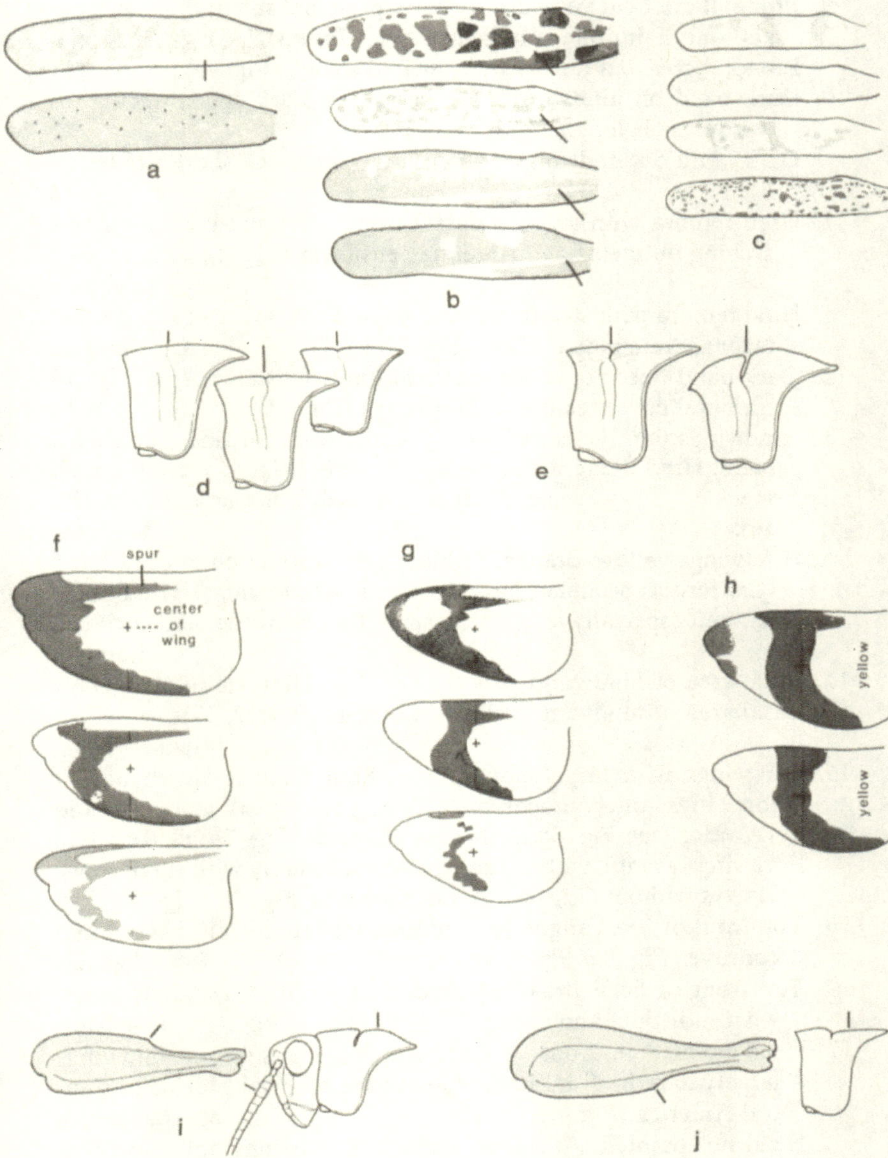

Fig. 7

9. Pronotal crest cut very shallowly by principal sulcus; in some species only faint line is indicated (Fig. 7*d;* Pls. 2, 3): **Arphia**
   Pronotal crest cut deeply by principal sulcus (Fig. 7*e*): **10**
10. Dark band on hindwings near periphery; dark spur directed toward base is long (Fig. 7*f;* Pl. 1): **11**
    Dark band on hindwings near center; spur much shorter (Fig. 7*g, h*): **12**
11. Hind femora with step on dorsal ridge. Pronotal crest high and arching on metazona. Antennae ensiform (Fig. 7*i;* Pl. 1): **Tomonotus**
    Hind femora with step on ventral ridge. Pronotal crest low on metazona. Antennae filiform (Fig. 7*j;* Pl. 1): **Leuronotina**
12. Dark band does not cover center of hindwings (Fig. 7*g*): **13**
    Dark band covers center of hindwings (Fig. 7*h*): **14**
13. Hindwings faintly colored with yellow, with indistinct outer dark band. Hind tibiae with long apical spurs (Fig. 45*d, e, f*). Small, speckled, hairy species from coastal California and Oregon (Pl. 14): **Microtes**
    Hindwings yellow, orange, or blue, with distinct outer dark band (California specimens have deep yellow hindwings). Hind tibiae without especially long apical spurs (not as in Fig. 45*d, e, f*) (Pl. 4): **Lactista**
14. Basal area of hindwings rosy red (Pl. 13): **Dissosteira pictipennis**
    Basal area of hindwings yellow or orange (Pls. 12, 13): **Spharagemon**
15. Forewings with large spots and without distinct crossbands; if forewings largely unicolorous, having pale streak along Cu_2 and 1A veins (see Fig. 8*p* for names of veins) (Fig. 7*b;* Pl. 9): **16**
    Forewings usually with darker crossbands, rarely with small spots; always without Cu_2 pale streak shown in Fig. 7*b* (Fig. 7*c*): **10**
16. Top front of head angulate in profile. fastigium well marked and concave (Fig. 8*a;* Pls. 7, 9): **Pardalophora**
    Top front of head broadly rounded in profile. Fastigium somewhat indistinct and convex (Fig. 8*b;* Pls. 7, 9): **Hippiscus**
17. Head pointed in front of eyes. Antennae ensiform. Hindwings largely dark in distal half, blue at base. Central Mexico to Central America (Fig. 8*e*, Pl. 1): **Machaerocera**
    Head not pointed. Antennae filiform. Hindwings not colored as above: **18**
18. Head large relative to pronotum and held distinctly higher than pronotum (Fig. 8*c*). Frontal ridge usually strongly grooved (Figs. 58–61; Pls. 20, 21, 22): **19**

Head small relative to pronotum. Top of head nearly in line with top of pronotum (Fig. 8*d*). Frontal ridge flat or shallowly grooved: **25**

19. Antennae ensiform. Male hindwings banded as in Fig. 8*f* (Pl. 20): **Psinidia**

Antennae filiform. Male hindwings not banded as in Fig. 8*f*: **20**

20. Cerci flattened, spoon-shaped in both sexes (Fig. 8*g;* Pl. 21): **Metator**

Cerci conical: **21**

21. Lateral pronotal carinae forming two thin white streaks. Area of lateral lobe adjacent to lateral carinae dark (Fig. 8*h;* Pl. 20): **Hippopedon**

Lateral pronotal carinae not forming thin, well-defined white lines. Area of lateral lobe adjacent to lateral carinae not particularly dark (Fig. 8*i, j, k, l*): **22**

22. Lower posterior margin of pronotum broadly rounded (Fig. 8*j;* Pl. 22): **Derotmema**

Lower posterior margin of pronotum angulate, pointed, or with a tooth (Figs. 8*i, k, l*): **23**

23. Small white spot or thumb-shaped marking near center of lateral lobe (Fig. 8*i*, Pl. 20): **Trachyrhachys**

Broad pale mark near center of lateral lobe (Fig. 8*k, l;* Pl. 21): **24**

24. Hindwings with outer dark band (Fig. 8*m;* Pl. 21): **Mestobregma**

Hindwings without outer dark band (Fig. 8*n;* Pl. 21): **Trepidulus**

25. Outer face of hind femora with strongly oblique dark band (Fig. 8*o;* Pl. 10): **Hadrotettix**

Outer face of hind femora banded or unbanded; posterior dark band, if present, never strongly oblique: **26**

26. Forewings without intercalary vein as shown in Fig. 8*p* (Plate 11): **Heliastus**

Forewings with intercalary vein (Fig. 8*p*): **27**

27. Forewings with broad crossbands and usually one narrower, paler, central band (Fig. 9*a*; Pl. 10): **Leprus**

Forewings variable, not like Fig. 9*a*: **28**

28. Forewings spotted or with pale streak along line dividing dorsal and lateral surfaces (Fig. 7*b*). Disk of pronotum roughly sculptured (Fig. 33*a–i*): **38**

Forewings not spotted or without distinct pale lines as in Fig. 7*b*. Disk of pronotum not sculptured as in Fig. 33*a–i*: **29**

29. Hind femora with broad dark band at narrowest part (Fig. 9*b;* Pls. 12, 13): **Eximacris**

Hind femora without dark band covering narrowest part: **30**

Fig. 8

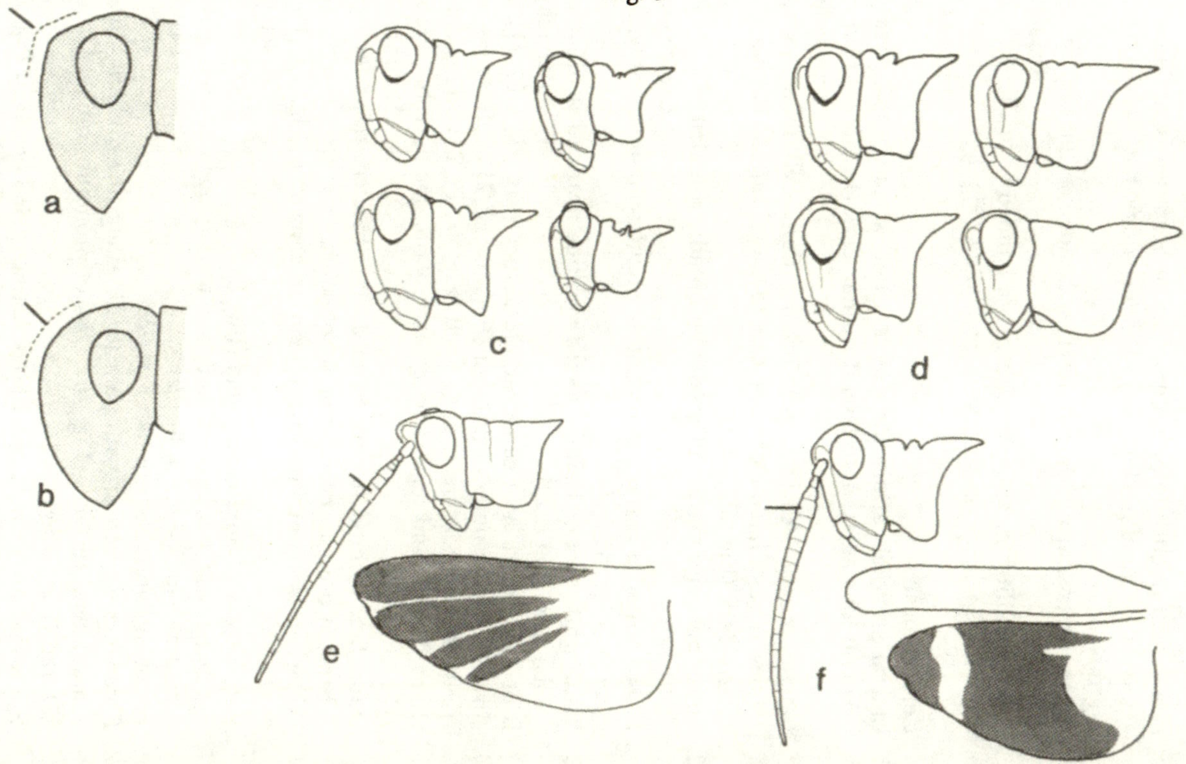

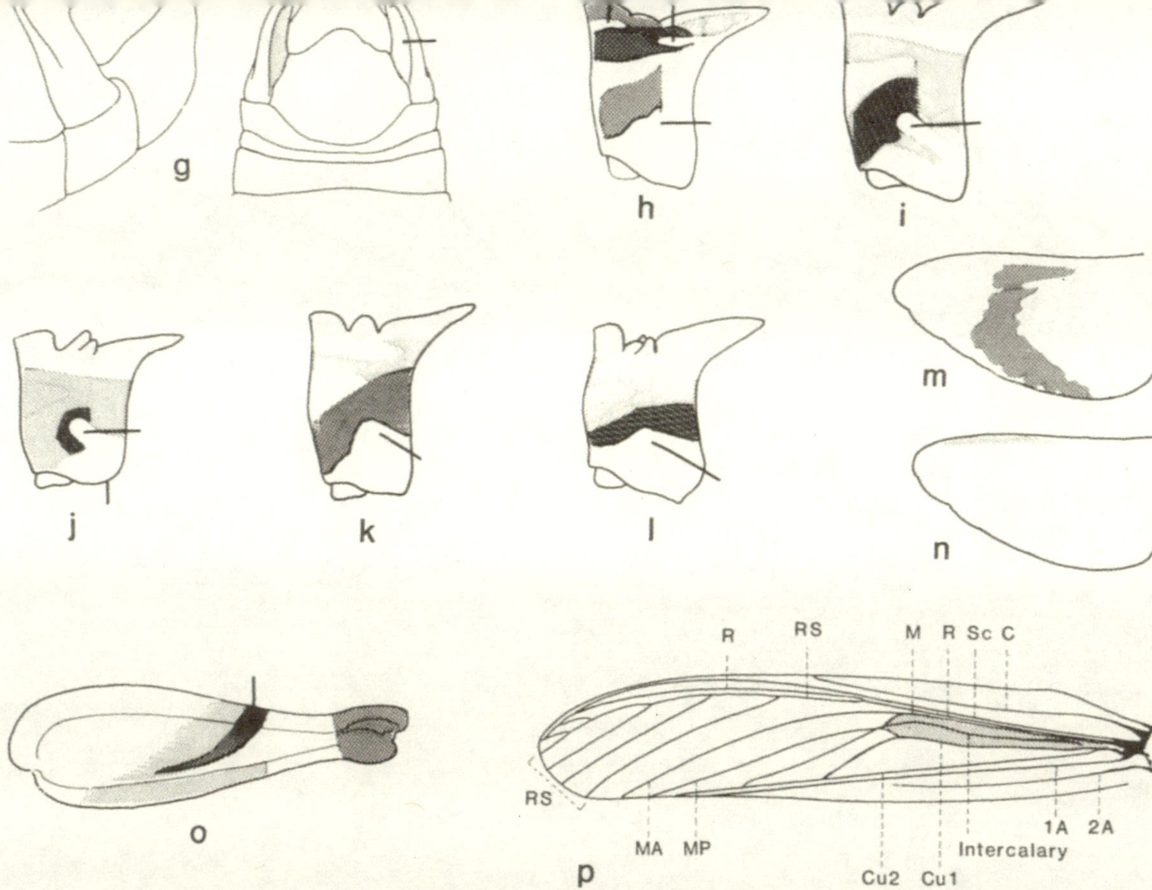
g
h
i
j
k
l
m
n
o
p
R
RS
M
R
Sc
C
RS
MA
MP
Cu2
Cu1
Intercalary
1A
2A

Fig. 9

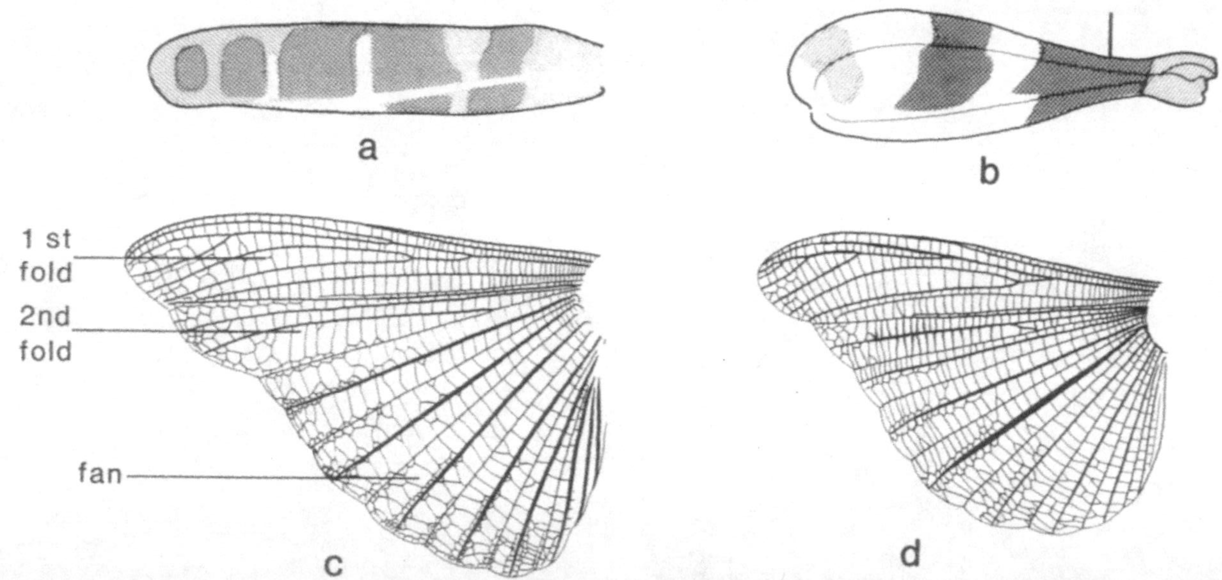

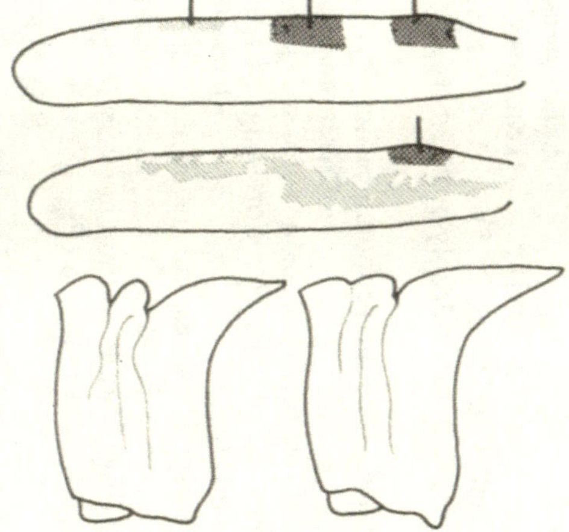

e

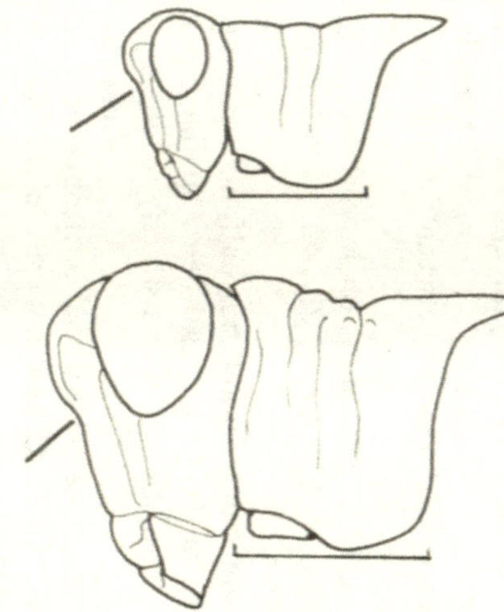

f

30. Fan of hindwings expanded; margin between fan and anterior fold concave (Fig. 9*c, d;* Pl. 19): **Circotettix**
Fan of hindwings not so expanded; margin between fan and anterior fold straight or convex: **31**
31. Small, hairy, heavily speckled. Apical spurs on hind fibiae long. Coastal regions of California (Fig. 45*d–f;* Pl. 14): **Microtes**
Not fitting above description: **32**
32. Dark crossbands on forewings only in anterior half, not reaching dorsal horizontal surface or crossing dorsal field. Median crest of pronotum well defined, distinctly or indistinctly trilobate in side view (Fig. 9*e;* Pl. 14): **Conozoa**
Dark crossbands on forewings reaching dorsal horizontal surface and usually crossing dorsal field. Median crest of pronotum variable, very low in many species: **33**
33. Frontal ridge of head concave in profile. Pronotum long in relation to height (Fig. 9*f;* Pl. 22): **Anconia**
Frontal ridge of head straight or convex in profile. Pronotum not long in relation to head: **34**
34. Found only on Cuba and other Caribbean islands (Pl. 14): **Sphingonotus**
Not occurring in Caribbean area (Pls. 15–18): **Trimerotropis**
35. Fastigium triangular and flat (Fig. 25*a–d;* Pl. 5): **Chortophaga**
Fastigium ovoid or elongate, shallow or deeply grooved (Fig. 25*e, f*): **36**
36. Forewings largely unbanded. Median pronotal carina forming narrow crest (Pl. 5): **Chimarocephala**
Forewings distinctly banded, especially along anterior margin: **37**
37. Fastigium very narrow and shallow (Fig. 27*i*). Forewings with one large and one small band (Pl. 6): **Shotwellia**
Fastigium variable, deep in some species, but never like Fig. 27*i*. Forewings with two to four bands (Pl. 6): **Encoptolophus**
38. Fastigium wider than long (Fig. 34*g*). Pronotum not very rough on prozona. Lophi of epiphallus not trilobate (Fig. 68*o*): **Sticthippus**
Fastigium longer than wide (Figs. 34*c–f*). Pronotum rough on prozona. Lophi of epiphallus trilobate (Fig. 34*l, n*): **39**
39. Inner face of hind femora blue and black with orange band at narrowest part (Pls. 7, 9): **Agymnastus**
Inner face of hind femora variable, not as above: **40**
40. Forewings with many large well-defined black spots (Pls. 8, 9): **Xanthippus**
Forewings with scattering of small dark spots, or fairly uniform in

color, or with several somewhat indistinct crossbands (Pls. 8, 9): **Cratypedes**

Genus MACHAEROCERA Saussure

RELATIONSHIP. Whether this genus belongs with the Oedipodinae or the Acridinae is problematical. Jago (1971) did not include it in his work on the Acridinae and Gomphocerinae, so I gave it little thought while preparing volume I of this series. Perhaps it should be included among the Acridinae, but its wing and leg coloration is more oedipodine in character. In some respects it reminds one of the oedipodine *Psinidia fenestralis,* but it has adopted more acridine features because it inhabits vegetation. In any event, the placement of this genus at the beginning of the Oedipodinae acts as a bridge, so to speak, between the Acridinae of the previous volume and this subfamily.

TAXONOMY. Six species have been described in this genus. *Oxycoryphus tibialis* Walker, *Machaerocera sumichrasti* Thomas, *Machaerocera obscura* Bruner, and *Machaerocera pacifica* Bruner were synonymized under *Machaerocera mexicana* Saussure by Hebard (1925). But Hebard was uncertain about the status of *Machaerocera magna.* He writes: "Study of the type of *M. magna* Bruner in the British Museum . . . leaves us uncertain as to whether it is an exceptionally large and broad female of *mexicana* or represents a distinct species." The status of *M. magna* remains in doubt. I am inclined to believe that other specimens would have shown up in collections if it were a distinct species. Unfortunately we only know that it was from "Durango or Sinaloa." I have treated it here as a synonym.

RECOGNITION. See under *M. mexicana.*

REFERENCES. Bruner 1904, Hebard 1925a.

Machaerocera mexicana Saussure Pl. 1

DISTRIBUTION. Forests of eastern and western Mexico south to Belize and Guatemala.

RECOGNITION. Fig. 10. A dark, slender insect with a rather pointed fastigium. Hindwings blue in basal area, dark in distal half to two-thirds. Antennae ensiform. Pronotum with disk and lateral lobes well defined on prozona. Lateral lobes with peculiar shape. Forewings without an intercalary vein. Forewings in lateral view becoming slightly wider at distal end. Body color variable: sometimes almost entirely black, sometimes brown and pale brown on dorsum. Hind femora banded most distinctly on upper face. Hind tibiae mostly blue. Body length to end of forewings 22–30 mm in males, 32–44 mm in females.

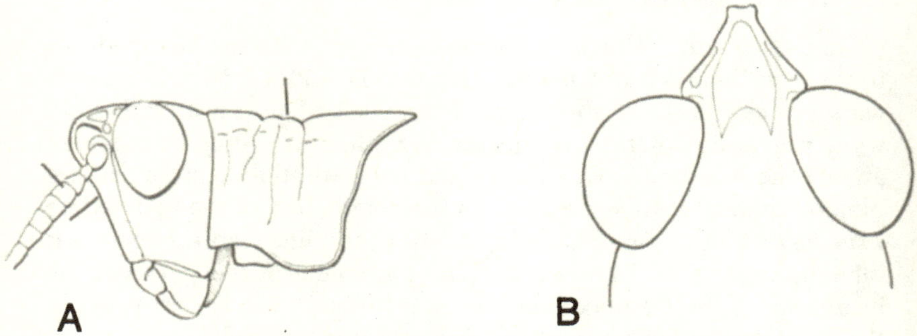

Fig. 10. *Machaerocera mexicana. a,* head and pronotum; note ensiform antennae, concave facial profile, and three cuts in the pronotal carina. *b,* top of head.

HABITAT. This species, like the Acridinae and Gomphocerinae, perches on shrubs and low trees along forest margins. I have found it mainly in rather lush mountain forests.

LIFE CYCLE. Adults throughout the year.

REFERENCES. Bruner 1904, Hebard 1925a.

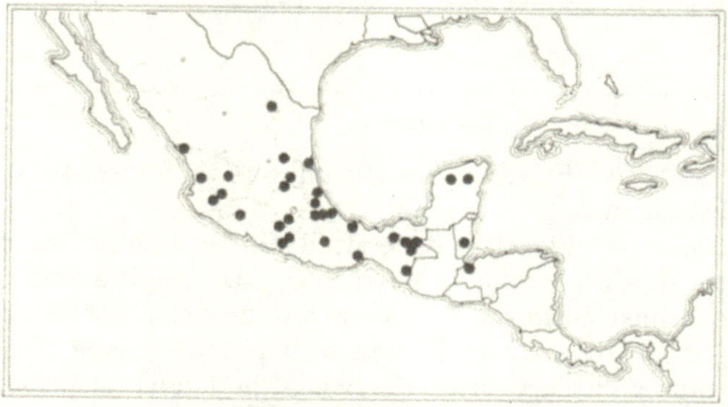

Machaerocera mexicana

Genus TROPIDOLOPHUS Thomas

RELATIONSHIP. This odd genus with one species does not seem closely related to any other North American genus. It resembles *Pyrgodera armata* from Asia in having a very high, toothed pronotal crest.

DISTRIBUTION. See *T. formosus.*

RECOGNITION. See *T. formosus.*

REFERENCES. Ball et al. 1942.

Tropidolophus formosus (Say) Pl. 1

DISTRIBUTION. Great Plains from Colorado south to northern Mexico and from Oklahoma and Texas to Arizona.

RECOGNITION. Fig. 6*a*. Very high, arching and serrated pronotal crest. Hindwings orange. Body green, usually with brown spots ringed with yellow, though many specimens lack such spots. Females sometimes pale brown. Hind tibiae with numerous spines. Female forewings not reaching end of abdomen. Body length to end of forewings 35–46 mm in males; to end of abdomen, 38–50 mm in females.

HABITAT. In Arizona *T. formosus* is found in dry, low-elevation grasslands and desert grasslands, where it feeds on low shrubs such as *Malvastrum* and *Sphaeralcea* (Ball et al. 1942). In west Texas Joern (1977) reports it feeding on *Croton texana.* Unlike most other Oedipodinae (except *Anconia integra*) this species lives primarily on vegetation that it matches closely in color. In west Texas I found it almost exclusively on vegetation.

BEHAVIOR. Males fly among bushes, flashing their orange hindwings and crepitating. Between flights they perch and produce O-stridulation with sound on both the up and the down stroke. Males approach females in a series of jerking motions. Immediately following stridulation they jerk their femora upward through 5–10 degrees. *T. formosus* is adapted to climbing about on vegetation, but its pair-forming and courtship behaviors are essentially similar to those of other Oedipodinae (Otte 1970). Females of *T. formosus* have short wings and are flightless.

LIFE CYCLE. Adults from mid-July to October.

REFERENCES. Ball et al. 1942, Otte 1970, Joern 1977.

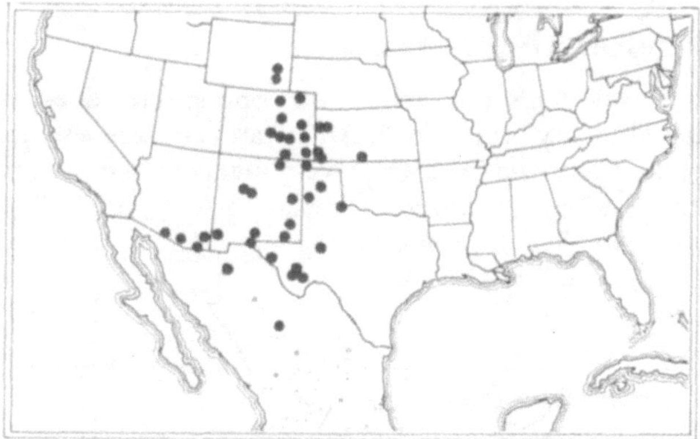

Tropidolophus formosus

Arphia Genus Group

Four genera are tentatively placed in this group: *Arphia, Leuronotina, Lactista,* and *Tomonotus.* This arrangement departs from the scheme envisaged by Rehn and Hebard, who placed *Arphia* with *Chortophaga, Chimarocephala,* and *Encoptolophus* under the Group Arphiae, based on similarities in the pronotal carinae. I am inclined to the view that *Arphia* is related most closely to *Leuronotina, Tomonotus,* and *Lactista,* which Hebard (1932) placed in the Group Lactistae. A relationship between *Arphia* and *Leuronotina* is seen in the forehead of *Leuronotina orizabae,* which is very similar to most *Arphia* species. A relationship with *Lactista* is suggested by the position of the arcuate groove of the fastigium, the presence of a single dorsal stripe on the forewings of some species in both genera, and the peculiar male–male signals in both genera (males strike the wings during femur-shaking, Otte 1970). *Tomonotus* differs from *Lactista* mainly in having a much higher pronotal crest. Fig. 11 shows the hypothetical phylogenetic relationships among the four genera (see Appendix 1 for discussion of characters).

RECOGNITION. Figs. 13–15. It is not simple to find characters shared by all members of this group. In general they can be characterized as follows. Transverse groove of fastigium located in back of middle. Pronotal crest cut by one transverse sulcus. Body brown to dark brown or blackish. Forewings with or without distinct crossbands; sometimes uniformly brown or gray. Male epiphallus with posterior lobes more or less unilobate. Hind tibiae rarely orange or red, usually bluish, sometimes brown to black, often banded with black.

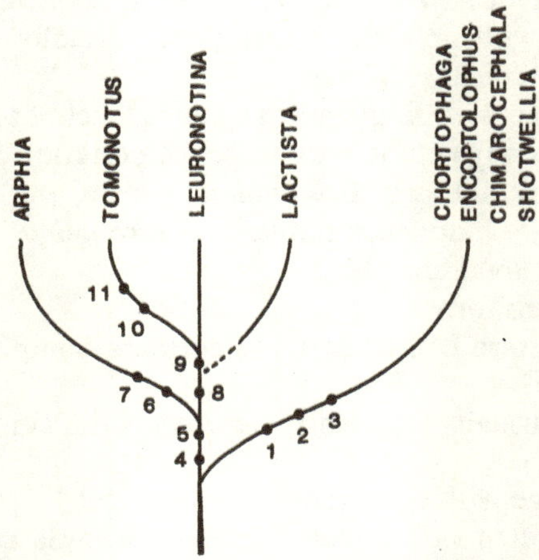

Fig. 11. Hypothetical phylogeny in Arphia and Chortophaga genus groups (see Appendix 1 for explanation).

• Identification of Genera

Arphia (widespread from central Mexican highlands to Alaska; Figs. 13, 14, 15; Pls. 2, 32)

1. Forewings without crossbands
2. Frontal ridge of head usually with two parallel grooves (or traces of them) above level of antennae
3. Fastigium usually with median carinula
4. Pronotal disk not as in *Leuronotina*
5. Dark band of hindwings variable, not shaped as in *Leuronotina*
6. Median pronotal crest variable in height; if very high, then without a deep cut
7. Antennae filiform

Leuronotina (southern Arizona to Central America; Fig. 19; Pl. 1)

1. Forewings usually spotted; spots sometimes aggregated into crossbands.
2. Frontal ridge with one or two grooves in dorsal third
3. Fastigium with or without median carinula
4. Pronotal disk rather strongly concave behind principal sulcus on either side of median ridge
5. Dark band of hindwings located near periphery and with long dark spur directed toward base (Pl. 1)
6. Median pronotal crest not high, distinctly cut by principal sulcus
7. Antennae filiform

Tomonotus (southern Arizona through western Mexico; Fig. 20; Pl. 1)

1. Forewings with or without crossbands; usually absent in pale brown or reddish, specimens
2. Frontal ridge with single groove above level of antennae
3. Fastigium elongate and with traces of posterior carinula
4. Pronotal disk not as in *Leuronotina*
5. Dark band of hindwings similar to *Leuronotina*
6. Median pronotal crest high
7. Antennae ensiform

Lactista (southwestern United States to northern South America; Figs. 21–23; Pl. 4)

1. Forewings usually with dark crossbands (indistinct in *gibbosus* and *humilis*)
2. Frontal ridge with one groove in dorsal third
3. Fastigium often with traces of median carinula at posterior end
4. Pronotal disk not as in *Leuronotina*
5. Dark band of hindwings not located near periphery as in *Leuronotina* and *Tomonotus*
6. Median pronotal crest variable in height, usually distinctly cut by principal sulcus
7. Antennae filiform

Genus ARPHIA Stål

DISTRIBUTION. This genus, with fifteen species, ranges over the grasslands of North America from the Atlantic to the Pacific and from Alaska and Canada to central Mexico. Several species (including *pseudonietana* and *conspersa*) are extremely wide ranging; others (*novissima, secreta,* and *townsendi*) have very restricted ranges. Only *secreta* may be adapted to desert conditions; the remainder inhabit open prairies, mountain grasslands, or grassy woodlands. Eastern species such as *granulata, xanthoptera, sulphurea,* and *simplex* often occur on eroded hillsides or in small woodland openings where grass is interspersed with patches of dry ground. Species such as *sulphurea* and *xanthoptera* doubtless greatly expanded their ranges in the eastern United States as the woodland was opened by farming. But pockets of these species probably existed there prior to disturbances by man in isolated hill prairies, cedar breaks, and drier hillsides.

Rarely do adults of more than one species of *Arphia* inhabit the same place at the same time. The species are either geographically or seasonally separated. *A simplex* and *sulphurea* are the only two species I have collected together as adults, but their distributions are largely nonoverlapping. Although several species have widely overlapping

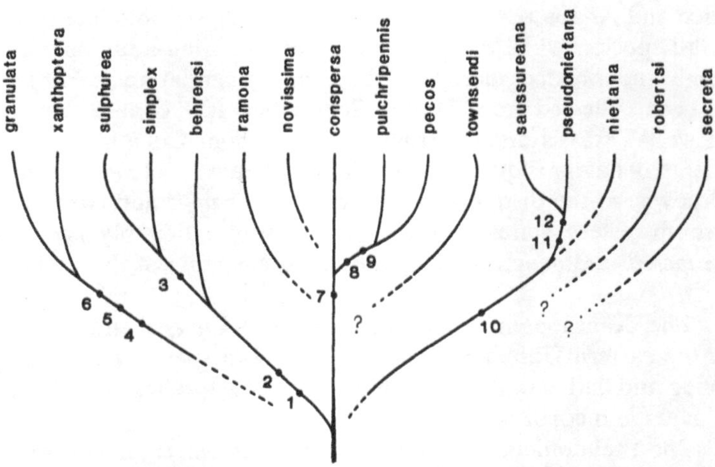

Fig. 12. Hypothetical phylogeny in *Arphia* (see Appendix 1 for explanation).

distributions (*conspersa* and *pseudonietana, xanthoptera* and *sulphurea*), the adults appear at different seasons.

TAXONOMY. The genus includes forty nominal species, but most of the names were subsequently synonymized, most by Hebard. I have examined all types that are not lost, as well as most large collections of this genus. The classification adopted here deviates only slightly from that of authors who considered only regional faunas. Of the forty names, only fifteen are considered valid. Four new species from Mexico are here added to the genus: *secreta, novissima, pecos,* and *robertsi.*

The types of *difficilis* Bruner (1905) and *fallax* Saussure (1884) are lost; these two names are therefore considered *nomina dubia. Arphia sordida* Osborn is a *nomen nudum,* for the species was not described and has no type.

The genus is tentatively subdivided into several species groups, which are admittedly rather poorly defined; in some cases I am unable to provide good diagnostic features (see Appendix 1 for discussion of characters).

1. The Sulphurea group contains *sulphurea, simplex, behrensi,* and perhaps *ramona* and *novissima.* The first three species overwinter as late instars, and adults are most abundant in spring and early summer.

2. The Conspersa group contains *conspersa, townsendi, pecos,* and *pulchripennis.* The geographic borders of the new species *pecos* (Texas) and of *pulchripennis* (California) remain uncertain; both species seem to grade morphologically into what appears to be *conspersa* in New

Mexico and Arizona. *A. pulchripennis* and *pecos* are both late summer and fall species, while *conspersa* is mainly a spring-adult species. In Arizona and Mexico specimens that appear similar to *pulchripennis* have been collected from May to September. It is certain that *pecos* from west Texas is distinct from *conspersa* from Colorado, for their courtship behavior is quite distinct. Mating behavior of *pulchripennis* is not known, so the distinction between *pulchripennis* and *conspersa* in the southwestern states requires further study. Possibly *pecos* and *pulchripennis* are geographically separated elements of the same species.

3. The Xanthoptera group contains *xanthoptera* and *granulata* from the eastern United States. Adults of both species appear in the summer and fall months; both have rounded foreheads and higher than average median pronotal crests.

4. The Pseudonietana group contains *pseudonietana, saussureana, nietana,* and possibly *ramona* and *novissima.* All are late summer or fall adult species. The hindwings have a bright red or orange basal area and the overall body color is dark gray to black.

5. The Secreta group contains only *secreta.* This species is rather distinct from all other *Arphia* in lacking a strong dark band on the hindwings and in possessing rather well-defined lateral pronotal carinae. Presently the species is represented by a single male from Coahuila, Mexico. Possibly the species should belong in a separate genus.

RECOGNITION. Figs. 13–15. *Arphia* species can be separated from other members of the genus group as follows. Forewings without crossbands, either unicolorous or spotted. Frontal ridge usually with two parallel grooves (or traces of them) dorsally. Principal sulcus forms a line where it crosses the median pronotal crest but does not cut it.

BIOLOGY. The members of this group live in a variety of habitats, from mountain meadows and forest openings (*nietana, conspersa*), to low-altitude thorn savannah (*simplex*), open short-grass prairies (*pseudonietana*), desert scrub (*pecos, secreta, pulchripennis*). Males, usually solitary, perform conspicuous flight displays to attract females (Otte 1970). The brightly flashing hindwings and accompanying snapping noises make a conspicuous display, which is especially dramatic when large numbers of males perform simultaneously. Between flights males produce various leg signals as they walk about on the ground searching for females.

REFERENCES. Bruner 1905.

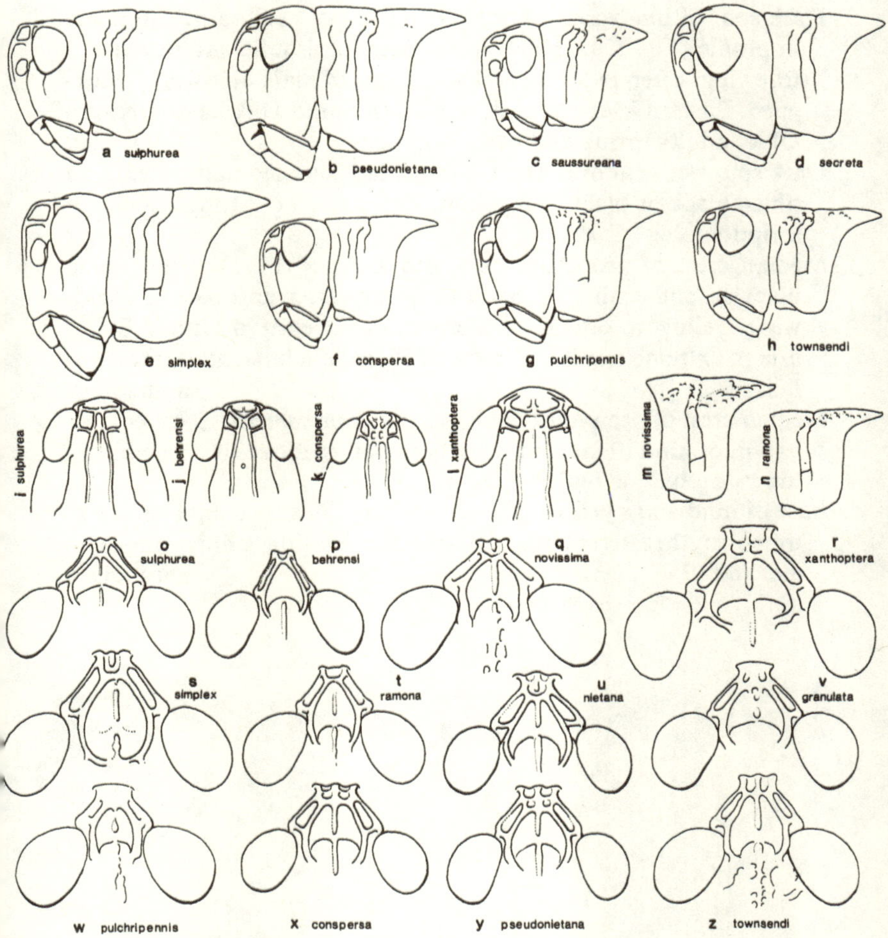

Fig. 13. *Arphia. a–h,* profiles of head and pronotum; *i–l,* frontal ridge and fastigium; *m, n,* lateral lobes; *o–z,* front of head, dorsal views.

▪ Key to Arphia Species

EASTERN UNITED STATES

1. Fastigium wider than long; frontal ridge very narrow between lateral foveolae (Fig. 13*a, i, o*). Widespread throughout eastern United States; spring and early summer adults: **sulphurea**

   Fastigium longer than wide or as long as wide; frontal ridge broader: **2**

2. Dark spur of hindwings narrow. Top front of head rather angulate in profile (Fig. 13*e*). Hind tibiae partly bluish. Sc and R veins of forewings often pale. Several anal veins of male hindwings thickened. Eastern Mexico north to eastern South Dakota and east to Ohio; adults spring and early summer: **simplex**

   Dark spur not as above. Top front of head rounded in profile. Hind tibiae black or black and yellow. Otherwise not fitting above description: **3**

3. Median crest of pronotum high and arching (Pl. 2). Hind tibiae blackish and with pale band in first quarter and base of hindwings yellow to orange. Widespread over eastern United States except extreme north and most of Florida; adults late summer to fall: **xanthoptera**

   Median crest of pronotum not high as in *xanthoptera*. Color combination of hind tibiae and wing base not as above (if tibiae black, hindwing base is red): **4**

4. Base of hindwings yellow. Males with one thickened anal vein. Extreme southeastern United States and Florida; adults midsummer to fall: **granulata**

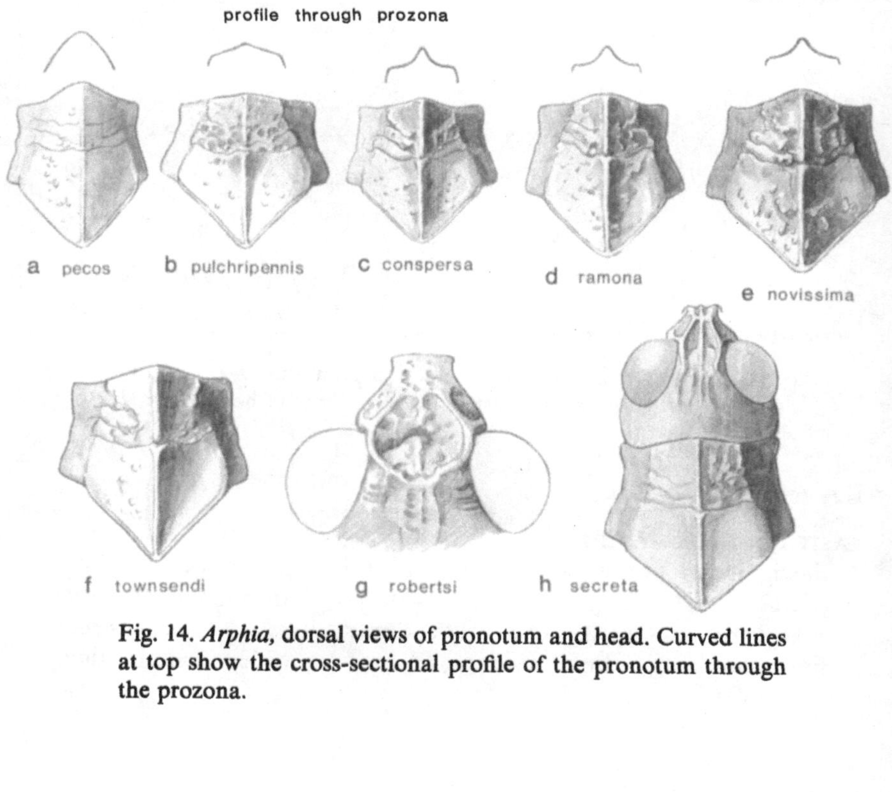

Fig. 14. *Arphia,* dorsal views of pronotum and head. Curved lines at top show the cross-sectional profile of the pronotum through the prozona.

Base of hindwings bright red. Male hindwings without thickened anal veins. In eastern United States known only in the extreme northwest; adults late summer to fall: **pseudonietana**

CALIFORNIA AND OREGON

1. Fastigium tapering to a point anteriorly (Fig. 13*p, t*). Frontal ridge between lateral foveolae narrow and without small median carinula. Hindwing base orange or yellow. California or southern Oregon: **2**

   Fastigium truncated anteriorly (Fig. 13*y, z*). Frontal ridge between lateral foveolae not as narrow and often with median carinula. Hindwing base variable: **3**

2. Hindwing base orange (rarely yellow east of the Sierra Nevada) and with narrow dark band. Hills and mountains in southern half of California; adults spring and early summer: **ramona**

   Hindwing base always yellow and with broader dark band. Hills

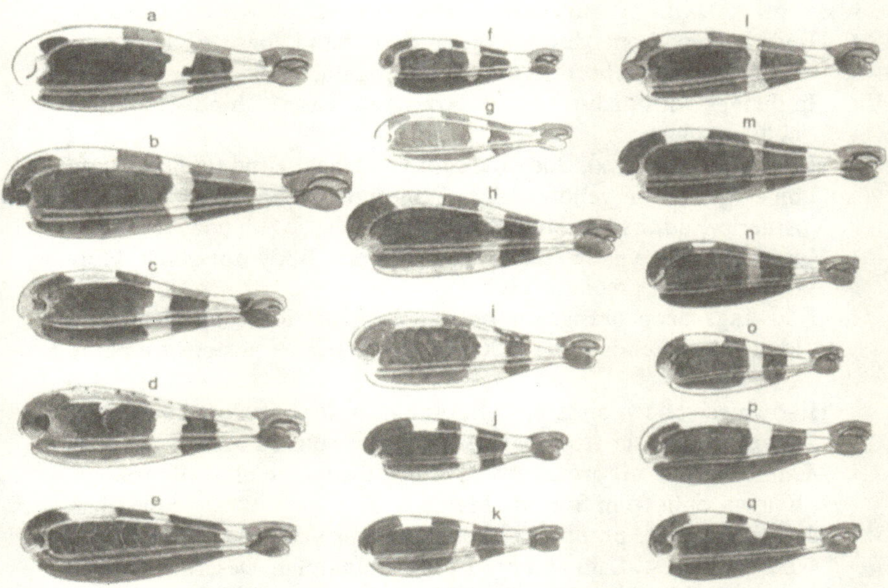

Fig. 15. *Arphia,* inner face of hind femur. *a, granulata; b, xanthoptera; c, sulphurea* (east Texas); *d, simplex* (Texas); *e, pseudonietana; f, saussureana; g, pulchripennis; h, pecos* (Texas); *i, conspersa* (Arizona); *j, conspersa* (Hidalgo); *k, simplex* (Hidalgo); *l, ramona; m, ramona* (darker specimen); *n, novissima; o, conspersa* (Durango); *p, q, nietana.*

and mountains of northern half of California and southern Oregon; adults spring and early summer: **behrensi**

3. Hindwing base orange. Top of frontal ridge between lateral foveolae not grooved and without distinct median carinula. Extreme southern California; adults midsummer to fall: **pulchripennis**

   Hindwing base yellow or red. Top of frontal ridge between lateral foveolae grooved or foveolate and with median carnula. Absent from southern California: **4**

4. Hindwing base bright red. Very widespread over western North America: **5**

   Hindwing base yellow, orange, or pinkish: **conspersa**

5. Abdomen light wood-brown or straw-colored. Hind tibiae usually partly blue. Body small. Northern half of California to Oregon; adults late summer to fall: **saussureana**

   Abdomen banded with dark brown or black, or entirely blackish. Hind tibiae blackish, without blue. Body large. Widespread over western North America; adults late summer and fall: **pseudonietana**

WESTERN CANADA TO NORTHERN MEXICO (EXCLUDING CALIFORNIA)

1. Hindwings without distinct outer dark band; base of hindwing pink. Known only from deserts of Coahila, Mexico: **secreta**

   Hindwings with distinct outer dark band; base of hindwings variable in color: **2**

2. Hindwings bright red. Body dark gray to black. Hind tibiae without blue, green, or yellow color. Widespread over western North America; adults late summer and fall: **pseudonietana**

   Hindwings pale yellow, orange, or pale red. Body not black. Hind tibiae yellow, green, or blue: **3**

3. Hindwings deep orange or red-orange; anterior colored area of hindwing extends same distance from base as posterior colored area: **4**

   Hindwings yellow, pink, or light orange; anterior colored area does not extend as far from base as posterior colored area: **5**

4. Median carina of pronotum forming distinct crest on prozona. Known only from Sonora, Mexico: **novissima**

   Median carina of pronotum indistinct on prozona, more like raised line than crest. Known only from Chihuahuan Desert: **pecos**

5. Greatest width of head is 0.90–0.96 times pronotal length in males. Hindwings very pale yellow. Known only from Chihuahua, Mexico: **townsendi**

   Greatest width of head is 0.80–0.84 times pronotal length. Hindwing disk pale yellow, yellow-orange, or red. Widespread over western North America; adults spring and early summer, occasionally late summer: **conspersa**

CENTRAL MEXICO

1. Hindwings pale blue: **robertsi**
   Hindwings yellow, orange, or red: **2**
2. Hindwing band narrow (see Pl. 3): **simplex**
   Hindwing band wider (see Pl. 3): **conspersa and nietana**

Arphia sulphurea **(Fabricius)** Pls. 2, 3

DISTRIBUTION. Eastern United States, from Atlantic coast to about 98° east and from near the Gulf Coast to southeastern Canada.

RECOGNITION. Figs. 13*a, i, o,* 15*c*. Base of hindwings almost always yellow, but in a few specimens from eastern Texas it is orange. Top front of head rather angulate. Fastigium wider than long. Frontal ridge between lateral foveolae very narrow. Lateral foveolae as high as long, or higher. Median carina of pronotum slightly to moderately arched (side view). Median carina not cut, or indistinctly cut by principal sulcus. Hind tibiae with a black band separating the first from the second third; distal two-thirds light brown to blackish. Body length to end of forewings 23–31 mm in males, 28–38 mm in females.

HABITAT. Dry upland pastures, woods margins, open woodlands with a grassy understory, and prairie remnants; generally inhabits well-drained places with patches of bare soil. In the southeastern United States it inhabits open pine woods with scrub oaks.

BEHAVIOR. Males have a conspicuous crepitation flight. Longer flights are up to ten feet or more and last up to one and a half seconds. Males usually fly in a straight line or in a low arc not more than three feet above the ground. Occasionally males fly a few inches straight up and return to their original perching place. Long flights are usually made by solitary males, while short flights are made in the presence of other individuals. On sunny days solitary males fly repeatedly, with intervals between flights as short as 15 seconds. Between flights they walk about, occasionally stridulating. Females are not known to crepitate. Courting males give three signals: one is a silent raising and lowering of the hind legs, another is a rapid vibration of the hind femora against the forewings, and the third is a series of whistling sounds produced by lowering the hind femora against the forewings (see Otte 1970 for detailed observations).

LIFE CYCLE. In most states this species overwinters in the last nymphal instar. In northern states adults appear in May and are gone by early July. In eastern Texas adults are present from April to August.

REFERENCES. Somes 1914, Blatchley 1920, Hubbell 1922b, Cantrall 1943, Coppock 1962, Otte 1970.

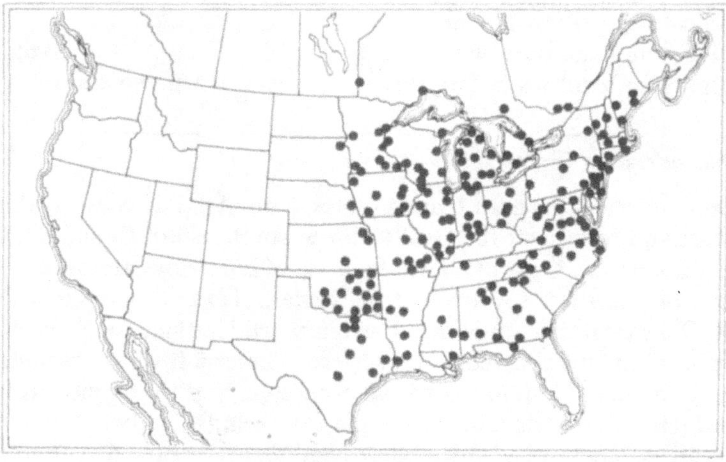

Arphia sulphurea

Arphia simplex Scudder Pls. 2, 3

DISTRIBUTION. Eastern Mexico nothward through Texas, Oklahoma, and Nebraska to southeastern South Dakota.

RECOGNITION. Figs. 13*e, s,* 15*d.* Base of hindwings yellow, yellow-orange, or orange. Dark band on hindwings has characteristic shape. Anal veins of hindwings thickened (not as in *sulphurea*). Top front of head rather angulate. Fastigium as long as wide or longer; in females it is sometimes wider than long. Lateral foveolae longer than high, often not clearly concave. Hind tibiae usually with a black band separating the first and distal two-thirds; second third usually bluish, but orange or reddish in Nuevo León and Hidalgo, Mexico. Sc and R veins of forewings often pale. Upper marginal area of hind femora usually distinctly banded. On specimens from Galeana, Nuevo León, the hindwing band is similar to that of *conspersa,* but the hind tibiae are red. Possibly *simplex* breaks up into several species in eastern Mexico.

In specimens from Chiapas the head is not so angulate in side view and the hind tibiae are largely greenish-blue; the dark spur on the dark band of the hindwing is wider and extends farther toward the base of the wing.

Body length to end of forewings 28–40 mm in males, 34–44 mm in females. Chiapas males 24–26 mm; females 26–30 mm.

HABITAT. Open fields, bunchgrass prairies, open grassy woodlands, and old fields. In Texas it shows a preference for timber margins. Often found in low-lying areas with knee-high grasses and at least small patches of bare ground. Isely (1935) found it associated with a heavy tangle of curly mesquite grass in northeastern Texas.

BEHAVIOR. Crepitation flights last up to six seconds and are usually in a straight line. A distinct loud clicking sound is produced, roughly twelve clicks per second at 85° F. Females have been observed flying toward males that have just crepitated. Courtship signals are similar to those of *sulphurea* (see Otte 1970 for detailed descriptions).

LIFE CYCLE. Adults are most abundant in spring and early summer. In eastern Kansas Smith (1981) collected nymphs from June through October and adults from April to June. The species probably overwinters as fourth instars. In Texas adults are most common in May and June, but some are present from April to December.

REFERENCES. Scudder 1875:514, Hebard 1925b, Isely 1937, Otte 1970, Smith 1981.

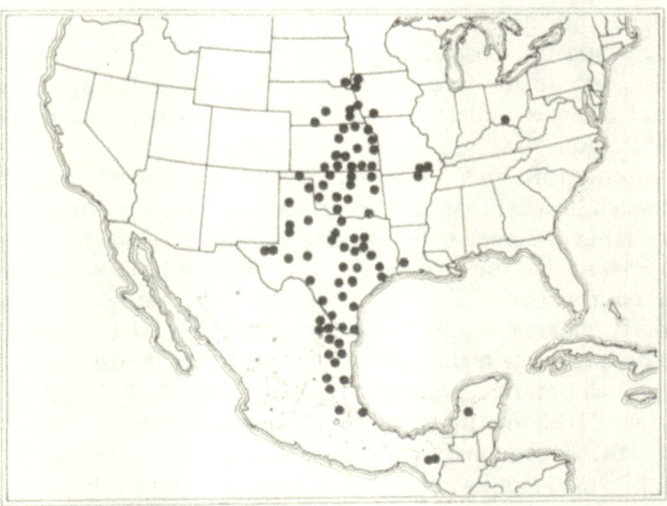

Arphia simplex

Arphia behrensi **Saussure** Pl. 2

TAXONOMY. Strohecker, Middlekauff, and Rentz (1968: 87) considered *behrensi* to be a subspecies of *conspersa*. Measurements of head dimensions, however, indicate that two taxa are involved (see Figs. 16,

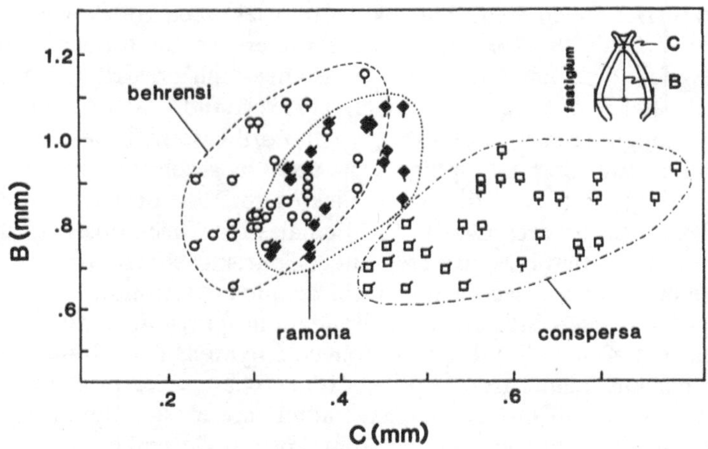

Fig. 16. *Arphia.* Scatter diagram showing separation of *ramona, behrensi,* and *conspersa.*

17). This species is similar to *sulphurea* in its wing coloration, head shape, and life cycle.

DISTRIBUTION. Coastal ranges and Sierra Nevada from central California north to the Oregon border. The Philadelphia Academy of Natural Sciences collection contains a single specimen bearing the label "Oregon."

RECOGNITION. Figs. 13*j, p,* 16, 17. Base of hindwings yellow; black band much broader than in *ramona.* Body dark brown or dark gray to blackish, but abdomen pale brown to yellowish. Head characters similar to *ramona,* but the species is smaller and can be separated from *ramona* on the basis of wing coloration. The fastigium is pointed anteriorly, making it quite distinct from *conspersa.* Hind tibiae blue in distal half; the blue region is usually preceded by a black or dark gray band, which in turn is preceded by a pale band. Body length to end of forewings 21–25 mm in males, 29–32 mm in females.

HABITAT. Open grassy woodlands and mountain meadows.

LIFE CYCLE. Adults are common from March to July; from collection data they appear to be most common in May.

REFERENCES. Saussure 1884, Bruner 1905, Strohecker, Middlekauff, and Rentz 1968.

Arphia ramona **Rehn** Pls. 2, 3

TAXONOMY. Strohecker, Middlekauff, and Rentz (1968: 87, 89) considered *ramona* to be a subspecies of *conspersa.* Measurements of

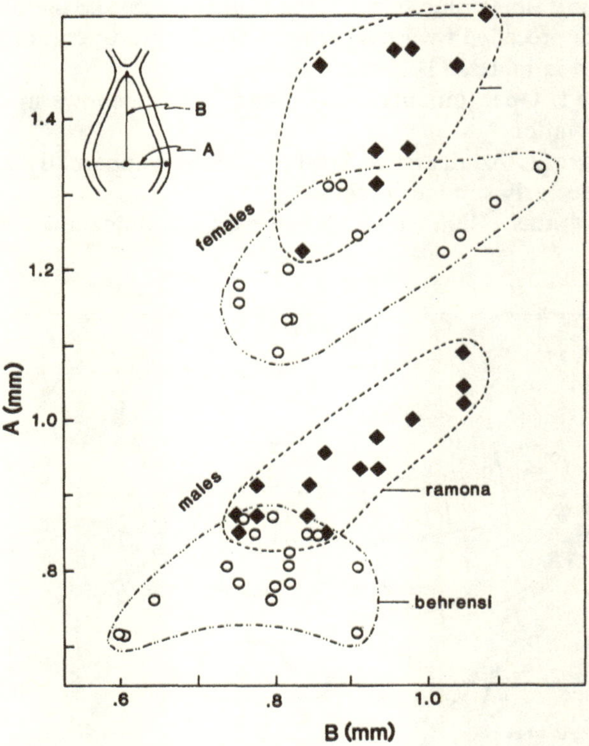

Fig. 17. *Arphia.* Scatter diagram showing separation of *ramona* and *behrensi.*

fastigial characters show quite clearly that *ramona* is a distinct species. This is especially evident since there is no overlap in fastigial dimensions where the two species overlap geographically. *A. ramona* is also quite distinct from *behrensi,* which Strohecker, Middlekauff, and Rentz also considered to be a subspecies of *conspersa.*

DISTRIBUTION. Southern California and Baja California.

RECOGNITION. Figs. 13*n, t,* 14*d,* 15*l, m,* 16, 17. Base of hindwings usually orange, but in several specimens from the Santa Cruz Mountains, California, and in a female from Ormsby, Nevada, it is deep yellow. Dark band relatively narrow. Body generally dark gray or dark brown to blackish, but abdomen wood-brown to yellowish. Most males with a pale median stripe on the folded forewings. Fastigium much narrower in front than in *conspersa,* and lateral foveolae more closely spaced. Median carina of promotum distinctly cut by principal sulcus. Hind tibiae blue through the distal two-thirds; the blue area is sometimes preceded by a pale band. Body length to end of forewings 25–31 mm in males, 33–41 mm in females.

HABITAT. Open mountain woodlands and meadows up to 10,000 feet in elevation.

LIFE CYCLE. Adults from April to October, apparently most common in the early summer months.

REFERENCES. Rehn 1902, Strohecker, Middlekauff, and Rentz 1968.

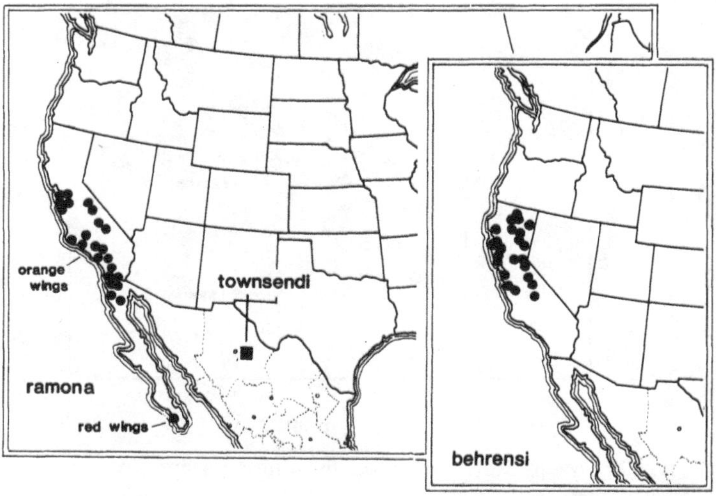

Arphia behrensi, A. ramona, and *A. townsendi*

Arphia novissima **Otte, n. sp.** Pls. 2, 3

DISTRIBUTION. Known only from two males collected by D. C. Rentz at Alamos, Sonora, Mexico.

RECOGNITION. Figs. 13*m*, 14*e*, 15*n*. Basal area of hindwings deep orange-red. This species differs from Mexican specimens of *conspersa* as follows: anterior colored area on anterior fold extends as far distally as colored area on the fan, and distal margin of colored area is well defined; fastigium wide relative to distance between lateral foveolae; median pronotal carina forms a distinct crest on the prozona; lower inner marginal area of hind femora with a single pale band (similar to *nietana*); hind tibiae dark gray-blue. Overall body color dark gray-brown. Body length to end of forewings 30–32 mm.

HABITAT. Not known.

LIFE CYCLE. The only two specimens known were collected in August.

SPECIMENS. Holotype male and paratype male, Alamos, Sonora, Mexico, August 12, 1960 (D. C. Rentz). Philadelphia Academy of Natural Sciences collection.

Arphia granulata Saussure — Pls. 2, 3

DISTRIBUTION. Florida and southeastern Georgia to southern Mississippi.

RECOGNITION. Figs. 13*w*, 15*a*. Basal area of hindwings always yellow; black band wider than in *sulphurea* and similar to that in *xanthoptera*. Head, pronotum, and wings usually uniformly brown or reddish brown, but occasionally quite strongly speckled. Frontal ridge much wider near dorsum than in *sulphurea*, and lateral foveolae widely separated. Top front of head rounded in profile, and lateral foveolae longer than high. Median pronotal carina moderately arched (in *xanthoptera* it is strongly arched). Hind tibiae strongly banded with yellow and black. Body length to end of forewings 27–33 mm in males, 30–37 mm in females.

HABITAT. Open pine woods, open palmetto scrub, grasslands. At Key West Rehn and Hebard (1914) reported it as common through the scrub and in open gumbo-limbo forest.

LIFE CYCLE. Adults have been collected between March and November, most in summer. Probably a few adults are present throughout the year.

REFERENCES. Saussure 1884, Rehn and Hebard 1905, 1912b, 1914, 1916, Otte 1970.

Arphia xanthoptera (Burmeister) — Pls. 2, 3

DISTRIBUTION. Most of eastern United States except the extreme north and most of Florida.

RECOGNITION. Figs. 13*l*, *r*, 15*b*. Base of hindwings yellow to deep orange; black band relatively broad. The species is most easily distinguished from its relatives in having a very high arching pronotal crest. Top front of head rounded in profile. Frontal ridge broad, truncated where it joins the fastigium; lateral foveolae well separated, as in *granulata*. Hind tibiae black but with a pale ring in the first quarter. Body length to end of forewings 31–38 mm in males, 36–46 mm in females.

HABITAT. Open sunny woodlands, pine-woods undergrowth, grassy fields, woods margins.

LIFE CYCLE. Probably egg-overwintering. Adults from July to October in northern states, July through November in southern states. Adults have been collected in January in Houston, Texas. In eastern Kansas Smith (1981) collected nymphs from June to August and adults from July through November.

REFERENCES. Blatchley 1920, Otte 1970, Smith 1981.

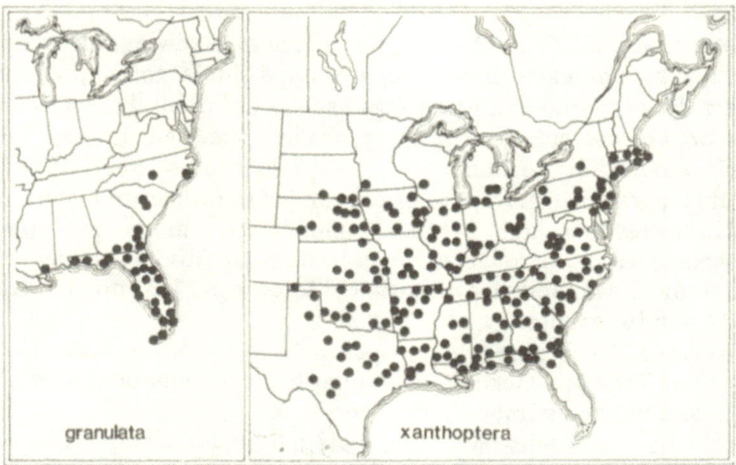

Arphia granulata and *A. xanthoptera*

Arphia conspersa Scudder Pls. 2, 3

TAXONOMY. The specific status of *pallidipennis* Bruner from the Mexico City area remains in doubt because only two specimens are known. This form was originally considered to represent a distinct species on the basis of hindwing color. Its head characters are similar to those of *conspersa.* Since some specimens of *conspersa* from Arizona, New Mexico, and Chihuahua, Mexico, also have rather pale greenish yellow wings, it is unlikely that *pallidipennis* will be considered distinct when intervening populations are studied. The type of *Arphia aberrans* is also pale-winged and is here treated as a synonym of *conspersa.* In its head and leg characters it is identical to *conspersa.* Furthermore, numerous specimens from New Mexico and Arizona with wing coloration intermediate between *aberrans* and *conspersa* exist in collections. *Arphia townsendi* was synonymized under *conspersa* by Hebard. But the type series of this species have distinctly larger heads than *conspersa.* Since head size appears to be constant throughout the range of *conspersa,* I believe that *townsendi* is probably distinct.

DISTRIBUTION. Most of western North America from central Mexico to Alaska.

RECOGNITION. Figs. 13*f, k, y,* 14*c,* 15*i, j, o.* Basal area of hindwings highly variable in color, from pale yellow to orange and reddish. Top front of head rounded in side view. Fastigium longer than wide and truncated at anterior end. Lateral foveolae often triangular. Dorsum of forewings often with a pale median stripe. Hind tibiae usually yellowish, but often with a greenish or bluish cast in the distal two-thirds. Abdomen yellowish to wood-brown. *A. conspersa* emerges in the late spring and early summer and so is not likely to be confused with the sympatric *pseudonietana.* In California it can be separated from *ramona* and *behrensi* on the basis of fastigial and foveolar shapes (see Figs. 16, 17). In Mexico this species can be separated from *novissima* and *nietana* on the basis of hindwing color and color pattern and in the banding of the hind femora. Body length to end of forewings 20–25 mm in males, 25–35 mm in females.

VARIATION. Yellow to reddish wings occur in all parts of the range, but regionally one color tends to predominate. On wing color variation in Colorado, Schennum and Willey (1979) write: "Several distinct monomorphic population centers occur, including southwestern and central Colorado, the Grand Mesa–West Elks region, and the western Great Plains. These centers are separated by relatively narrow intergrade zones." The problem of gene flow among populations has been extensively investigated by Willey and Willey (1967a, b). See also Hebard 1937: 361 for a discussion of color variation.

BEHAVIOR. Crepitation flights and courtship displays are similar to those of *sulphurea* (Willey and Willey 1969, Otte 1970). This species feeds mainly on grasses and sedges.

HABITAT. Short-grass prairies, forest clearings, and brushland openings; common in mountain meadows; seems to prefer sandy or gravelly soil. The altitudinal range of this species is quite wide, from 1,000 to 4,000 m above sea level in Colorado (Alexander and Hilliard 1969).

LIFE CYCLE. In the mountain states and Great Plains this species overwinters in late nymphal stages. In western Canada Brooks (1958) reports adults from May to September, but they are most common from May to July; the life cycle there may be completed in one or two years. In eastern Kansas, Smith (1981) collected nymphs from June through October and adults from April to June; here the species overwinters as fourth and fifth instars. In southern New Mexico, Arizona, and Mexico adults occur from May to at least late fall.

REFERENCES. Scudder 1875b, Strohecker, Middlekauff, and Rentz 1968, Scoggan and Brusven 1972, Smith 1981, and other references cited above.

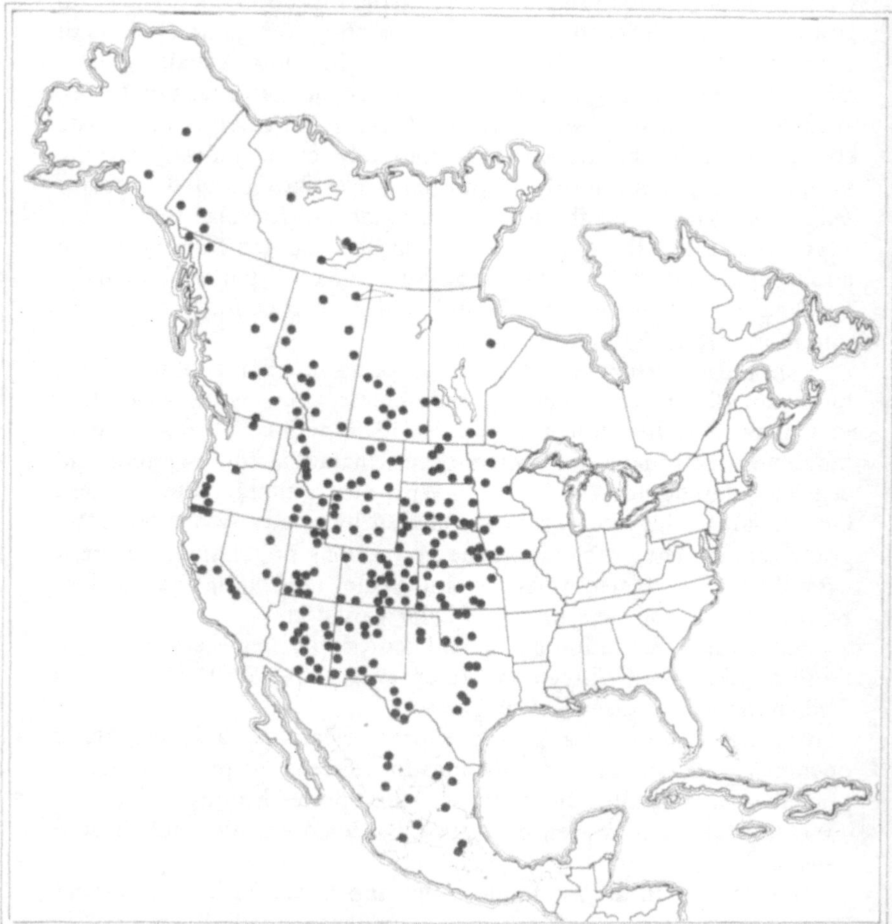

Arphia conspersa

***Arphia pecos* Otte, n. sp.** Pls. 2, 3

TAXONOMY. Specimens of this species were studied by Scudder (1876) and are represented in the type series under the name *A. teporata* Scudder. However, Hebard (1929: 333 n) chose as the lectotype a male from southern Colorado and then synonymized the name under *conspersa.* The Texas specimens of Scudder's type series therefore belonged to an unnamed species, here named *pecos.* Some Texas specimens of this species were determined by Hebard as *A. aberrans* Bruner. The latter name is based on a single pale-winged female from southern Arizona, which I consider to belong to *A. conspersa.*

The courtship behavior of specimens collected along the Pecos River in 1970 is very different from that described for *conspersa* (Otte 1970). *A. pecos* males do not stridulate like *conspersa* or other *Arphia* species as they approach females; instead they shake their hind legs, striking them against the forewings. This signal is used only in male–male encounters in other *Arphia* species.

DISTRIBUTION. Western Texas to northern Mexico.

RECOGNITION. Figs. 14*a,* 15*h.* Basal area of hindwings deep orange to orange. Anterior color area even with posterior area. Top front of head rounded in lateral profile. Carina in anterior part of fastigium poorly developed or obsolete (usually well defined in *conspersa*). Profile of pronotum through prozona largely convex (concave in *conspersa*). Region between RS veins has more than two rows of cells (usually only two rows of cells in *conspersa*). Hind tibiae dark blue to pale blue or greenish in distal two-thirds. In specimens whose courtship behavior was observed, the tibiae are dark blue in the distal two-thirds. Body length to end of forewings 30–40 mm in males, 36–43 in females.

HABITAT. Along the Pecos River I found this species inhabiting grassy patches with bare ground among rocks and shrubbery.

LIFE CYCLE. Adults from June to August. The holotype and paratypes were collected in August.

SPECIMENS. Four males, two females along the Pecos River at U.S. 90, Val Verde County, Texas, August 12, 1970 (Otte). Two males, Pecos River at Palmdale, Val Verde County, August 18, 1970 (Otte). One male, 9 mi east of Pleasanton, Texas, July 2, 1925. One male, two females, "Mexico" (Scudder's specimens). One male, Ceballos, Durango, Mexico, 8,800 ft, July 22, 1938 (Roberts). Two males, one female, Nuevo León, Mexico, 15 mi east of Monterrey, July 27, 1938. Five males, four females, Mamulique Pass, Nuevo León, July 11, 1936 (Roberts). One male, Higueros, between Monterrey and Saltillo, Coahuila, July 15, 1936 (Roberts). All specimens in the Philadelphia Academy of Natural Sciences collection.

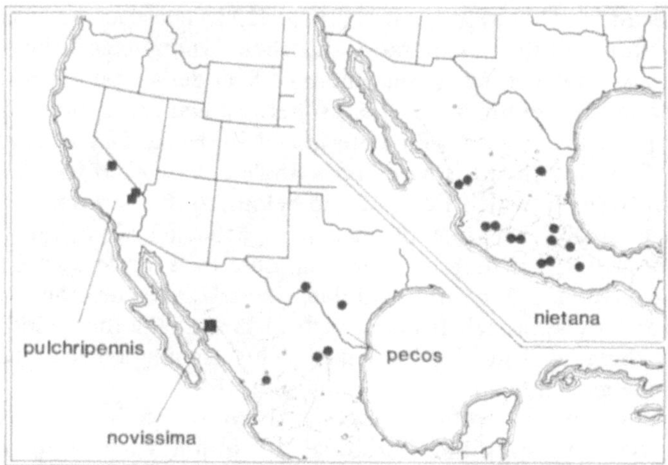

Arphia pecos, A. novissima, A. pulchripennis, and *A. nietana*

***Arphia nietana* (Saussure)** pls. 2, 3

TAXONOMY. I have compared this species with *truculenta* Rehn and *aurantipennis* Bruner. Except for sight differences in hindwing coloration, the type of the latter appears to be largely identical to specimens of *nietana.* Fastigial and facial characteristics are very similar, and both have orange hind tibiae. Because *truculenta* is also largely indistinguishable from *nietana,* I am here treating the name as a junior synonym.

DISTRIBUTION. Oaxaca to Durango, Mexico. Rehn's type specimen of *A. truculenta* (here synonymized) is from Guadalajara, Jalisco.

RECOGNITION. Figs. 13*v*, 15*p*, *q*. Basal area of hindwings orange to red orange. As in *pseudônietana,* hind femora have one inner pale band. This species differs also from *conspersa* and *simplex* in having only one pale band on the lower inner marginal area of the hind femora (similar to *novissima* and *pseudonietana*). Fastigium longer than wide and with a median carinula. Upper end of frontal ridge with small foveolar depressions and usually with a small median ridge. Hind tibiae orange or bluish.

In the northern part of its range *nietana* is very similar to *pseudonietana,* for here the abdomen and hind tibiae are black (abdomen brown, tibiae pale in southern part of range). But the wing coloration remains distinctive. In Durango state the two species have not been found to coexist at the same localities.

Body length to end of forewings 27–31 mm in males, 29–32 mm in females.

HABITAT. Mountain grasslands and forest glades.

LIFE CYCLE. Adults throughout the summer months.

REFERENCES. Saussure 1884, Rehn 1904a.

Arphia pulchripennis Bruner Pls. 2, 3

TAXONOMY. This species is very similar to *conspersa* and may eventually be shown to be identical with it (see discussion of *Arphia* taxonomy).

DISTRIBUTION. Known only from southern California.

RECOGNITION. Figs. 13*g, x,* 14*b,* 15*g.* Hindwings orange, dark band relatively narrow. Head: fastigium longer than wide and without a median carinula; lateral foveolae not entirely concave and rather lumpy; upper side of frontal ridge lumpy, mostly convex. Median pronotal carina low, forming a broad ridge. Hind tibiae brown at base, then whitish through most of the first quarter, then blue through the distal two-thirds. Body length to end of forewings 26.0–30.5 mm in males; 32–36 mm in females.

HABITAT. Not reported in literature; probably lightly wooded hillsides.

LIFE CYCLE. Southern California adults were collected from July 7 to September 1.

REFERENCES. Bruner 1905, Strohecker, Middlekauff, and Rentz 1968.

Arphia townsendi Bruner Pls. 2, 3

DISTRIBUTION. Known only from Colonia García, Chihuahua, Mexico. Bruner also records this species from the San Francisco Mountains of Arizona, but the paratypic specimen is clearly a member of *conspersa.*

RECOGNITION. Figs. 13*h, a′.* Basal area of hindwings pale yellow with a slight greenish tinge. This species differs from *conspersa* principally in the size of the head. In *townsendi* males the greatest head width is 0.90–0.96 times the pronotal length; in *conspersa* males it is 0.80–0.84 times the pronotal length. Fastigium truncated at front, with or without a median carinula in front half. Upper end of frontal ridge convex and with a median carinula. Body gray-brown, lateral lobes characteristically marked. Dorsal field of folded forewings mostly pale. Hind tibiae blue in distal three-quarters; blue band preceded by a pale band. Body length to end of forewings 23–26 mm in males (females not known).

HABITAT. Not reported in literature; probably similar to *conspersa.*

LIFE CYCLE. No collection date is given on the specimens from Colonia García.

REFERENCES. Bruner 1905.

***Arphia pseudonietana* (Thomas)** Pls. 2, 3

TAXONOMY. Larger, darker specimens from the southwestern states have traditionally been called *A. pseudonietana crassa.* The transition to such conditions is gradual, however, so recognition of a subspecies is of dubious value.

DISTRIBUTION. Widespread over western North America from Jalisco, Mexico, north to British Columbia, and from western to midwestern United States and Ontario, Canada.

RECOGNITION. Figs. 13*b, z,* 15*e.* Disk of forewings nearly always bright red to deep red and ringed with a heavy black band. A few yellow-winged specimens have been collected in Minnesota. In the southwestern states the black band becomes noticeably broader. Body often speckled gray, but usually dark gray-brown to black. Abdomen dark brown to black. Dorsum of closed forewings rarely with a pale median stripe. Disk of pronotum sometimes pale, straw-colored. Hind tibiae brown to black, but sometimes with a pale ring in the proximal quarter. Hind tibiae with a single pale band on inner face (as in *nietana*). Body length variable; specimens from the southwestern states are the largest. Length to end of forewings 30–41 mm in males, 32–47 mm in females.

BEHAVIOR. Crepitation flights are very conspicuous and can be heard at a distance of a hundred meters or more. Most flights are in a straight line and not more than four feet above the ground. Courtship signals are similar to those of *sulphurea* and *conspersa* (Otte 1970).

HABITAT. Open prairies and forest and brushland openings, especially where there are sizable patches of bare ground.

LIFE CYCLE. Egg-overwintering; adults from July to September in northern states and into November in southwestern states.

REFERENCES. Thomas 1870, Caudell 1903b, Hebard 1935b, Cantrall 1943, Otte 1970, Scoggan and Brusven 1972.

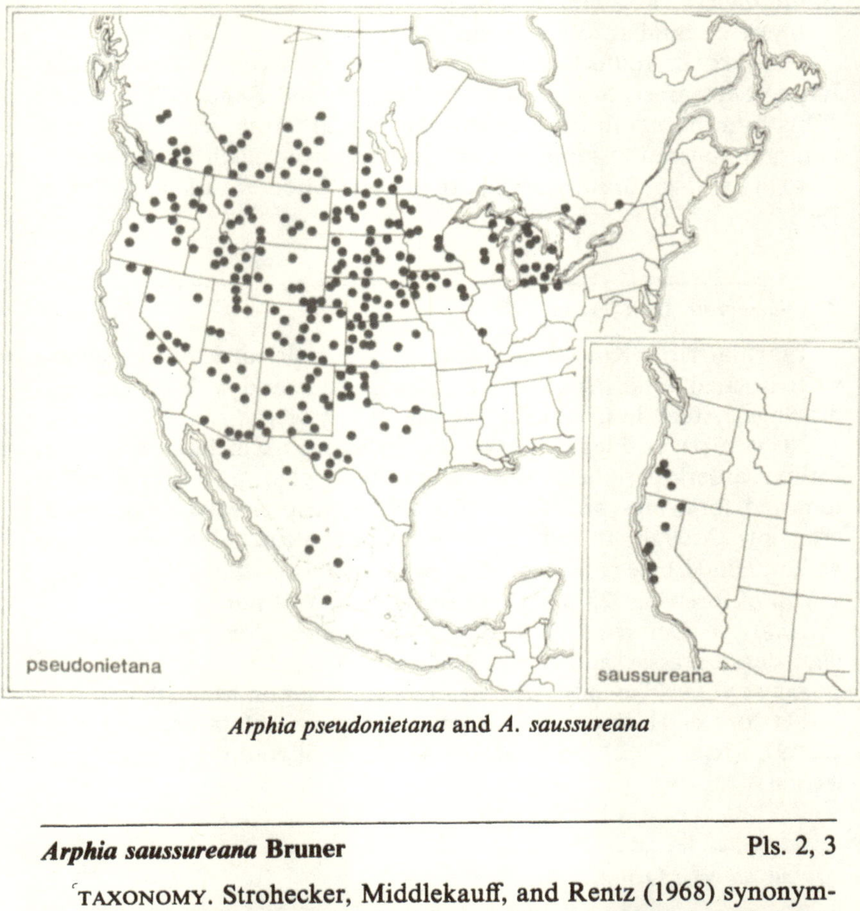

Arphia pseudonietana and *A. saussureana*

Arphia saussureana **Bruner** Pls. 2, 3

TAXONOMY. Strohecker, Middlekauff, and Rentz (1968) synonymized this species under *pseudonietana* without explanation. Because body size, abdominal coloration, and tibial coloration are quite different in the two forms, I consider them to belong to different species.

DISTRIBUTION. Northern California and Oregon.

RECOGNITION. Figs. 13*c*, 15*f*. Basal area of hindwings red. Body, except abdomen, usually dark brown or dark gray, sometimes brown or gray with numerous darker speckles. Pronotal disk occasionally straw-colored. Abdomen wood-brown or yellow (blackish or dark brown in *pseudonietana*). Hind femora with two pale bands on inner face. Hind tibiae often bluish in distal two-thirds, but sometimes yellowish or greenish, rarely brownish; distal extremity of hind tibiae always dark brown or blackish. Body length to end of forewings 21–28 mm in males; 24–32 mm in females.

HABITAT. Similar to *pseudonietana.*

LIFE CYCLE. Adults from May to September, usually collected from July to September. Strohecker, Middlekauff, and Rentz (1968) write: "This form occurs in the western part of California very late in the year and apparently supplants [*A. behrensi*] over much of its range."

REFERENCES. Bruner 1890, Strohecker, Middlekauff, and Rentz 1968.

Arphia secreta Otte, n. sp. Pls. 2, 3

DISTRIBUTION. Known only from a single male collected in a somewhat isolated valley, surrounded by mountains and with an internal drainage system, in Coahuila, Mexico.

RECOGNITION. Figs. 13*d,* 14*h*. Hindwings pink in basal area and without a dark peripheral ring. Overall body color pale gray; pronotum and forewings with faint longitudinal stripes. Median pronotal crest low throughout and cut by principal sulcus. Abdomen straw-yellow. Hind tibiae pale greenish blue in distal two-thirds. Body length to end of forewings 25 mm; hind femur length 12 mm.

HABITAT. Desert on valley floor. This area is dominated by *Atriplex,* desert grasses, and *Larrea* bushes.

LIFE CYCLE. The only specimen was collected on May 20.

SPECIMENS. Holotype male, Cuatrocienegas, Coahila, Mexico, May 20, 1975 (R. C. Rice). Philadelphia Academy of Natural Sciences collection.

Arphia robertsi Otte, n. sp. Pl. 3

DISTRIBUTION. Known only from the vicinity of Zitácuaro, Michoacán, Mexico.

RECOGNITION. Fig. 14*g*. This is the only *Arphia* with blue hindwings and an indistinct outer dark band. Forewings quite noticeably mottled in some specimens and with a pale dorsal streak in some. Frontal ridge convex. Fastigium almost round, with arcuate groove near center and with a median ridge. Pronotal carina distinct but low. Pronotal disk sometimes with a faint X-shaped marking when viewed from above. Hind femora sometimes with faint oblique dark bands on outer face. Inner face with two pale bands and with dark bands becoming blue ventrally. Hind tibiae almost entirely blue. Abdomen yellowish. Body length to end of forewings about 25 mm in males, about 30 mm in females.

HABITAT. Not known.

LIFE CYCLE. Eight females and two males were collected in March; one male was collected in February. This species appears to reach the adult stage early in the year, which may explain why it has not been collected more frequently in the heavily collected regions of central Mexico.

SPECIMENS. The holotype male, two paratype males, and eight paratype females were all collected by M. J. D. White at Zitácuaro, Michoacán, Mexico, on March 13, 1938. Another paratype male was collected by W. F. Chamberlain 20 mi east of Zitácuaro on February 11, 1948.

Genus LEURONOTINA Hebard

DISTRIBUTION. *Leuronotina* ranges northward from Costa Rica through the mountains of western Mexico to the Santa Rita Mountains of southern Arizona. In western Mexico two species live in the mountains but their distributions are still poorly known. The mountain ranges north and south of the Durango-Mazatlán road have not been explored and may reveal other members of this genus. Two species are found in these mountains, *ritensis* in the north and *philorites* in the south.

RECOGNITION. Figs. 18, 19. Disk of pronotum forms a concave depression immediately behind the principal sulcus. Forewings spotted (*orizabae, ritensis*) or rather dark (*philorites*), sometimes with crossbands. Hindwings pink or orange in basal area (but see variation in *orizabae*). Dark band on hindwings is near the periphery with a long spur directed toward base of wing. Hind femora bluish on inner face. Lower ridge on hind femora becomes suddenly narrower near the center.

REFERENCES. Rehn 1910.

• Identification of Leuronotina Species

orizabae (Fig. 19*c*)

1. Fastigium wider than long
2. Hindwings pink in basal area; rarely almost colorless
3. Hind tibiae with broad distal black band

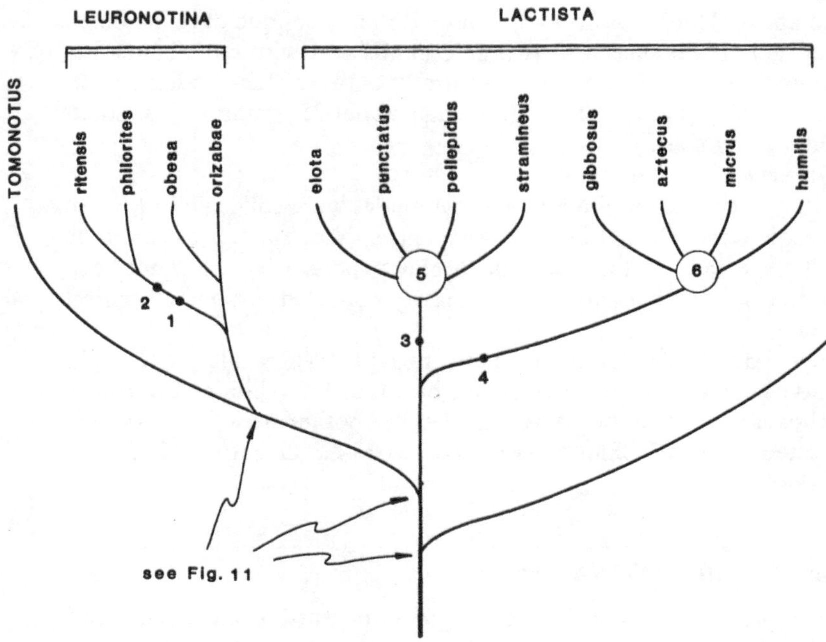

Fig. 18. Hypothetical phylogeny in *Tomonotus, Leuronotina, Lactista,* and *Arphia* (see Appendix 1 for explanation).

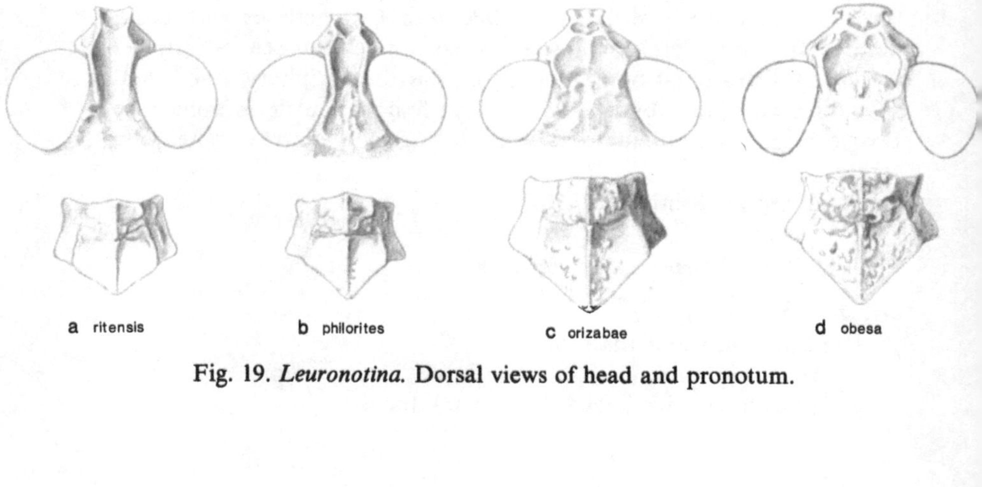

Fig. 19. *Leuronotina.* Dorsal views of head and pronotum.

obesa (Fig. 19*d*)
1. Fastigium wider than long
2. Hindwings yellow
3. Hind tibiae blue, unbanded

philorites (Fig. 19*b*)
1. Fastigium much longer than wide
2. Hindwings reddish to orange in basal area
3. Hind tibiae blue distally

ritensis (Fig. 19*a*)
1. Fastigium much longer than wide
2. Hindwings orange in basal area
3. Hind tibiae banded black and blue

Leuronotina orizabae (Saussure) Pl. 1

DISTRIBUTION. Central Mexico to Costa Rica.

RECOGNITION. Fig. 19*c*. Hindwings deep pink, very occasionally transparent. Overall body color light gray-brown and strongly speckled. Head: frontal ridge never deeply grooved, sometimes flat, usually shallowly grooved and with a low median carinula dorsally, foveolate at junction with fastigium. Fastigium wider than long and with an undulating surface, including a low median carinula in the anterior part. Pronotum: median carina low, cut only by principal sulcus. Disk of prozona strongly raised relative to concave anterior part of metazona. Lateral carinae deeply cut by transverse sulci. Disk rough with many tubercles; posterior angle less than 90 degrees. Forewings usually

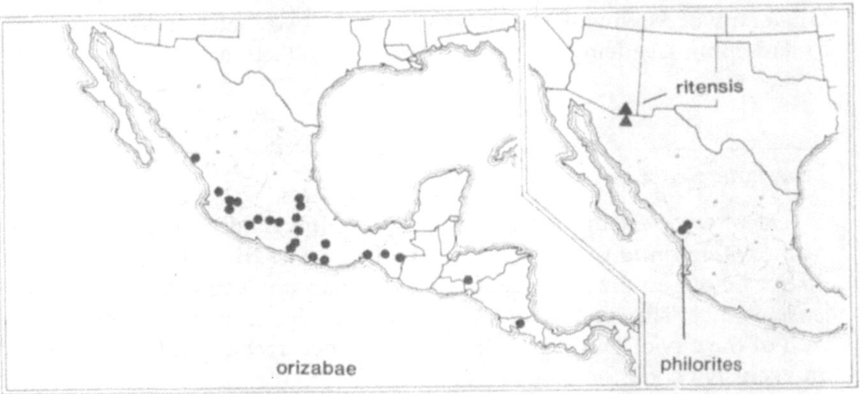

Leuronotina orizabae, L. ritensis, and *L. philorites*

with two or three dark crossbands. Hind femora with two dark bands in central two-thirds, darker on upper face; lower carina becoming suddenly narrower between the second and last third. Hind tibiae banded with black. Abdomen is straw-yellow or ivory-yellow. Body length to end of forewings 21–25 mm in males, 25–29 mm in females.

HABITAT. Open, thinly grassed, gravelly areas.

LIFE CYCLE. Adults are probably found during all seasons. They have been collected in March in Costa Rica and in July in central Mexico.

REFERENCES. Saussure 1884, Hebard 1932.

Leuronotina obesa Otte, n. sp. Pl. 1

DISTRIBUTION. Known only from Aguascalientes, Mexico.

RECOGNITION. Fig. 19*d*. Hindwings yellow, dark spur of band wider than in *orizabae*. Most similar to *orizabae* in overall appearance. Forewings with a narrow, somewhat interrupted band above base of hind femora. Fastigium wider than long (*orizabae* similar). Hind tibiae blue, without black bands (*orizabae* has black bands). Body length to end of forewings 24–26 mm in males; 28 mm in only female. Fastigium wider than long (like *orizabae* and unlike *philorites* and *ritensis*).

HABITAT. Mountain grassland. The only known specimens were collected at 8,200 feet.

LIFE CYCLE. The only adults were collected on March 9.

SPECIMENS. Holotype male, five paratype males, and one paratype female are all from Sierra Fria, 40 miles south of Rincón de Romos, Aguascalientes, Mexico (Cantrall). Holotype and two males in the University of Michigan Museum of Zoology, two males and female in Philadelphia Academy of Natural Sciences collection.

Leuronotina ritensis (Rehn) Pl. 1

TAXONOMY. Rehn placed this species in the genus *Scirtetica* along with *S. marmorata* from the eastern United States. But the two species do not belong in the same genus or even the same genus group. *L. ritensis* is most similar to *philorites,* a new species from the mountains west of the city of Durango, Mexico. These two species probably share an exclusive common ancestry (Fig. 18).

DISTRIBUTION. Presently known only from between 5,000 and

8,000 feet on the Santa Rita Mountains in southern Arizona and from Sierra de los Ajos, at the head of Canyon de Evans, Sonora, Mexico.

RECOGNITION. Fig. 19*a*. Body strongly speckled with pale green and black, resembling a lichen-encrusted rock. Basal area of hindwings orange; outer dark band continuous, without a transparent window at distal end. Forewings without distinct crossbands. Inner faces of hind femora are bluish. Hind tibiae banded bluish and black. Body length to end of forewings 21–23 mm in males, 26–28 mm in females.

HABITAT. Lichen-encrusted rocks in mountains.

REFERENCES. Rehn 1910.

Leuronotina philorites Otte, n. sp. Pl. 1

DISTRIBUTION. This species is known only from the mountainous region southwest of Durango, Mexico.

RECOGNITION. Fig. 19*b*. Hindwings reddish to orange. Eyes relatively large, fastigium very narrow. Hind tibiae blue in distal two-thirds. Dorsal field of folded forewings with or without a pale, narrow median band. Overall body color dark gray-brown to blackish. Lateral pronotal carinae form two very distinct lumps on the prozona.

HABITAT. Above 6,000 feet in openings in mountain woodlands.

LIFE CYCLE. All five specimens were collected in late August.

SPECIMENS. Mexico, 142 mi southwest of the city of Durango on Highway 40, August 26, 1963 (Grant and Howard), one male, five females, Philadelphia Academy of Natural Sciences collection. Mexico, on Highway 40, 63 mi east of its junction with Highway 15, August 28, 1963 (Grant and Howard), one male, one female, Philadelphia Academy collection. "El Ranchero," 7.6 mi northeast of Revolcadero, Durango, Mexico, at 7,620 feet, August 26, 1961 (Cantral and Cohn no. 44), one male, one female, University of Michigan Museum of Zoology.

Genus TOMONOTUS Saussure

RANGE. Southern Arizona south to Guerrero and Oaxaca, Mexico.

RECOGNITION. Figs. 18, 20. The genus differs from *Leuronotina* and *Lactista* mainly by its high median pronotal crest. Fastigium long and roughly pear-shaped in outline. Antennae ensiform. Upper ridge of hind femora becoming abruptly narrower at two-thirds point. Forewings with or without faint crossbands (*mexicanus* usually with bands;

ferruginosus specimens that resemble dry leaf litter usually homogenously colored). Hindwing patterning is very similar to *Leuronotina*.

REFERENCES. Bruner 1905, Hebard 1932.

• Identification of *Tomonotus* Species

ferruginosus (Sierra Madre Occidental to southern Arizona; Fig. 20*a*)

1. Hindwings orange
2. Forewings often without dark crossbands
3. Fastigium relatively long

mexicanus (central volcanic plateau of Mexico south to Oaxaca; Fig. 20*b*)

1. Hindwings pink
2. Forewings usually with dark crossbands
3. Fastigium relatively short

Tomonotus ferruginosus Bruner — Pl. 1

DISTRIBUTION. Southeastern Arizona to Durango, Mexico.

RECOGNITION. Fig. 20*a*. Antennae slightly ensiform. Base of hindwings orange, bordered by a narrow dark band. Overall body color variable, especially among males. Some males are rather strongly speckled on side of pronotum and with four dark bands on upper face of hind femora. Other males are unicolorous wood-brown. One male from southern Arizona has thin pale streaks along wing angle. Fe-

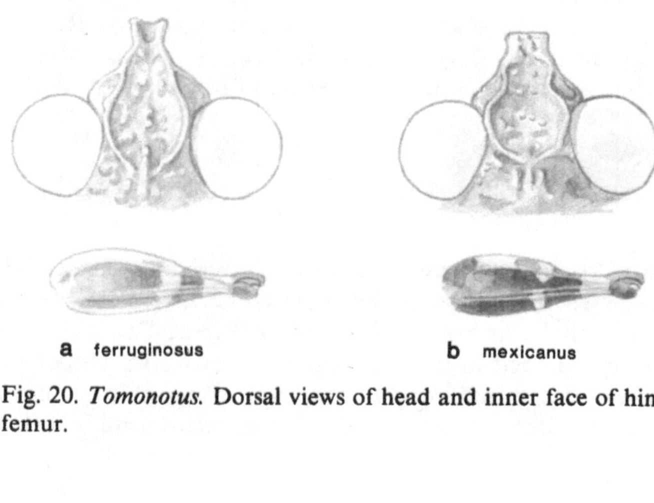

Fig. 20. *Tomonotus.* Dorsal views of head and inner face of hind femur.

males usually uniformly dead-leaf brown. Pronotal crest very high, almost half as high as the lateral lobe. Forewings without dark crossbands. Posterior end of forewings usually obliquely truncated. Hind tibiae ranges from yellowish to bluish in distal two-thirds. Abdomen straw-yellow to wood-brown. Body length to end of forewings 25–29 mm in males, 33–40 mm in females.

HABITAT. Oak woodlands. Ball et al. (1948) write: "The color very closely resembles that of dry oak leaves, and it is usually among them."

LIFE CYCLE. In Arizona nymphs overwinter, and adults are most common from April to August.

REFERENCES. Bruner 1905, Ball et al. 1942.

Tomonotus mexicanus **Saussure** Pl. 1

DISTRIBUTION. Morelia, Michoacán, and Guerrero, Mexico.

RECOGNITION. Fig. 20*b*. Hindwings translucent pink. Body color variable, from dark speckled gray to uniform woodbrown. Unlike *ferruginosus,* this species almost always has dark crossband in middle of forewings. In dark specimens hind tibiae are strongly banded with black and ivory.

HABITAT. No published records but probably similar to that of *ferruginosus.*

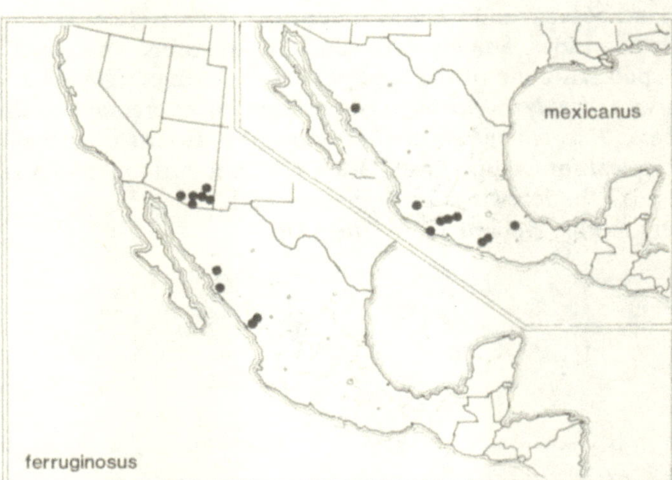

Tomonotus ferruginosus and *T. mexicanus*

LIFE CYCLE. Adults have been collected in March, July, August, and December.

REFERENCES. Hebard 1932.

Genus LACTISTA Saussure

TAXONOMY. Eight species are included in the genus. Two (*aztecus* and *micrus*) were previously in the genus *Platylactista.* That genus is here synonymized because Hebard's diagnosis (1932: 250) does not adequately separate the species from *Lactista,* relying on subtle differences in the height of the pronotal crest and on pronotal proportions. Possibly *pellepidus* and *stramineus* do not represent distinct species.

DISTRIBUTION. Southern California and southern Arizona to Venezuela.

RECOGNITION. Figs. 18, 21, 22, 23. This genus differs from *Tomonotus* in lacking a high, deeply cut pronotal crest and in having filiform antennae. Unlike *Tomonotus* and *Leuronotina,* in *Lactista* the hind femora do not narrow suddenly. It differs from *Arphia* in having forewing crossbands, in lacking a double groove on the frontal costa, and in possessing (most species) a small tooth along the lateral lobe margin.

BIOLOGY. The members of this genus are found on eroded or disturbed ground and occasionally along beaches (*elota*) through much of lowland Mexico and Central America. They are often found along gravelly roadsides, among Aztec and Mayan ruins, and other open sunny spots. In none of the species I have observed (*gibbosus, elota, punctatus, pellepidus, stramineus*) do solitary males crepitate as they do in *Arphia, Encoptolophus,* and *Chortophaga.* Courtship behavior is rather typical for Oedipodinae (Otte 1970), but during femur-shaking, males strike the forewings as *Arphia* males do.

REFERENCES. Hebard 1932, Otte 1970.

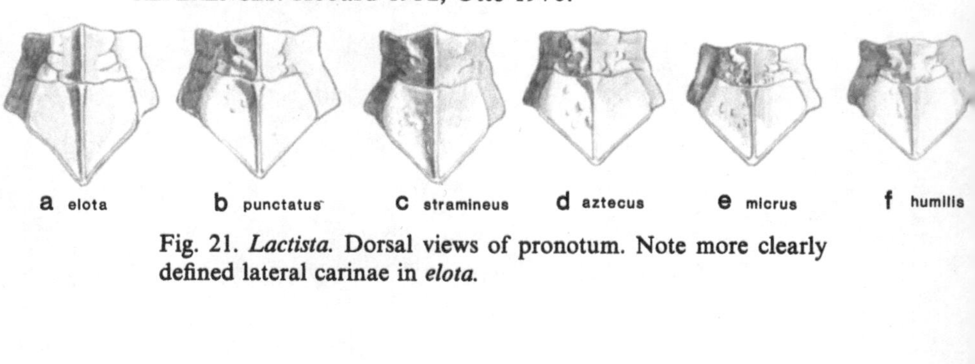

Fig. 21. *Lactista.* Dorsal views of pronotum. Note more clearly defined lateral carinae in *elota.*

■ Key to Lactista Species

1. Lower posterior margins of lateral lobes rounded (or with at most a very small projection) Fig. 23*a–d:* **2**
   Lower posterior margins of lateral lobes angulate and with distinct projection or tooth (Fig. 23*e–h*): **5**
2. Hindwings yellow: **3**
   Hindwings greenish or bluish: **4**
3. Inner faces of hind femora with one pale or ivory crossband. Forewings without crossbands. Southern Arizona, southern California, Baja California: **gibbosus**
   Inner faces of hind femora with two or three pale crossbands. Forewings usually with crossbands. Widespread: **aztecus**

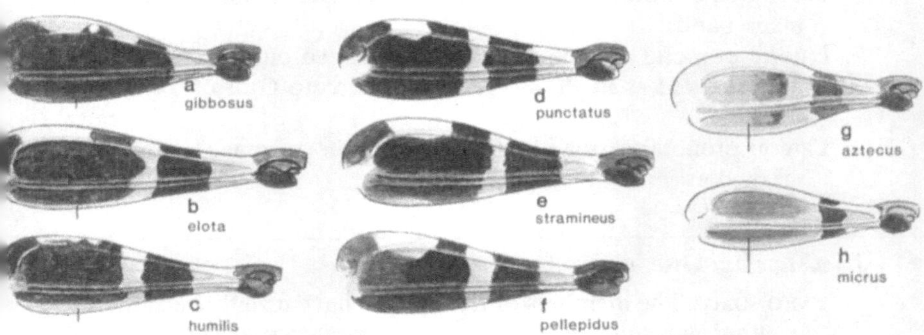

Fig. 22. *Lactista.* Inner face of hind femur.

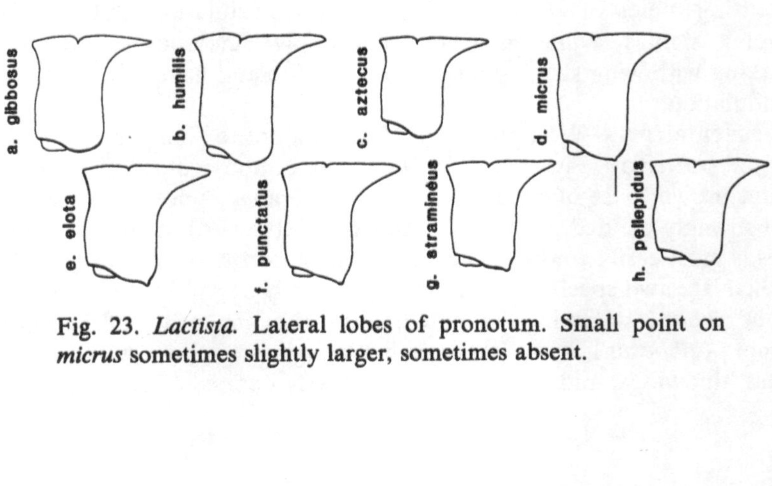

Fig. 23. *Lactista.* Lateral lobes of pronotum. Small point on *micrus* sometimes slightly larger, sometimes absent.

4. Lateral field of forewings with two distinct dark crossbands, anterior one directly above base of hind femur. Lateral lobes often slightly pointed (Fig. 23*d*): **micrus**
   Lateral field of forewings with faint dark bands or none and always without dark band above base of hind femur. Lateral lobes always rounded: **humilis**
5. Basal area of hindwings orange-red. Lateral lobes with straight lower edge (Fig. 23*f*). Eastern Mexico from Tamaulipas to Veracruz: **punctatus**
   Basal area of hindwings light orange, yellow, pale blue, or greenish. Lateral lobes with a more sinuous lower edge (Fig. 23*e, g, h*). Western Mexico to central America and Yucatán: **6**
6. Basal area of hindwings pale blue, pale green or transparent. Hind tibiae with two black bands. Chiapas and Campeche to Venezuela: **stramineus**
   Basal area of hindwings yellow or orange. Hind tibiae without two black bands: **7**
7. Lateral pronotal carinae rather well indicated on prozona. Yellow area marked as in Pl. 4. Western Mexico to Gulf of Tehuantepec: **elota**
   Lateral pronotal carinae largely obsolete on prozona. Yellow area marked as in Pl. 4. Yucatán peninsula: **pellepidus**

Lactista elota **Otte, n. sp.** Pl. 4

TAXONOMY. The members of this species have usually been identified as either *punctatus* or *oslari.* I first suspected that *punctatus* and *elota* were distinct when I found (Otte 1970) that male courtship behavior is quite different in the two species. In my 1970 paper, *elota* was called *punctatus,* and *punctatus* was called *Lactista* (near *punctatus*). Courting males of *elota* produce ordinary stridulation and femur-tipping signals, while *punctatus* males have incorporated femur-shaking with wing-striking into their courtship and have lost ordinary stridulation.

DISTRIBUTION. Western Mexico from Sonora to Chiapas.

RECOGNITION. Figs. 21*a,* 22*b,* 23*e.* Overall body color light to medium brown; base of hindwings yellow to orange; sides of forewings not strongly banded. Hind tibiae bluish to yellow or brown. This species is most easily confused with *gibbosus* in northwestern Mexico, although the two species are largely separated geographically. It differs from *gibbosus* in having two pale bands on the inner face of the hind femora (*gibbosus* has one band), in having opaque yellow, not translucent, hindwings, and in having more clearly defined lateral pronotal carinae on the prozona. It differs from *punctatus* as follows: lateral carinae of prozona more clearly defined, lower posterior corner of lateral lobes less pointed, and lower margin not as straight. Body length to end of forewings 21–26 mm in males, 26–31 mm in females.

HABITAT. Bare eroded areas such as dry hillsides, roadsides, and beaches.

LIFE CYCLE. Adults have been collected from July to October.

REFERENCES. Otte 1970.

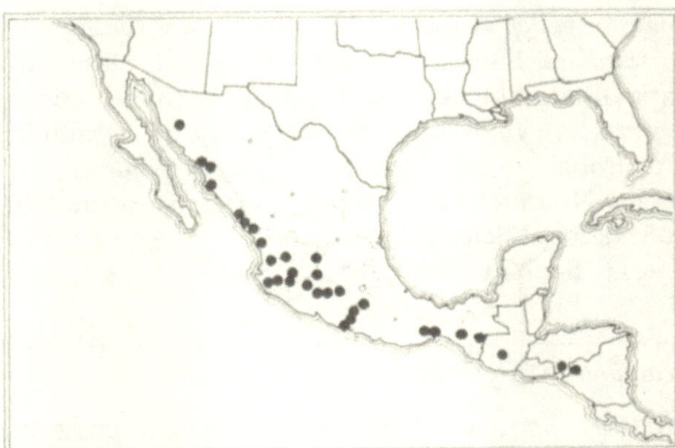

Lactista elota

Lactista punctatus (Stål) Pl. 4

DISTRIBUTION. Eastern Mexico.

RECOGNITION. Figs. 21*b*, 22*d*, 23*f*. Basal area of hindwings usually deep orange, rarely yellow. This species can be distinguished from *pellepidus* as follows: shape of colored area of hindwings is different, lateral pronotal lobe more pointed posteriorly, and lower margin nearly straight. It can be distinguished from *elota* as follows: lateral pronotal carinae more or less obsolete on the prozona, especially in the most constricted part of the disk; lateral pronotal lobe more pointed on lower posterior corner, and lower margin straighter; overall body color gray-brown, and forewings usually with three distinct crossbands. Body length to end of forewings 21–24 mm in males; 27–33 mm in females.

BEHAVIOR. Courtship and agonistic behavior of this species is quite distinct from that of *elota*. Otte (1970) incorrectly referred to this species as *Lactista* (near *punctatus*).

HABITAT. Bare patches of ground in sparsely vegetated places such as riverbanks, roadsides, and exposed hillsides.

LIFE CYCLE. Adults have been collected from June to December.
REFERENCES. Bruner 1905, Otte 1970.

Lactista pellepidus Saussure Pl. 4

DISTRIBUTION. Yucatán peninsula; two specimens bear Costa Rica labels.

RECOGNITION. Figs. 22*f*, 23*h*. Hindwings yellow in basal area. Distinguished from the very similar *stamineus* by color of hindwings and of hind tibiae; in *pellepidus* hind tibiae are blue through most of their length, in *stramineus* they are banded with black. Differs from *punctatus* in shape of lateral lobes and of colored area of hindwings. Body length to end of forewings 22–24 mm in males, 26–31 mm in females.

HABITAT. Probably open, sparsely vegetated ground.

LIFE CYCLE. No dates are given on specimens in the Philadelphia Academy of National Sciences collection.

REFERENCES. Saussure 1884, Bruner 1905.

Lactista stramineus (Erichson) Pl. 4

DISTRIBUTION. Southern Mexico and Yucatán peninsula to Colombia, Venezuela, and Netherlands Antilles.

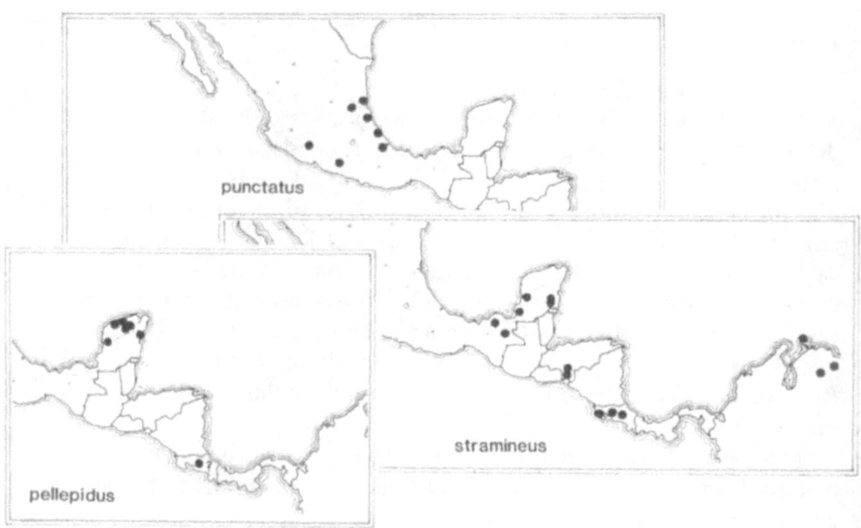

Lactista punctatus, L. pellepidus, and *L stramineus*

RECOGNITION. Figs. 21*c*, 22*e*, 23*g*. Basal area of hindwings very pale greenish to pale yellow. Forewings with two dark crossbands in first half and a less distinct one in distal half. This species differs from the most similar species, *pellepidus*, in coloration of hind tibiae, which are banded black at base (especially on inner face), then ivory, then blackish, then yellowish or pale greenish, then blackish (mostly blue in *pellepidus*). Body length to end of forewings 22–25 mm in males, 28–33 mm in females.

HABITAT. Sparsely vegetated.

LIFE CYCLE. In Costa Rica adults have been collected throughout the year.

REFERENCES. Rehn 1940.

***Lactista aztecus* (Saussure)** Pl. 4

DISTRIBUTION. Widespread over the southwestern United States and the arid regions of northern Mexico.

RECOGNITION. Figs. 21*d*, 22*g*, 23*c*. Basal area of hindwings pale to deep yellow and with a strong peripheral dark band. Overall body color is highly variable: pale gray or cream to dark gray or dark brown; occasionally strongly speckled, blue-gray, or reddish; color usually resembles pevailing ground color. Frontal ridge of males is shallowly grooved and continues onto fastigium without much interruption; that of females is more or less flat above bases of antennae. Fastigium longer than wide. Pronotum: median carina very low, sometimes only

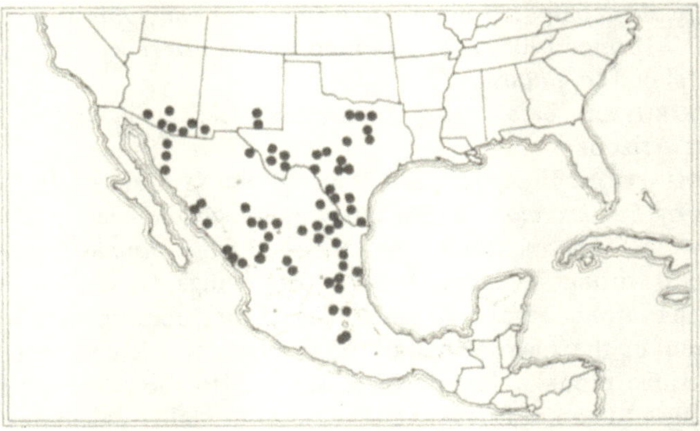

Lactista aztecus

a raised line on the metazona; disk rough on prozona, extremely wide posteriorly; posterior angle less than 90 degrees. Forewings usually with one strong crossband opposite dark band on hind femur. Hind femora with a single dark band between second and last third and with a pale band at narrowest part. Hind tibiae pale blue to greenish yellow in distal three-quarters.

HABITAT. Open ground with sparse vegetation. This species depends on camouflage for protection. It usually remains motionless until nearly touched.

LIFE CYCLE. In Arizona the species overwinters in egg or nymphal stages. Adults from March to November (Ball et al. 1942). In west Texas adults from June to October (Tinkham 1948).

REFERENCES. Hebard 1932, Ball et al. 1942, Tinkham 1948, Otte 1970.

Lactista gibbosus Saussure — Pl. 4

TAXONOMY. *Lactista oslari* Caudell is here for the first time treated as a synonym of *gibbosus.* Arizona specimens of *oslari,* including the holotype, are essentially identical to *gibbosus* from southern California and Baja California, except for a small difference in average size. The key of Strohecker, Middlekauff, and Rentz (1968) for distinguishing the two species is not useful—the hind tibiae are blue and the hindwings are translucent yellow in both nominal species. Caudell also provides no information on how the two could be distinguished. He writes: "Professor Bruner says this is near to, but distinct from, *gibbosus.* He also compared it with *pellepidus* and *punctatus* and found it different." Bruner 1905: 168, says merely that *oslari* seems to be very closely related to *gibbosus.*

DISTRIBUTION. Baja California, southern California, and eastern to southern Arizona.

RECOGNITION. Figs. 22*a,* 23*a.* This species is replaced in Sonora, Mexico, by *Lactista elota,* a previously unrecognized species, also with yellow wings. *Gibbosus* can be distinguished from *elota* as follows: lateral lobes without a tooth; base of hindwings translucent yellow (rather than opaque yellow); inner face of hind femora with a single pale band in the narrowest part (two bands in *elota*); lateral pronotal carinae more or less obsolete anterior to the principal sulcus. Body length to end of forewings 18–27 mm in males; 24–30 mm in females.

HABITAT. Dry, bare ground with scanty vegetation, often on exposed hillsides and along roadsides.

REFERENCES. Saussure 1884, Caudell 1905, Rehn and Hebard 1910, Strohecker, Middlekauff, and Rentz 1968, Otte 1970.

Lactista micrus **Hebard** Pl. 4

DISTRIBUTION. Known only from the state of Oaxaca, Mexico.

RECOGNITION. Figs. 21*e*, 22*h*, 23*d*. Hindwings faintly blue in basal two-thirds and with a brown peripheral band that does not extend along the posterior margin. Overall body color pale wood-brown. Frontal ridge large, flattened above the median ocellus, but hollowed at the upper end. Ridge between the frontal ridge and the fastigium clearly defined (not as in *aztecus*). Fastigium with two anterior and two posterior depressions. Vertex with two depressions behind fastigium. Disk of pronotum very much as in *Leuronotina orizabae* but not as rough; posterior angle about 90 degrees (less than 90 degrees in *aztecus*). Forewings with two dark crossbands, one above the base of the hind femora and the other opposite the dark bands of the hind femora. Hind femora with one strong dark band just anterior to narrowest part. Hind tibiae yellowish, brownish at distal end. Body length of female (to end of forewings) 23 mm.

HABITAT. Probably bare dry ground.

LIFE CYCLE. Adults were collected in March, June, and August.

REFERENCES. Hebard 1932.

Lactista humilis **Hebard** Pl. 4

DISTRIBUTION. States of Puebla, Guerrero, and Morelos, Mexico.

RECOGNITION. Figs. 21*f*, 22*c*, 23*b*. Basal area of hindwings is very pale translucent green or blue, occasionally colorless and has a poorly defined outer dark band. This species differs from other *Lactista* species in having the lower posterior margin of the lateral lobes rounded and in lacking a small point there. Lower, inner marginal area of hind femora is pale in the half or third that borders the lower carinula. Hind tibiae yellow with brown at the distal end. Forewings usually with one fairly distinct dark crossband at halfway point. Body length to end of forewings 18–22 mm in males; 23–28 mm in females.

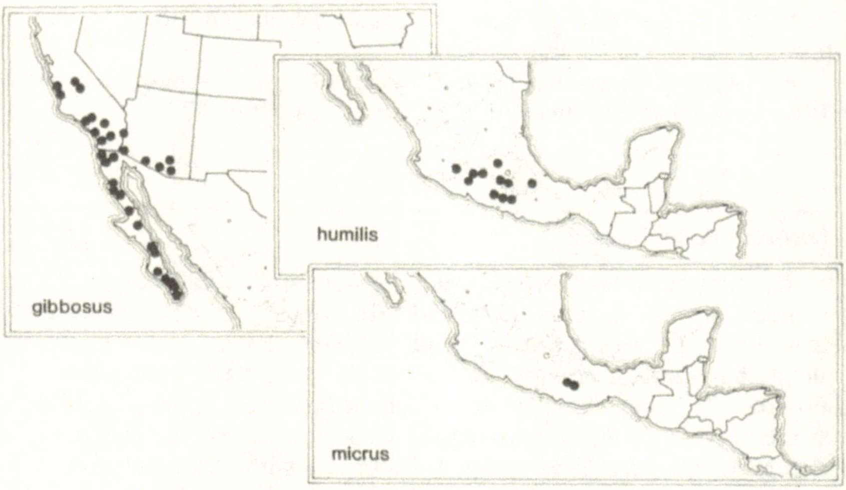

Lactista gibbosus, L. humilis, and *L. micrus*

LIFE CYCLE. Adults have been collected in May, June, September and December.

REFERENCES. Hebard 1932.

Chortophaga Genus Group

Four genera are included in the group; these are *Chortophaga, Chimarocephala, Encoptolophus,* and *Shotwellia* (Fig. 24). The first three have always been considered closely related; in his checklist Hebard placed them in the Group Arphiae. In this work I have placed *Arphia* elsewhere (see Arphia genus group). *Shotwellia* resembles several *Encoptolophus* species (*subgracilis, pallidus*) sufficiently to include the genus in this group. Also the male aedeagus of *S. isleta* is almost identical to those of *Chortophaga* and *Encoptolophus.*

In all four genera the hindwings lack dark coloration or a distinctly defined dark band; the leading edge of the hindwings has a dark thickened region; the forewings have one to three distinctive dark bands along the leading edge (sometimes indistinct in *Chortophaga viridifasciata* and *C. mendocino*); the hind tibiae are usually blue (except in *Shotwellia* and most specimens of *Chortophaga viridifasciata*). On the basis of frontal ridge shape and hind femur shape, *Chortophaga* and *Chimarocephala* make up one subgroup, and *Encoptolophus* and *Shotwellia* another.

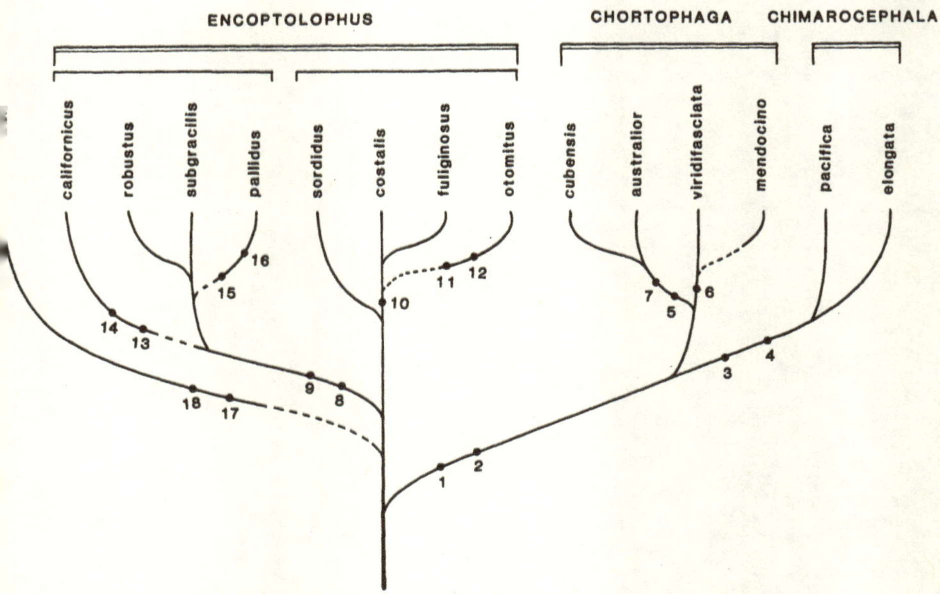

Fig. 24. Hypothetical phylogeny in *Encoptolophus, Chortophaga,* and *Chimarocephala* (see Appendix 1 for explanation).

• Identification of Genera

Chortophaga (Fig. 25; Pl. 5)

1. Fastigium strongly triangulate and flat
2. Distance between front of face and front of eyes relatively long
3. Forewings variable but not as strongly banded as in *Encoptolophus*
4. Frontal ridge shallowly grooved
5. Median carina of pronotum not deeply cut

Chimarocephala (Fig. 25; Pl. 5)

1. Fastigium elongate and deep
2. Distance between front of face and front of eyes long
3. Forewings scarcely banded
4. Frontal ridge deeply grooved
5. Median carina of pronotum cut shallowly to deeply

Encoptolophus (Figs. 28, 29, 30, 31; Pl. 6)

1. Fastigium variable but not strongly triangular; narrow or wide, shallow or deep
2. Distance between front of face and front of eyes not long
3. Forewings with two to four distinct bands
4. Frontal ridge moderately to strongly grooved

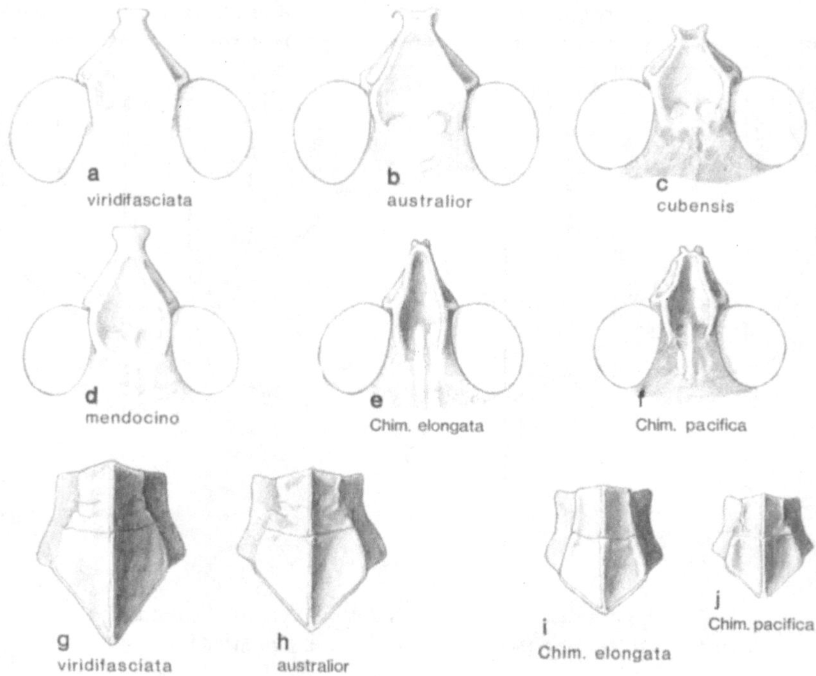

Fig. 25. *Chortophaga* and *Chimarocephala:* head and pronotum.

5. Median carina of pronotum distinctly cut when ridge is well developed

Shotwellia (Fig. 28; Pl. 6)

1. Fastigium narrow and shallow
2. Distance between front of face and front of eyes short
3. Forewings with one large and one small dark band
4. Frontal ridge shallowly grooved
5. Median ridge of pronotum obsolete

Genus CHORTOPHAGA Saussure

TAXONOMY. The genus can be divided into two species groups. The Viridifasciata group includes *viridifasciata* and *mendocino:* These species have narrower and more protruding frontal ridges, convex or flat at the level of the antennae; they lack strong dark marks on the lateral field of the forewings and on the upper marginal areas of the hind femora; and their hind tibiae are dirty yellow to dark brown. The Australior Group includes *australior* and *cubensis.* These two species have strong dark markings on the sides of the forewings and upper portions of the hind femora, a frontal ridge that is slightly concave at the level of the antennae, and bluish tibiae.

DISTRIBUTION. The genus is distributed widely over North America from Canada to Costa Rica. Only locally abundant, it is usually found in rather lush grassy fields and pastures.

RECOGNITION. Fig. 25. Members of this genus are often sexually dimorphic in color. Males are usually brownish and rarely greenish. Females are usually greenish and occasionally brownish. The frequency of brown and green morphs varies with the lushness of the habitat. In hot and drying vegetation brown morphs predominate; in green and moist vegetation green morphs are more common. The genus differs from *Chimarocephala* in the shape of the fastigium and pronotal disk. It differs from *Encoptolophus* in the shape of the fastigium and the degree of forewing banding.

REFERENCES. See references under each species.

• Identification of Chortophaga Species

viridifasciata (widespread except far west and Florida; Figs. 25*a, g,* 26)

1. Disk of pronotum strongly acute-angulate along posterior margins
2. Lateral field of forewings without strong dark marking directly above base of hind femora
3. Upper face of hind femora not strongly banded and without strong triangular dark mark in central part
4. Frontal ridge not concave at level of antennae
5. Hind tibiae dirty yellow or brown

mendocino (north central coastal California; Fig. 25*d*)

1. Posterior margins of pronotal disk form almost right angle
2. Lateral field of forewings as in *viridifasciata*
3. Upper face of hind femora as in *viridifasciata*
4. Frontal ridge not concave at level of antennae
5. Hind tibiae dark brown

australior (Florida and southern Georgia; Figs. 25*b, h,* 26)

1. Disk of pronotum with posterior margins usually forming almost right angle
2. Lateral field of forewings with strong dark marking directly above base of hind femora
3. Upper face of hind femora with three dark bands (including knee), middle band distinctly triangular
4. Frontal ridge slightly concave at level of antennae

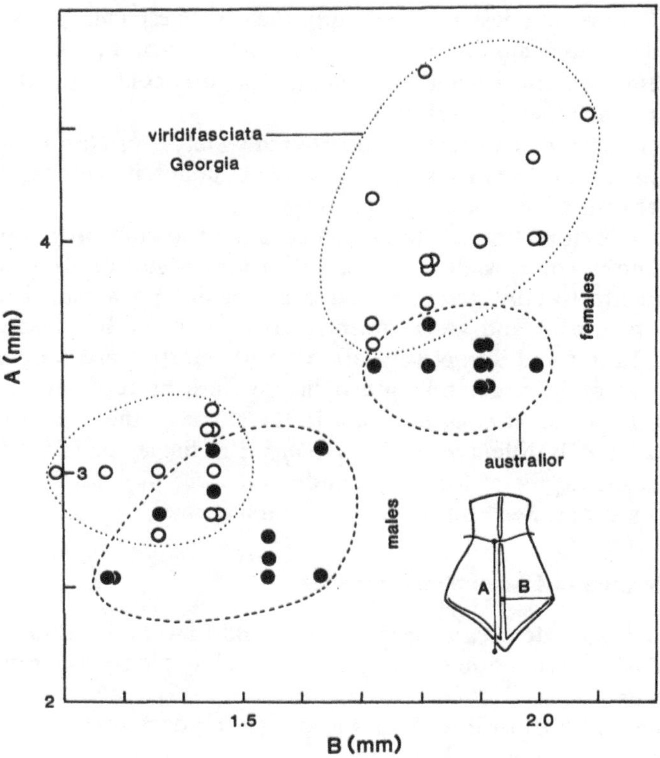

Fig. 26. *Chortophaga.* Scatter diagram showing separation of *viridifasciata* and *australior.*

5. Hind tibiae usually bluish or blue-green

cubensis (Cuba, Bahamas; Fig. 25*c*)

1. Disk of pronotum with distinct swelling beneath median carina of prozona; posterior angle as in *australior*
2. Lateral field of forewings as in *australior*
3. Upper face of hind femora as in *australior*
4. Frontal ridge slightly concave at level of antennae
5. Hind tibiae usually bluish

***Chortophaga viridifasciata* (DeGeer)** Pl. 5

TAXONOMY. Bruner's *Chortophaga meridionalis* is here treated as a synonym of *viridifasciata.* Bruner (1905) mentions differences in size, wing length, and median pronotal carina between the two species. But in each category *meridionalis* falls within the range of variation in *viridifasciata.*

DISTRIBUTION. Extremely widespread over the eastern two-thirds of North America, replaced by *australior* in southern Georgia and Florida.

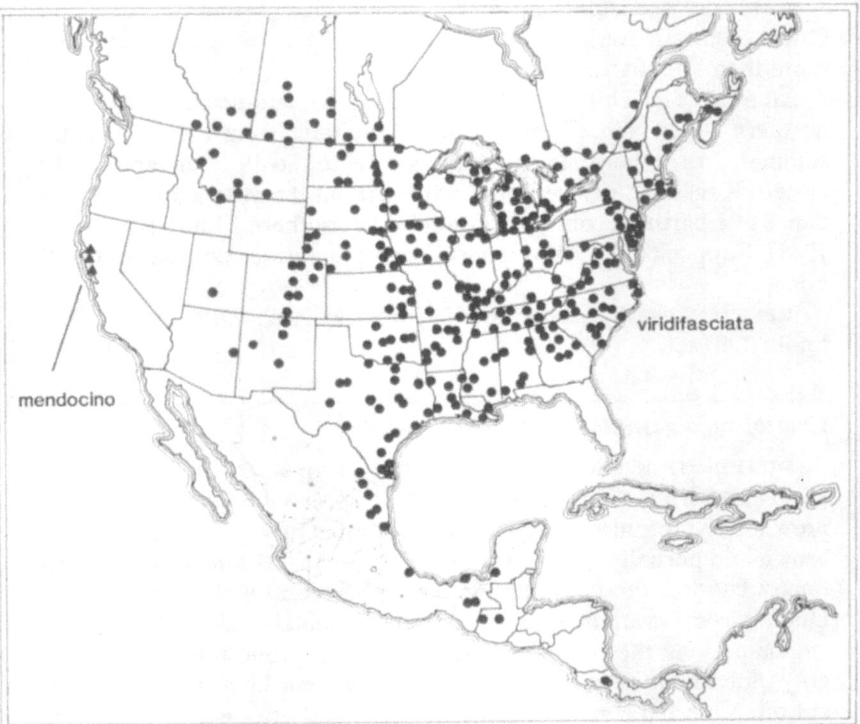

Chortophaga viridifasciata and *C. mendocino*

RECOGNITION. Figs. 25*a*, *g*, 26. Body color variable from gray-brown to bright green. Brown males are much more abundant than green, but green females are usually much more abundant than brown ones. In green pastures brown females are rare or absent (Otte 1972). Hindwings are pale yellow or greenish yellow at base, turning sooty gray-brown in central part. Posterior angle of pronotal disk is strongly acute. Disk of pronotum of brown individuals lacks the distinct X-shaped marking present in *australior*. Forewings do not have strong dark bands along the front margin. Upper marginal area of hind femora is usually not strongly banded and lacks a strong triangular dark mark in central part. Body length to end of forewings 23–30 mm in males, 28–38 mm in females.

BEHAVIOR. Pair formation and courtship were studied by Otte (1970). The principal difference between this species and *C. australior* is that the latter does not seem to stridulate while approaching females.

HABITAT. Roadsides, meadows, pastures, lawns, and moist swales. Often occurs in fresh spring growth. Usually does not live in grass more than a foot tall.

LIFE CYCLE. This species overwinters as late-instar nymphs in northern states, and adults are most abundant from spring to early summer. In southern states adults can be found in later months. In eastern Kansas adults were found throughout the year with some evidence of a partial second generation in the southern Flint Hills (Smith 1981). Coppock (1962) also reported two generations per year in Oklahoma.

REFERENCES. Blatchley 1920, Coppock 1962, Otte 1970, 1972, Smith 1981.

Chortophaga australior Rehn and Hebard — Pl. 5

DISTRIBUTION. Florida and southern Georgia.

RECOGNITION. Figs. 25*b, h,* 26. This species also displays green-brown polymorphism. Males are predominantly gray-brown; both brown and partially green females are abundant. Hindwings are indistinctly banded, the base almost clear, midsection with a dusky area. Differs from *viridifasciata* in that the pronotal disk is less acute-angulate along the posterior margin. Disk of pronotum in brown or gray individuals always has a pale X-shaped marking. Posterior margin often has dark markings. Median pronotal crest not as high as in *viridifasciata.* Forewings (folded position) almost always with a distinct dark mark along the lower edge directly above the basal part of the hind femora. Leading edge of forewings usually has three or four distinct dark bands. Upper marginal area of hind femora usually has three dark bands (including knee band) in the distal two-thirds; band nearest the middle is triangular. Body length to end of forewings 21–27 mm in males, 29–37 mm in females.

BEHAVIOR. Flight crepitation is similar to that of *viridifasciata.* Both males and females crepitate. Pair formation is assumed to be similar (Otte 1970). As males approach females during courtship, they tip their femora, but they do not stridulate, unlike *viridifasciata.*

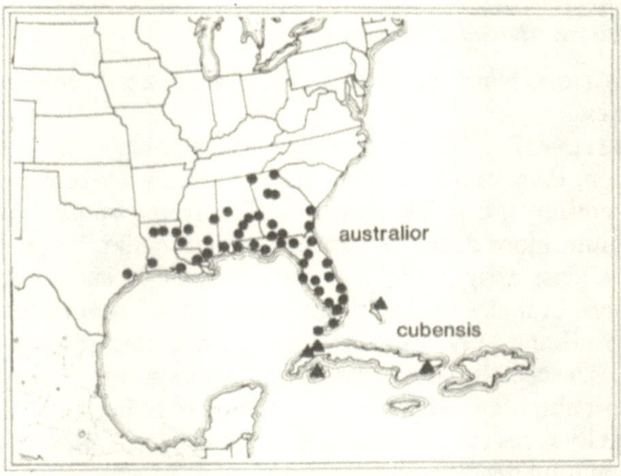

Chortophaga australior and *C. cubensis*

HABITAT. Short-grass fields, along roads, in uncultivated spots around towns, and in sandy fields overgrown with short grasses.

LIFE CYCLE. Adults are present throughout the year.

REFERENCES. Rehn and Hebard 1910, Otte 1970.

Chortophaga cubensis (Scudder) **Pl. 5**

DISTRIBUTION. Cuba, Isle of Pines, and Bahamas.

RECOGNITION. Fig. 25*c*. Superficially similar to *australior*, but the hindwings are entirely clear, the prozona has a distinct swelling beneath the median carina, and the fastigium is shorter. All males in the Philadelphia Academy of Natural Sciences collection are of the brown phase, and most females are of the green phase. As in *australior*, brown-phase individuals have distinct pale X-shaped markings on the pronotal disk. Brown- and green-phase individuals of both sexes have a strong triangular dark marking on the dorsum of the hind femora. Body length to end of hindwings 20–24 mm in males, 26–28.5 mm in females.

HABITAT. Not reported in literature; probably similar to that of *australior*.

LIFE CYCLE. Adults of Bahamas specimens were collected in January; Cuba adults, September and January. Isle of Pines specimens were collected in August.

REFERENCES. Scudder 1875c, 1876a, Saussure 1884.

Chortophaga mendocino Rentz — Pl. 5

DISTRIBUTION. North coast of California in Mendocino and Sonoma counties.

RECOGNITION. Fig. 25*d.* Disk of hindwings pale greenish yellow in basal portion, then smoky brown. Males are all gray-brown, females are almost entirely green. This species differs from *viridifasciata* as follows: fastigium more deeply hollowed; posterior margins of pronotal disk form almost a right angle; body length of to end of forewings 18.5–20.0 mm in males; to end of abdomen, 25–28 mm in females.

HABITAT. Rentz (1977) notes that *C. mendocino* prefers exposed grassy slopes, especially those with a southern exposure. The grasshoppers are most abundant where there is a mixture of herbaceous vegetation, especially forbs such as plantain (*Plantago* spp.), filaree (*Erodium* spp.), and buttercups (*Ranunculus* spp.).

LIFE CYCLE. Adults from February to May.

REFERENCES. Rentz 1977.

Genus CHIMAROCEPHALA Scudder

RANGE. The genus is known only from California.

RECOGNITION. Fig. 25*e, f, i, j.* Males are small and dark, females often pale green. Fastigium elongate and deep. Distance between front of face and front of eyes is usually long. The frontal ridge is deeply grooved. Forewings are scarcely banded.

REFERENCES. Hebard 1931b, Strohecker, Middlekauff, and Rentz 1968, Rentz 1977.

Chimarocephala pacifica (Thomas) — Pl. 5

TAXONOMY. Coastal and northern populations have usually been called *C. pacifica pacifica,* while southern populations have usually been called *C. pacifica incisa.* The configuration of the median crest distinguishes the two forms. I have not used the subspecies category, because intermediate conditions make it difficult to place many specimens in one form or the other.

DISTRIBUTION. Widespread in the southern two-thirds of California.

RECOGNITION. Fig. 25*f, j.* Males: small, dark brown or gray-brown, sometimes blackish. Disk of hindwings is mostly clear, but faintly greenish yellow at base and turning smoky at apex. Frontal ridge is deeply grooved. Fastigium is deeply grooved and has a median carinula at posterior end. Median pronotal crest cut or not cut by principal sulcus. Lateral pronotal carinae are well developed. In some populations, usually the more northern, they are continuous from front to rear; in more southern populations they consist of two ridges that do not quite meet near the principal sulcus. Upper face of hind femora has an oblique dark band near the middle. Body length to end of forewings 15.5–20.0 mm.

Females: Body green to gray-brown, either mostly green or mostly brown. Occasionally green with brown forewings; rarely with varying amounts of red pigmentation on body. Forewings vary from being shorter than the hind femora to considerably longer. Most females in the Philadelphia Academy collection, from the northern part of the range, are green and with forewings not extending beyond ends of hind femora. In San Diego County most females are gray-brown with wings extending beyond the ends of the hind femora. Female hind leg markings are similar to those of males, but the oblique dark band is often missing in green females. Greatest body length is 20–29 mm.

HABITAT. Mountain meadows, roadsides, hillsides, and open woodlands.

LIFE CYCLE. Adults are more common in the spring. At higher altitudes adults have been collected in July.

REFERENCES. Scudder 1875a, Caudell 1906, Rehn and Hebard 1909, Hebard 1931b, Strohecker, Middlekauff, and Rentz 1968. Thomas 1973b.

Chimarocephala elongata **Rentz** Pl. 5

DISTRIBUTION. Central California, in the Central Valley and its periphery.

RECOGNITION. Fig. 25*e, i.* Disk of hindwings is smoky, translucent, basal portion very weakly yellow. Body gray-brown or green to dark brown. Forewings often with a pale streak along the Cu_2 and 1A veins. Upper face of hind femora with weak oblique dark band that does not cross the upper marginal area. Lateral pronotal carinae are distinct and continuous from front to back margin and are cut by principal sulcus. Hind tibiae are gray to pale brown. Female forewings extend beyond ends of femora and abdomen. Body length to end of forewings 21–22 mm in males, about 28 mm in females.

HABITAT. Rentz (1977) reports that the species is restricted to alkaline areas bordering small streams, where it lives among salt grass (*Distichlis spicata*).

LIFE CYCLE. Adults from March to May.

REFERENCES. Rentz 1977.

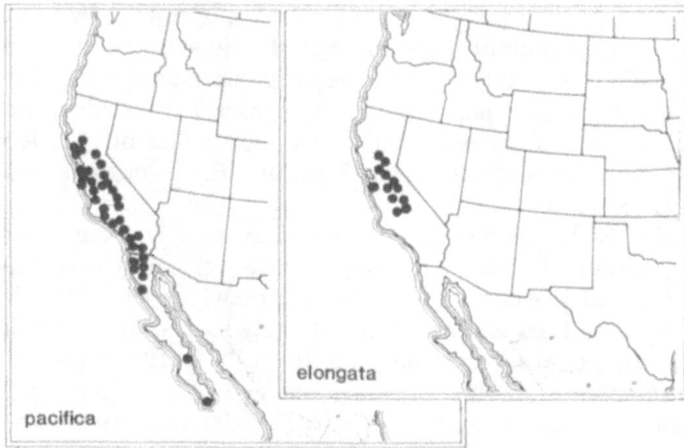

Chimarocephala pacifica and *C. elongata*

Genus ENCOPTOLOPHUS Scudder

RANGE. Widespread from southern and eastern Canada to central Mexico.

TAXONOMY. Several species groups can be recognized; one includes *sordidus, costalis, fuliginosus,* and *otomitus;* another includes *subgracilis, pallidus,* and *robustus;* a third contains only *californicus.*

In this work *californicus* Bruner is moved to this genus from *Chimarocephala,* and *costalis* is treated as a separate species rather than as a subspecies of *sordidus.* Studies of mating behavior in *sordidus* and *costalis* establish that these are two different species (Otte 1970). I have been unable to define the subspecies *E. subgracilis texensis* either morphologically or geographically and have therefore not treated it as a distinct entity.

RECOGNITION. Figs. 24, 27–30. Hindwings without strong bands. Forewings with two to four dark crossbands (sometimes absent in very pale *E. pallidus*). Fastigium somewhat triangular, usually with a small median ridge in posterior part. Frontal ridge grooved. Median pronotal ridge variable; distinctly cut when it is well developed. Hindwings without a radial sector vein. Region of hindwings between R and front margin often darkened and bordered by thickened veins (R, Sc, and C) (not so pronounced in *pallidus* and *californicus*).

BIOLOGY. habitat preference is quite variable within the genus—from thickly grassed and weedy old fields in the northeast (*sordidus*), to short-grass prairies (*costalis*), thorn scrub and desert grasslands

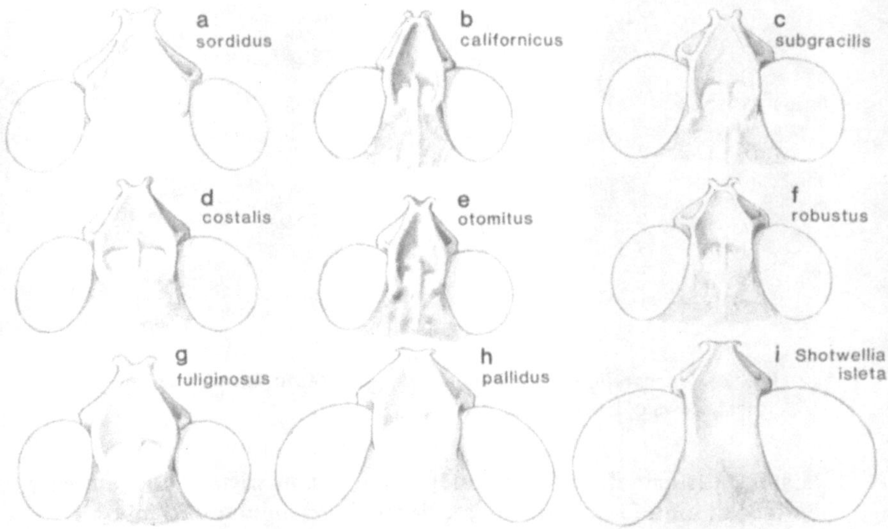

Fig. 27. *Encoptolophus* and *Shotwellia* fastigium shapes.

(*subgracilis*), and (*pallidus*). Mating behavior is known in only two species: in both *costalis* and *sordidus* solitary males crepitate frequently, and judging from similarities in wing venation, *otomitus* and *fuliginosus* probably behave similarly. The wing venation of *robustus, subgracilis, pallidus,* and *californicus* is less specialized (Fig. 29*d, e, f*). *Subgracilis* does not crepitate spontaneously, and it is likely, judging from the hindwings, that none of the four latter species crepitates.

REFERENCES. Scudder 1875, Bruner 1905.

▪ Key to Encoptolophus Species

CALIFORNIA AND BAJA CALIFORNIA

1. Overall body color extremely pale, often straw-yellow or sandy yellow. Median pronotal crest very low, lateral carinae very poorly defined. Deserts of southern California and Nevada: **pallidus**

   Overall body color gray-brown or brownish. Distinct median pronotal crest and more clearly indicated lateral carinae: **2**

2. Lateral carinae of pronotum continuous and uninterrupted in region of sulci (Fig. 28*c*). Upper face of hind femora with small dark marking or none at all. Fastigium deeply hollowed (Fig. 27*b*): **californicus**

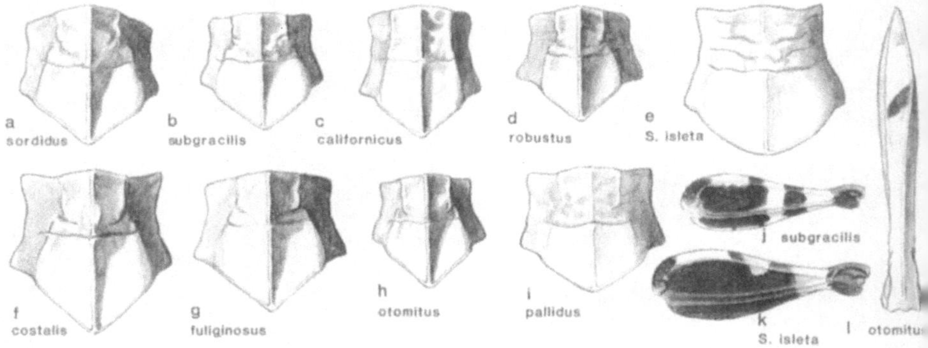

Fig. 28. *Encoptolophus* and *Shotwellia.* Pronotum, inner face and dorsal views of hind femur.

Lateral carinae of pronotum interrupted in constricted zone. Upper face of hind femora has a large, usually triangular dark marking. Fastigium not as above: **3**

3. Forewings barely extend beyond ends of hind femora. Head about as broad as pronotum (top view). Lateral pronotal carinae distinct (Fig. 28*d*). Extreme southwestern California and Baja California: **robustus**

Forewings extend about one head width or more beyond hind femora. Head appears distinctly narrower than thorax (top view). Lateral pronotal carinae very indistinct, especially on metazona (Fig. 28*b*): **subgracilis**

MEXICO

1. Forewing venation unspecialized (Fig. 29). Greatest head width usually less than 1.20 times greatest width of pronotal disk. Forewings extend one head width or more beyond hind femora. Posterior angle of pronotal disk more than 90 degrees and middle dark band on upper face of hind femora is triangular: **subgracilis**

Forewing venation specialized (Fig. 29). Greatest head width more than 1.20 times greatest width of pronotal disk. Forewings usually extend less than one head width beyond hind femora: **2**

2. Upper face of hind femora with small oblique dark band near center (Fig. 28*l*). Fastigium deeply hollowed and narrower (Fig. 27*e*): **otomitus**

Upper face of hind femora either with triangular band near center or without dark band. Fastigium shallow and broader (Fig. 27*d, g*): **3**

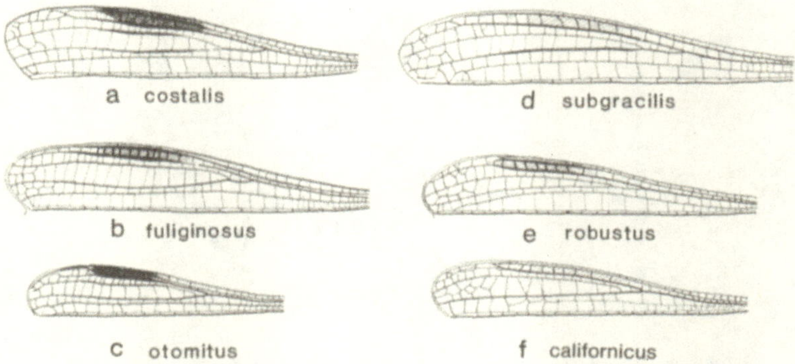

Fig. 29. *Encoptolophus.* Anterior fold of hindwings, showing enlarged cells in species on left.

3. Forewing venation as in Fig. 29*a*. Pronotal disk usually with pale X-shaped marking (except in green and pale brown specimens). Pronotal disk usually with dark markings on metazona: **costalis**

   Forewing venation as in Fig 29*b*. Pronotal disk without pale X-shaped marking and without markings on metazona: **fuliginosus**

UNITED STATES (EXCEPT CALIFORNIA)

1. Fastigium wider than long. Hind tibiae mostly brown to blackish. Pronotal disk often with short black marking on metazona. Northeastern quarter of United States and Ontario: **sordidus**

   Fastigium longer than wide. Hind tibiae bluish or gray. Pronotal disk not as above: **2**

2. Posterior angle of pronotal disk 90 degrees or less. Pronotal disk usually with dark triangles or dark streaks on metazona. Forewings extend less than one head width beyond hind femora. Hindwings with enlarged cells in first fold (Fig. 29*a*): **costalis**

   Posterior angle of pronotal disk more than 90 degrees. Pronotal disk without dark markings. Forewings usually extend one head width or more beyond hind femora. Hindwings without enlarged cells in first fold (Fig. 29*d*): **subgracilis**

Encoptolophus sordidus **(Burmeister)** Pl. 6

DISTRIBUTION. Northeastern United States and Ontario from Kansas and Nebraska to New England.

RECOGNITION. Figs. 27*a*, 28*a*. Disk of hindwings largely transparent, but with a faint touch of yellow-green near the base, becoming

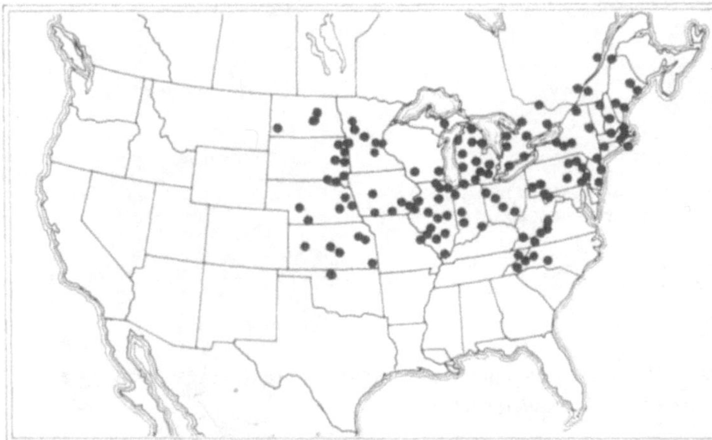

Encoptolophus sordidus

smoky brown or gray-brown in distal third or half. Body dark gray-brown with distinct black markings on side. Fastigium wider than long and with a low median carinula in posterior half (longer than wide in other United States species). Pronotum: disk often with a pale X-shaped marking, and posterior margin with a row of black spots (X-shaped marking is present also in *costalis*). Lateral lobe with a large shiny, blackish area located mainly on the prozona. Forewings: banded on the lateral field, usually with two larger dark bands, one directly above the base of the hind femora, the other above the middle section of the hind femora; posterior third largely blackish. Hind femora: upper face usually with four bands (including knees), those nearest the middle being triangular and black; outer face indistinctly banded, but knee and band preceding it rather distinct. Hind tibiae: dark brown in distal two-thirds (blue in *costalis*), with a distinct pale ring separating the first and second thirds, dark brown to blackish at the base. Venter of abdomen wood-brown with transverse black bands; subgenital plate shiny black. Body length to end of forewings 22–28 mm in males, 27–33 mm in females.

BEHAVIOR. Solitary adult males make crepitation flights, often landing on vegetation, where they remain until the next flight. Courting males produce three acoustical signals: (1) kicking the ends of the forewings with the hind tibiae, (2) vibrating the hind femora against the forewings, and (3) producing short bursts of vibration by rubbing the hind femora against the forewings.

HABITAT. According to Hebard (1925), in South Dakota "it appears late in the season and, unlike *E. costalis,* prefers waste weedy areas rather than prairie or plains conditions." Weedy fields, heavily overgrown pastures and woods margins, often in areas with knee-high grasses and weeds. Blatchley (1920) writes: "It frequents only dry upland timothy and clover meadows, blue-grass pastures, roadsides, etc. When living in woodland pastures it frequents sunny spots, seldom alighting in the shade when flushed."

LIFE CYCLE. Late summer and fall adults; egg-overwintering. In Kansas Smith (1981) recorded nymphs from June through August and adults from July through October.

REFERENCES. Blatchley 1903, 1920, Hebard 1925b, Otte 1970.

Encoptolophus costalis Scudder — Pl. 6

TAXONOMY. *E. costalis* has long been considered a subspecies of *sordidus.* There are three reasons for believing that it is distinct: (1) courtship behavior in the two forms is quite different (Otte 1970); (2) the two forms are sympatric in some places (for example, Lane County, Kansas), suggesting that they do not interbreed; and (3) they have different habitats (Hebard 1925b).

DISTRIBUTION. Rocky Mountain and Great Plains states, from central Mexico to western Canada.

RECOGNITION. Figs. 27*d,* 28*f,* 29*a,* 30. Very similar to *sordidus* but differing as follows: fastigium longer than wide in males; median pronotal carina lower; pronotal disk patterns different—usually with a more distinct X-shaped marking, and metazona with triangular dark markings or more linear black stripes; dorsal field of forewings very often with pale lines running along the angle and converging posteriorly; sometimes entire dorsal field is pale (never pale in *sordidus*); hind tibiae bright blue in distal two-thirds. Venter of abdomen variable: sometimes straw-yellow, sometimes straw-yellow banded with black. In central Mexico this species is most similar to *fuliginosus.* It differs from that species as follows: disk of pronotum usually has pale X-shaped marking and usually has posterior FDI or traces of them; posterior width of pronotal disk 1.63–2.00 times the narrowest width; front of pronotal disk slightly wider than region immediately posterior.

Some females are largely green; these retain the dark markings on the lateral field of the forewings but lack dark pigmentation on the pronotal disk and on the upper face of the hind femora. Venter of abdomen does not have dark bands. Body length to end of forewings 16–24 mm in males; 22–31 mm in females.

BEHAVIOR. Male courtship signaling is quite distinct from that of *sordidus*. Males vibrate the hind femora against the forewings and produce a chirping stridulation when the hind femora are lowered and scraped against the forewings (Otte 1970).

HABITAT. Prairies, open grasslands and, in Texas, open savanna woodlands. Ball et al. (1942) report: "A common species in the grasslands of Arizona from Douglas to Ashfork and Lupton. Upper and Lower Sonoran zones. Feeds to a large extent on range grasses but is also important at times in cultivated areas, especially in alfalfa fields." In Mexico I collected this species on a lightly wooded hillside.

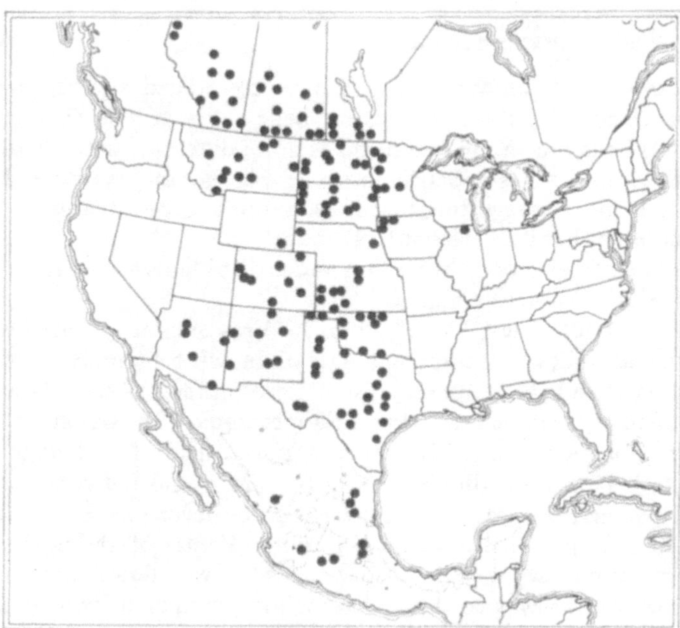

Encoptolophus costalis

LIFE CYCLE. Late summer and fall adults in the north. In Texas Isely (1937) reported that adults and nymphs could be found in all seasons, with a peak in abundance from September to December. In central Mexico adults have been collected from April to September.

REFERENCES. Scudder 1862, Somes 1914, Hebard 1925b, 1934a, 1934b, Ball et al. 1942, Otte 1970.

Encoptolophus fuliginosus Bruner Pl. 6

DISTRIBUTION. Known from Chihuahua to Oaxaca, Mexico.

RECOGNITION. Figs. 27*g*, 28*g*, 29*b*, 30. Hindwings largely transparent but very slightly yellowish at extreme base and very slightly smoky in distal third. In the Philadelphia Academy of Natural Sciences series, the following color forms are represented: (1) face and side of body blackish, dorsum brown (males); (2) like 1, but dorsum green (males); (3) side of body speckled brown and black, dorsum brownish (female); (4) face and side of body greenish, with blackish marks on metazona of lateral lobes, and dorsum green. This species differs from *costalis* as follows: front of pronotal disk is narrower than region immediately posterior to it; posterior width of disk is 1.40–1.73 times the narrowest width; disk lacks a pale X-shaped marking and FDI. In all color forms the hind femora are somewhat banded on outer face and upper marginal area; there is no distinct triangular dark mark in middle section of the latter. It differs from *subgracilis* in having relatively short forewings extending not more than one head width beyond the hind femora. Body length to end of forewings 20–23 mm in males, 22–28 mm in females.

HABITAT. Not known.

LIFE CYCLE. Adults in Philadelphia Academy collection were collected from August to October.

REFERENCES. Bruner 1905.

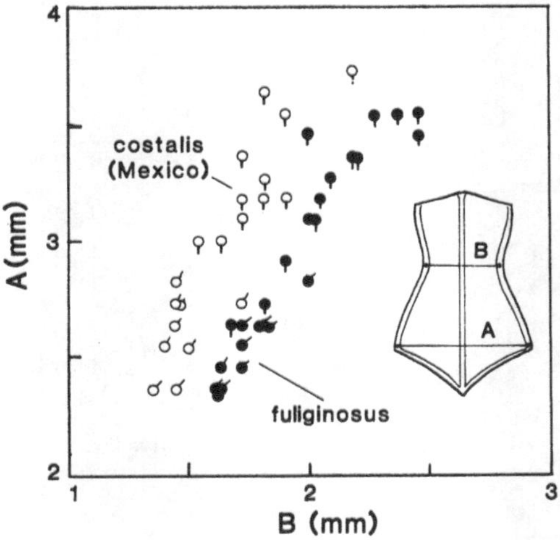

Fig. 30. *Encoptolophus.* Scatter diagram showing separation of *costalis* and *fuliginosus.*

***Encoptolophus otomitus* (Saussure)** Pl. 6

DISTRIBUTION. Central Mexico.

RECOGNITION. Figs. 27*e*, 28*h*, 29*c*. Hindwings entirely transparent. Overall body color light grayish brown or green, with distinct dark areas on lateral lobes and lateral field of forewings. All males in the Philadelphia Academy of Natural Sciences collection are gray-brown. Females are either brown or green. Green-phase females have lateral lobes unmarked or with distinct black markings. Head: fastigium is more deeply hollowed and narrower than in *fuliginosus*. Pronotum: disk lacks a distinct X-shaped marking (present in *costalis*). Disk lacks FDI (usually present in *costalis*). Median carina is well developed, lateral carinae are interrupted in constricted part; lateral lobes often have a distinctive dark area, but these markings are faded or absent in some individuals. Hind femora have an oblique dark mark on upper marginal area (absent in *costalis* and *fuliginosus*). Wings extend less than one head width beyond end of hind femora (more than one head width in *subgracilis*). Dorsal field of forewings is usually lighter than lateral field and sometimes with narrow pale lines along the wing angle. Hind tibiae bluish, yellowish, or brownish. Body length to end of wings 15.0–18.5 mm in males, 21–23 mm in females.

LIFE CYCLE. Adults have been collected in March, May, and November.

REFERENCES. Hebard 1932.

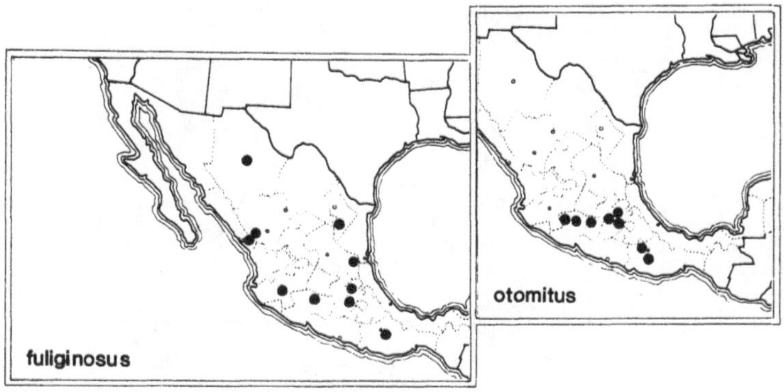

Encoptolophus fuliginosus and *E. otomitus*

COLOR PLATES

PLATE 1

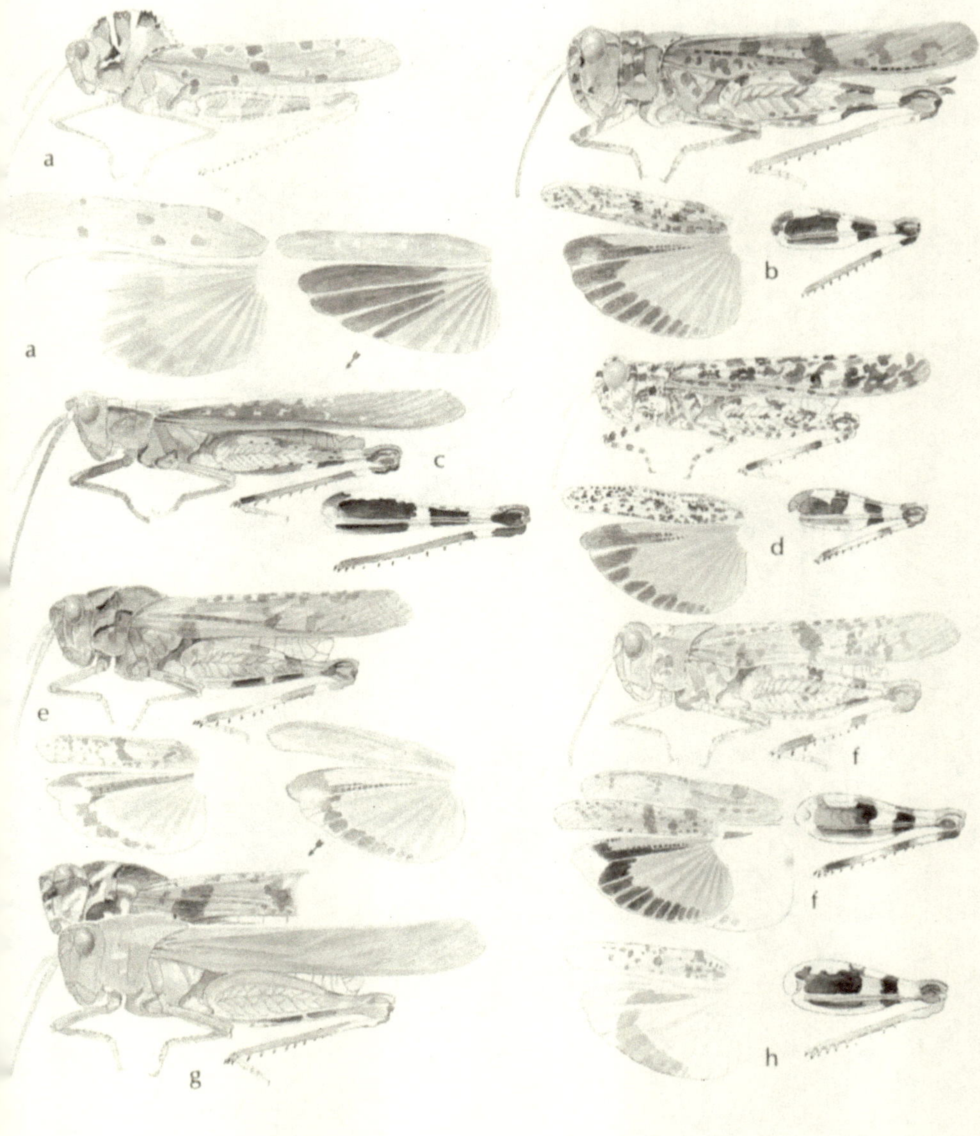

a. *Tropidolophus formosus;* b. *Leuronotina philorites;* c. *Machaerocera mexicana;* d. *Leuronotina ritensis;* e. *Tomonotus mexicanus;* f. *Leuronotina orizabae;* g. *Tomonotus ferruginosus;* h. *Leuronotina obesa*

PLATE 2

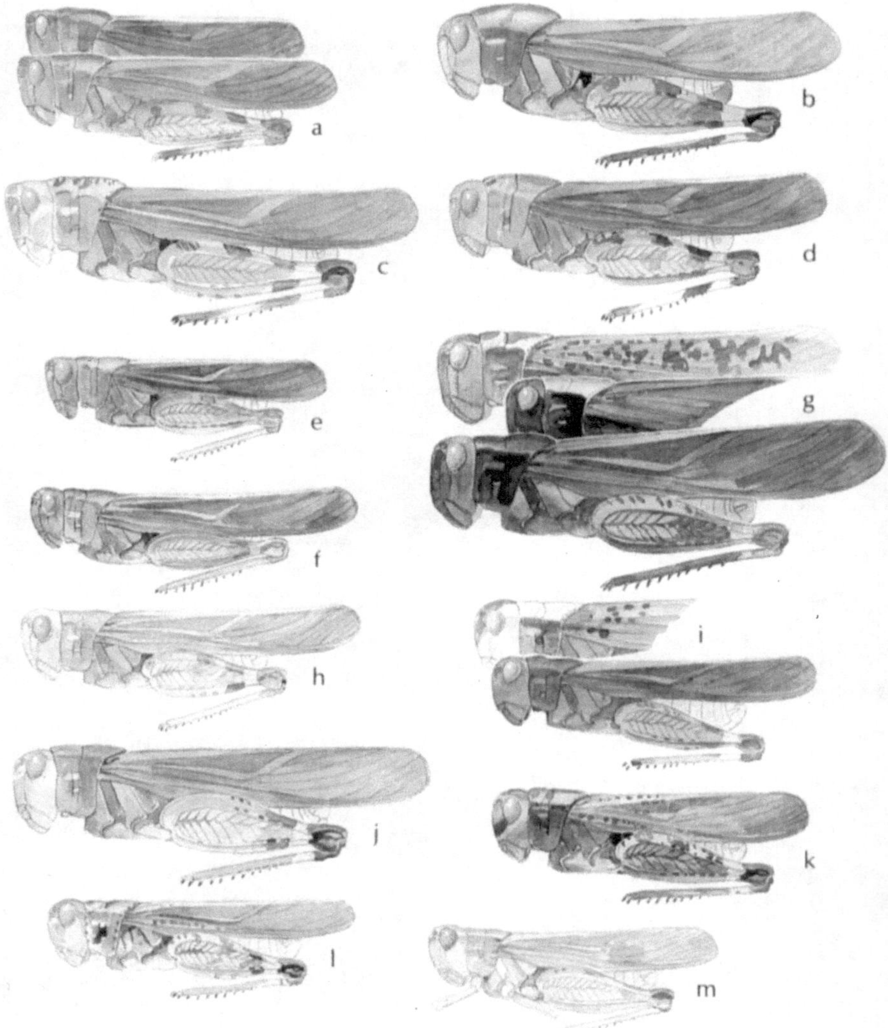

a. *Arphia sulphurea;* b. *Arphia xanthoptera;* c. *Arphia simplex;* d. *Arphia granulata;* e. *Arphia behrensi;* f. *Arphia ramona;* g. *Arphia pseudonietana;* h. *Arphia conspersa;* i. *Arphia saussureana;* j. *Arphia pecos;* k. *Arphia nietana;* l. *Arphia townsendi;* m. *Arphia secreta*

PLATE 3

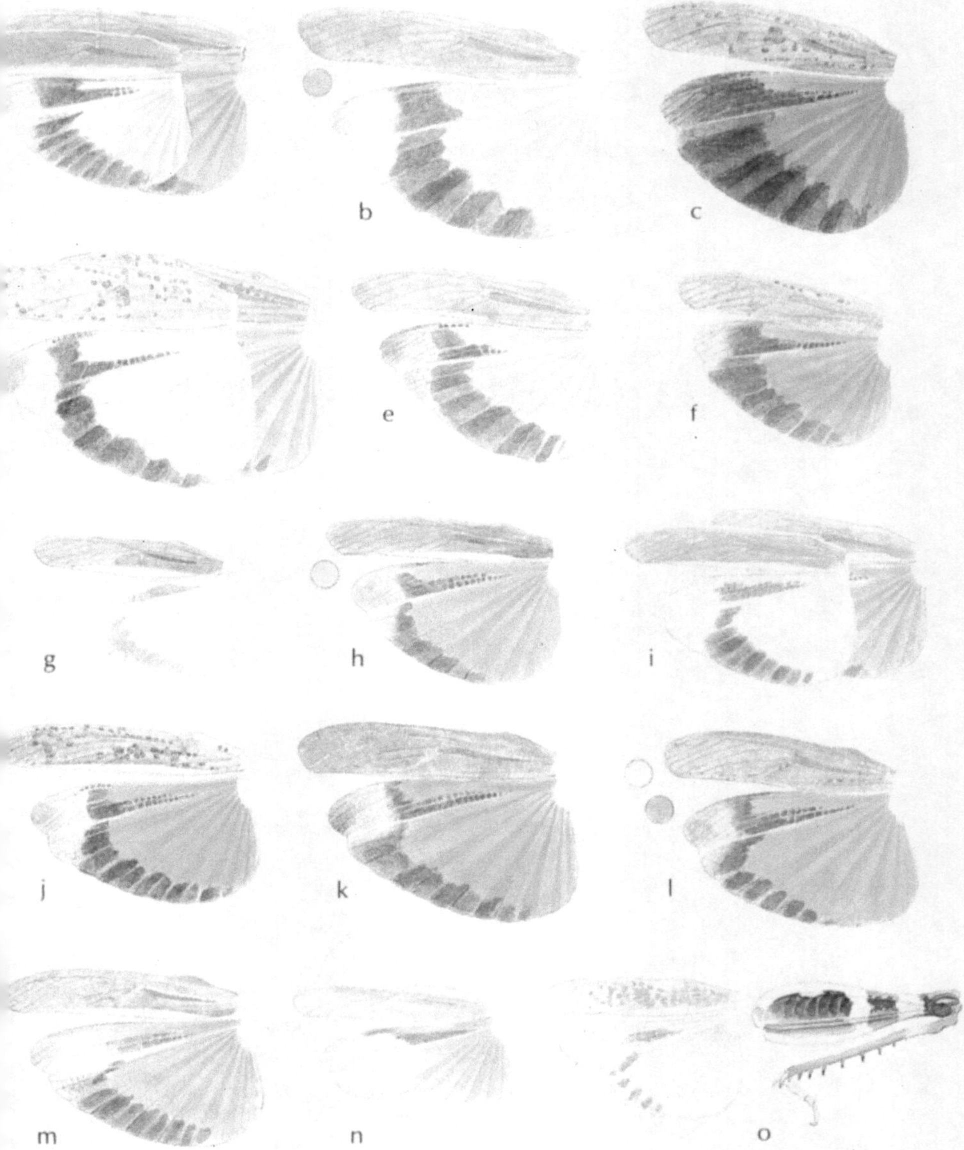

a. *Arphia sulphurea;* b. *Arphia xanthoptera;* c. *Arphia pseudonietana;* d. *Arphia simplex;* e. *Arphia granulata;* f. *Arphia saussureana;* g. *Arphia townsendi;* h. *Arphia ramona;* i. *Arphia conspersa;* j. *Arphia novissima;* k. *Arphia pecos;* l. *Arphia nietana;* m. *Arphia pulchripennis;* n. *Arphia secreta;* o. *Arphia robertsi*

PLATE 4

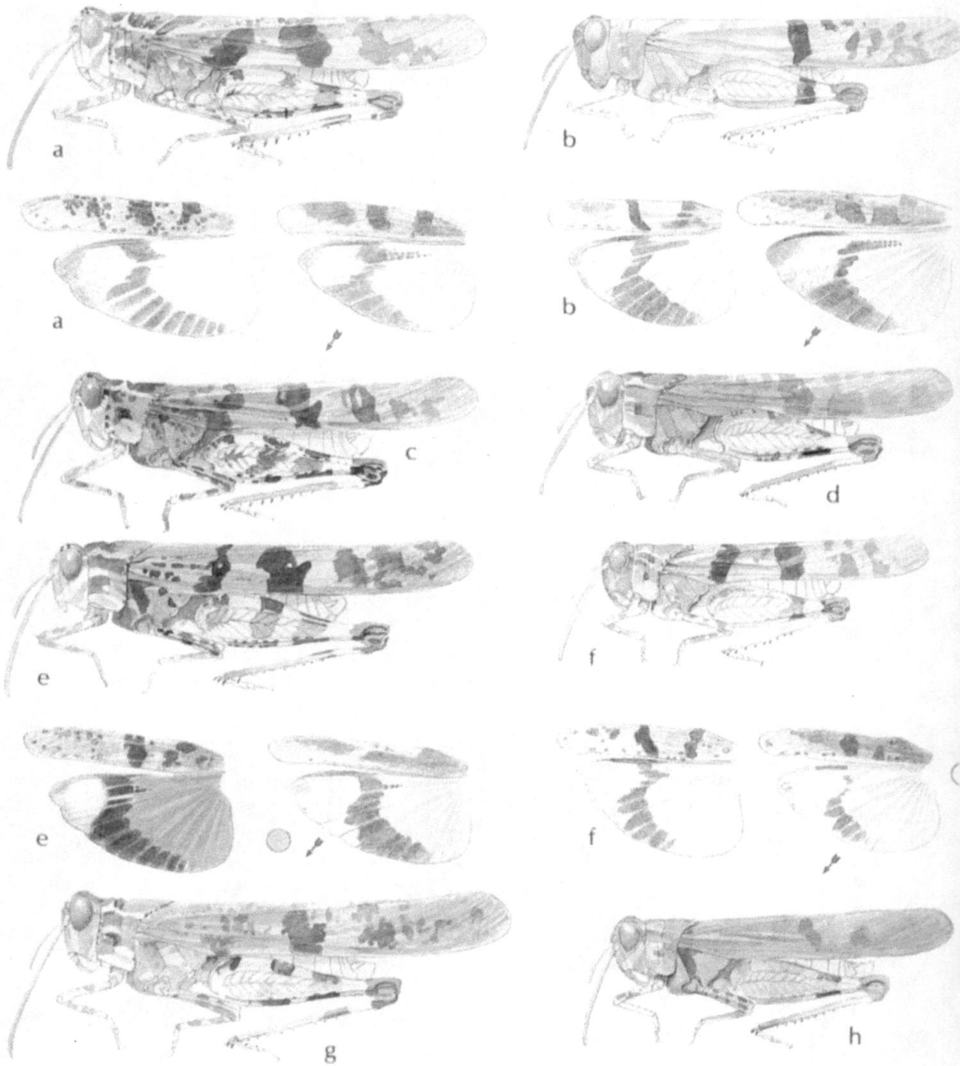

a. *Lactista stramineus;* b. *Lactista aztecus;* c. *Lactista pellepidus;* d. *Lactista gibbosus;* e. *Lactista punctatus;* f. *Lactista micrus;* g. *Lactista elota;* h. *Lactista humilis*

PLATE 5

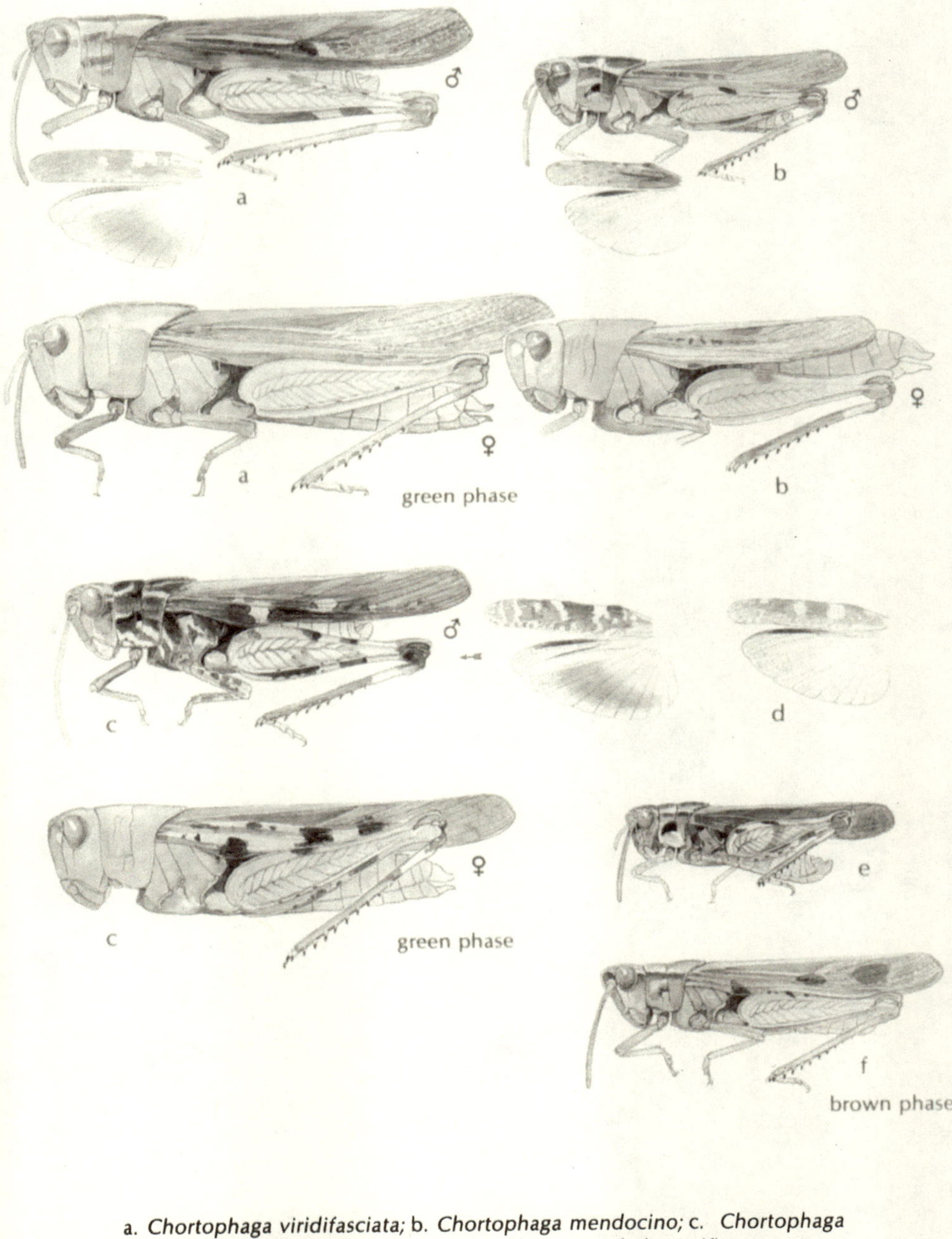

a. *Chortophaga viridifasciata;* b. *Chortophaga mendocino;* c. *Chortophaga australior;* d. *Chortophaga cubensis;* e. *Chimarocephala pacifica;* f. *Chimarocephala elongata,* brown phase.

PLATE 6

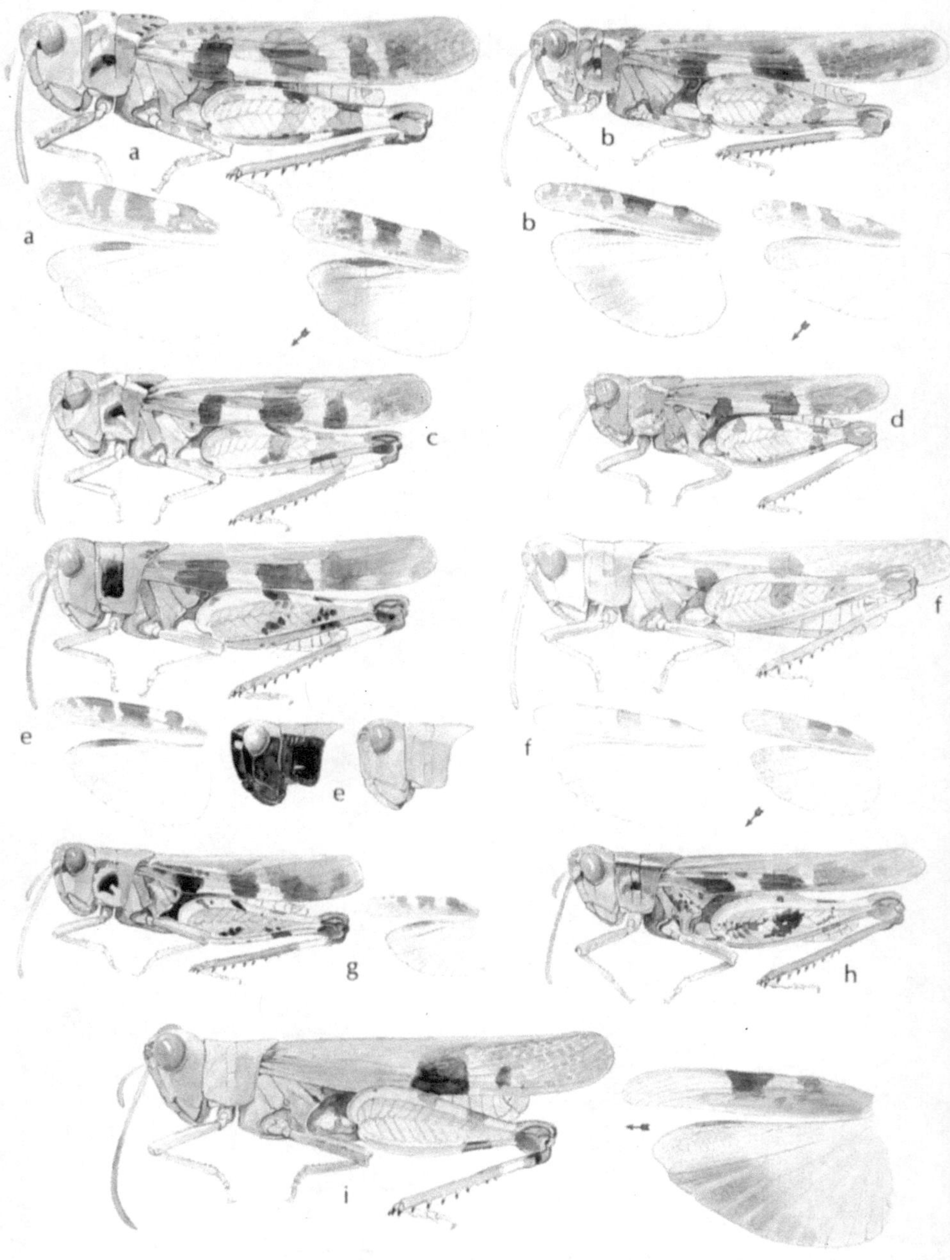

a. *Encoptolophus sordidus;* b. *Encoptolophus subgracilis;* c. *Encoptolophus costalis;* d. *Encoptolophus robustus;* e. *Encoptolophus fuliginosus;* f. *Encoptolophus pallidus;* g. *Encoptolophus otomitus;* h. *Encoptolophus californicus;* i. *Shotwellia isleta*

PLATE 7

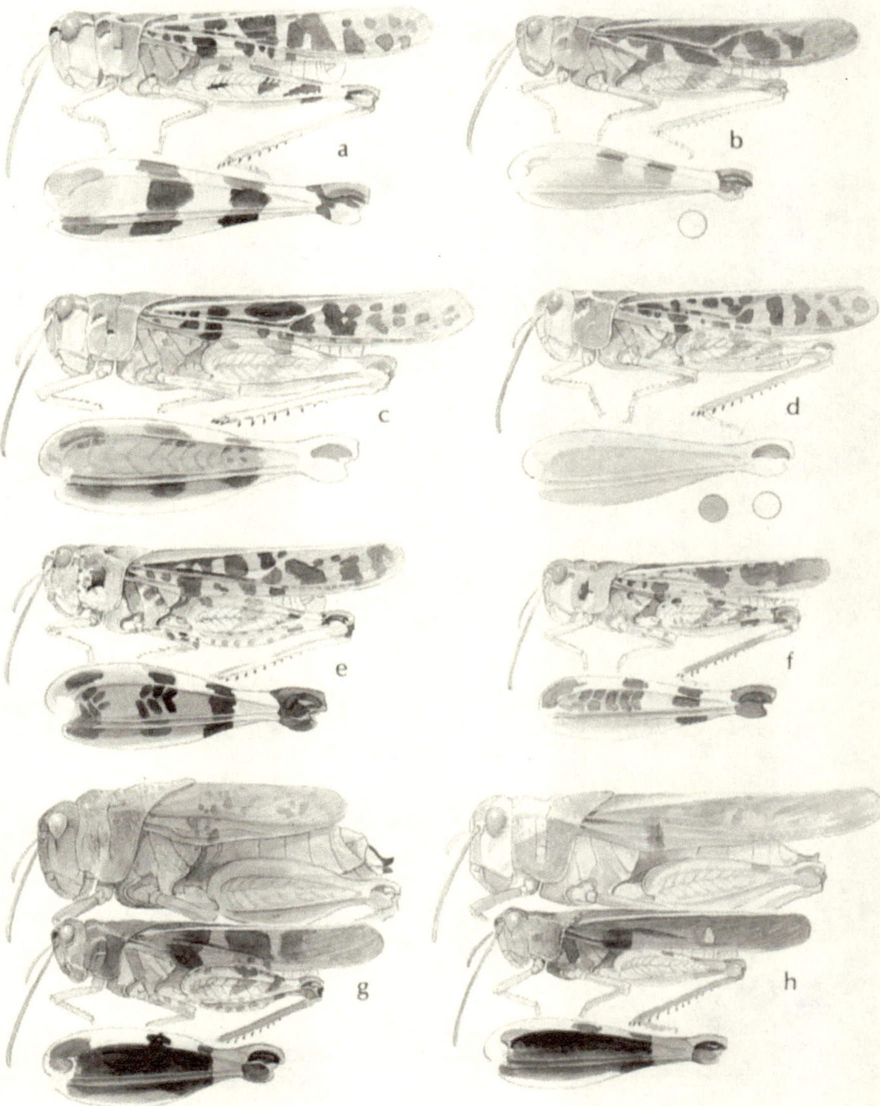

a. *Hippiscus ocelote;* b. *Pardalophora apiculata;* c. *Pardalophora saussurei;* d. *Pardalophora haldemani;* e. *Pardalophora phoenicoptera;* f. *Sticthippus californicus;* g. *Agymnastus ingens;* h. *Agymnastus venerabilis*

PLATE 8

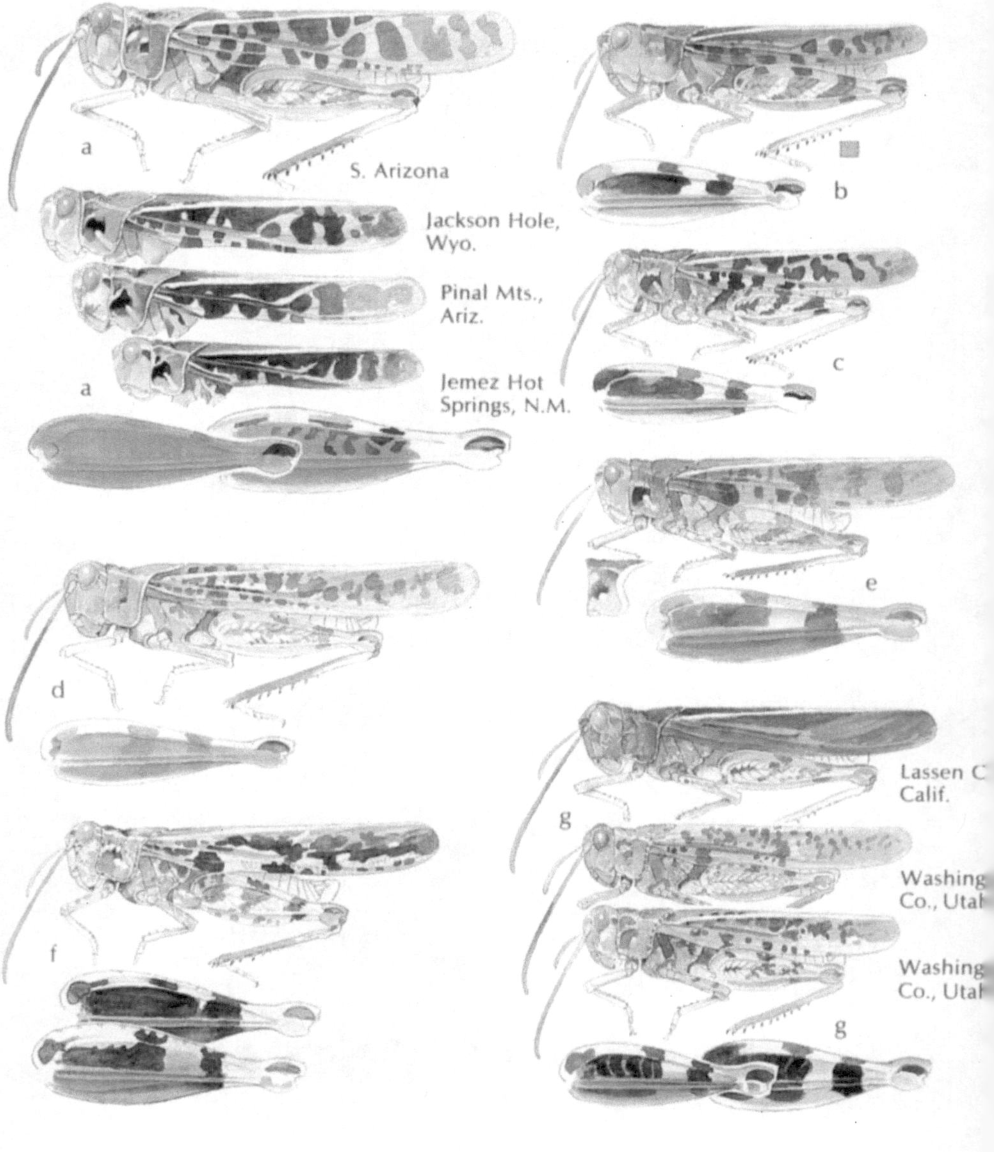

a. *Xanthippus corallipes;* b. *Xanthippus aquilonius;* c. *Xanthippus brooksi;* d. *Xanthippus montanus;* e. *Cratypedes neglectus;* f. *Xanthippus olancha;* g. *Cratypedes lateritius*

PLATE 9

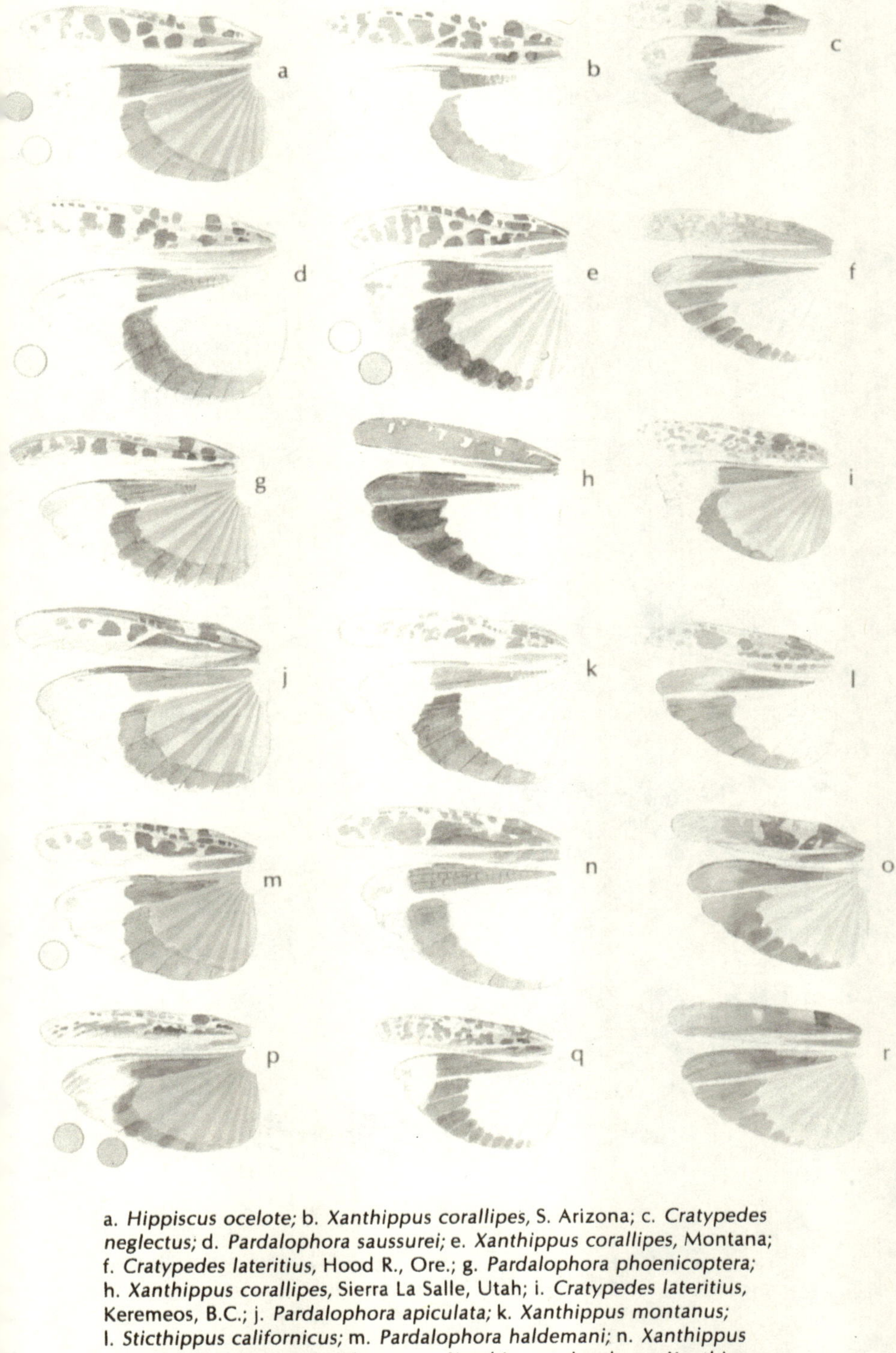

a. *Hippiscus ocelote;* b. *Xanthippus corallipes,* S. Arizona; c. *Cratypedes neglectus;* d. *Pardalophora saussurei;* e. *Xanthippus corallipes,* Montana; f. *Cratypedes lateritius,* Hood R., Ore.; g. *Pardalophora phoenicoptera;* h. *Xanthippus corallipes,* Sierra La Salle, Utah; i. *Cratypedes lateritius,* Keremeos, B.C.; j. *Pardalophora apiculata;* k. *Xanthippus montanus;* l. *Sticthippus californicus;* m. *Pardalophora haldemani;* n. *Xanthippus montanus;* o. *Agymnastus ingens;* p. *Xanthippus olancha;* q. *Xanthippus brooksi;* r. *Agymnastus venerabilis*

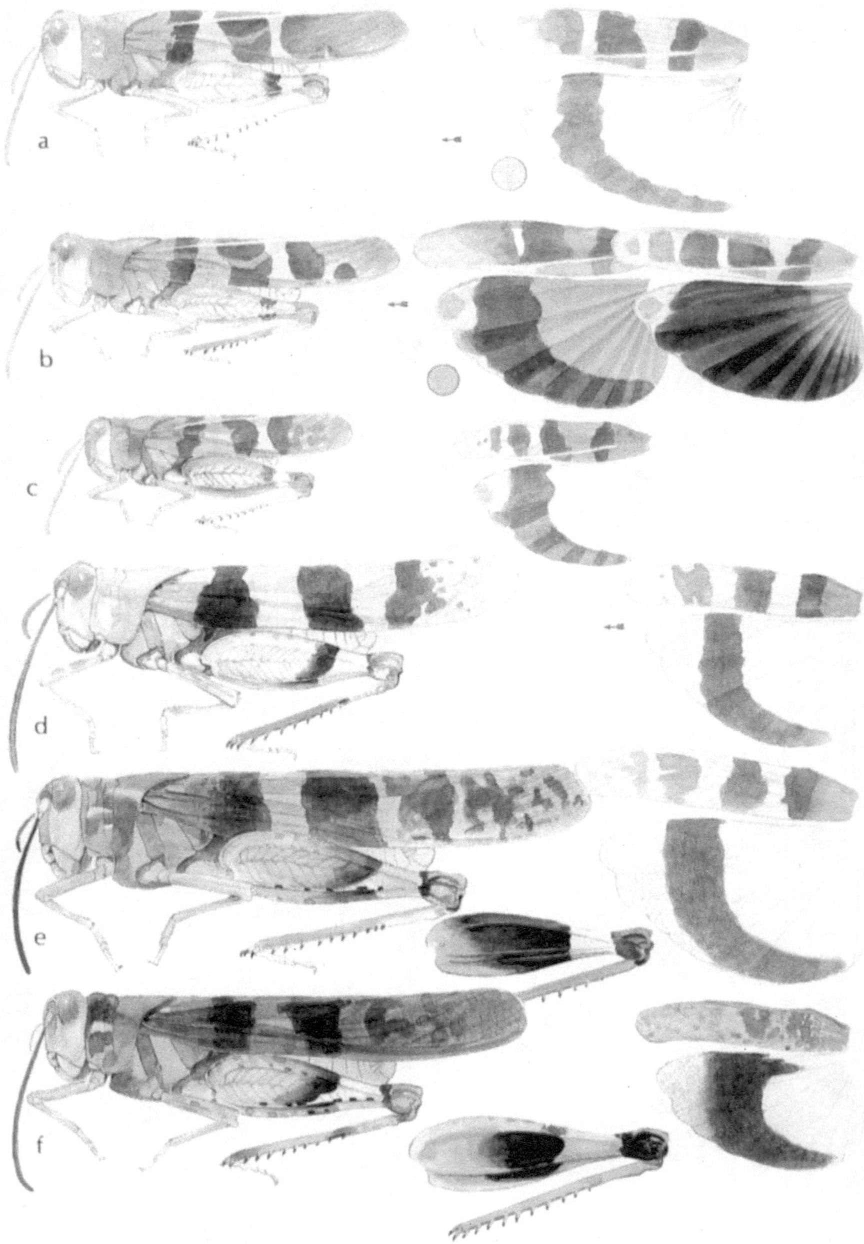

a. *Leprus wheeleri;* b. *Leprus intermedius;* c. *Leprus elephas;* d. *Hadrotettix trifasciatus;* e. *Hadrotettix magnificus;* f. *Hadrotettix scotodes*

PLATE 11

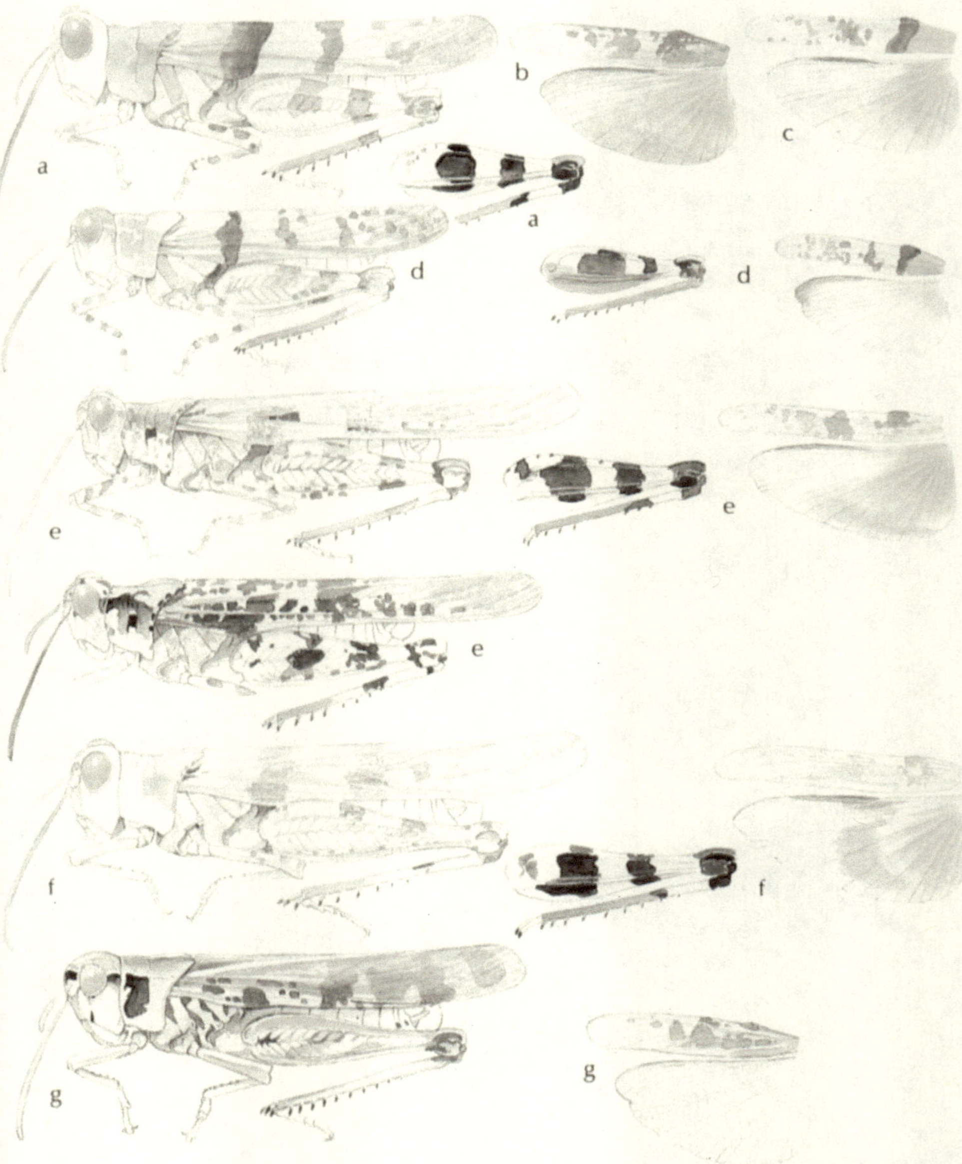

a. *Heliastus benjamini;* b. *Heliastus aztecus;* c. *Heliastus benjamini;* d. *Heliastus infrequens;* e. *Heliastus sumichrasti;* f. *Heliastus subroseus;* g. *Camnula pellucida*

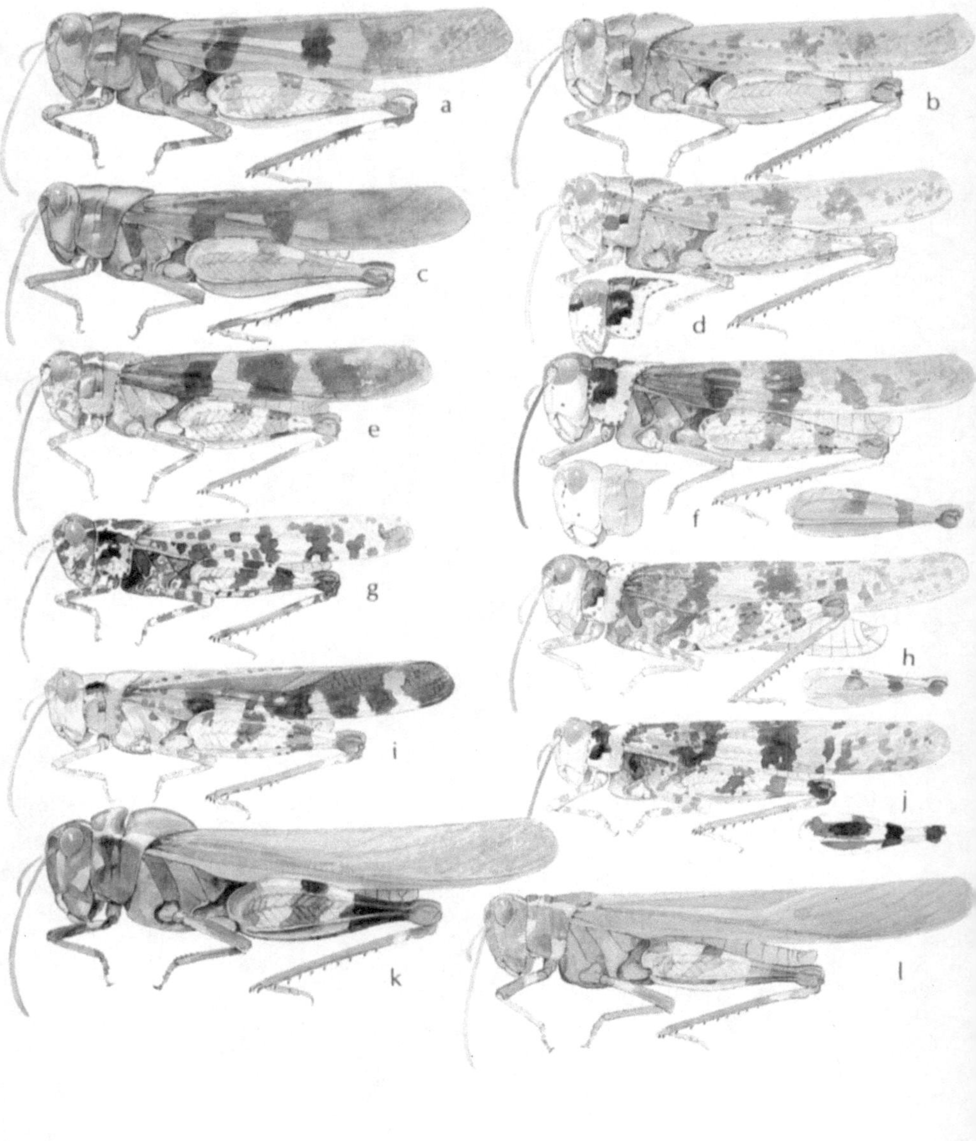

a. *Spharagemon bolli;* b. *Spharagemon cristatum;* c. *Spharagemon crepitans;* d. *Spharagemon collare;* e. *Spharagemon saxatile;* f. *Spharagemon equale;* g. *Spharagemon marmorata marmorata;* h. *Spharagemon campestris;* i. *Spharagemon marmorata picta;* j. *Spharagemon bunites;* k. *Eximacris superbum.* l. *Eximacris phenax*

PLATE 13

a. *Dissosteira carolina;* b. *Dissosteira longipennis;* c. *Dissosteira spurcata;* d. *Dissosteira pictipennis,* shown twice size; e. *Eximacris superbum;* f. *Spharagemon bolli;* g. *Spharagemon crepitans;* h. *Spharagemon collare;* i. *Spharagemon cristatum;* j. *Spharagemon equale;* k. *Spharagemon campestris;* l. *Spharagemon bunites;* m. *Spharagemon marmorata marmorata;* n. *Spharagemon marmorata picta*

PLATE 14

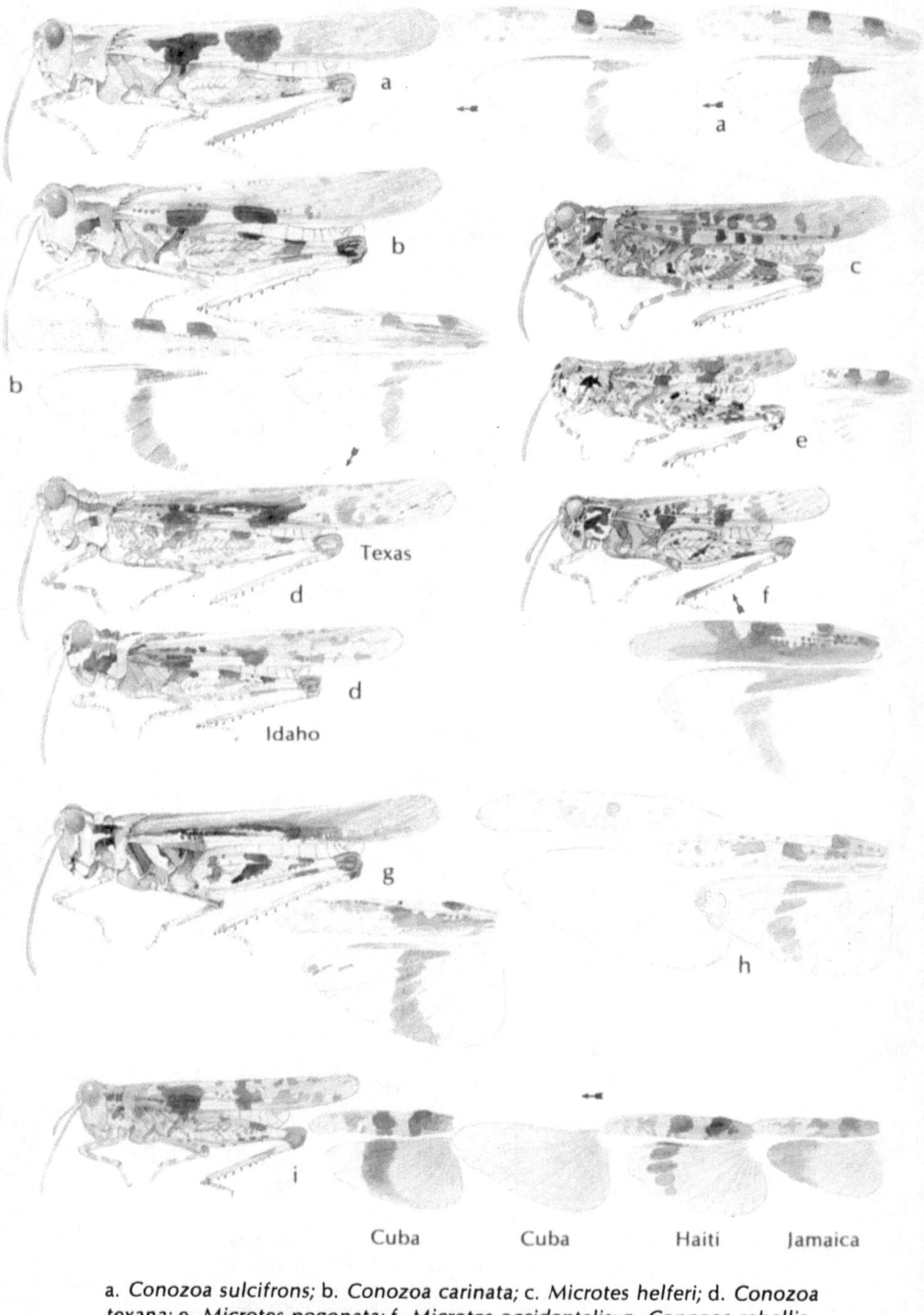

a. *Conozoa sulcifrons;* b. *Conozoa carinata;* c. *Microtes helferi;* d. *Conozoa texana;* e. *Microtes pogonata;* f. *Microtes occidentalis;* g. *Conozoa rebellis;* h. *Conozoa hyalina;* i. *Sphingonotus haitensis*

PLATE 15

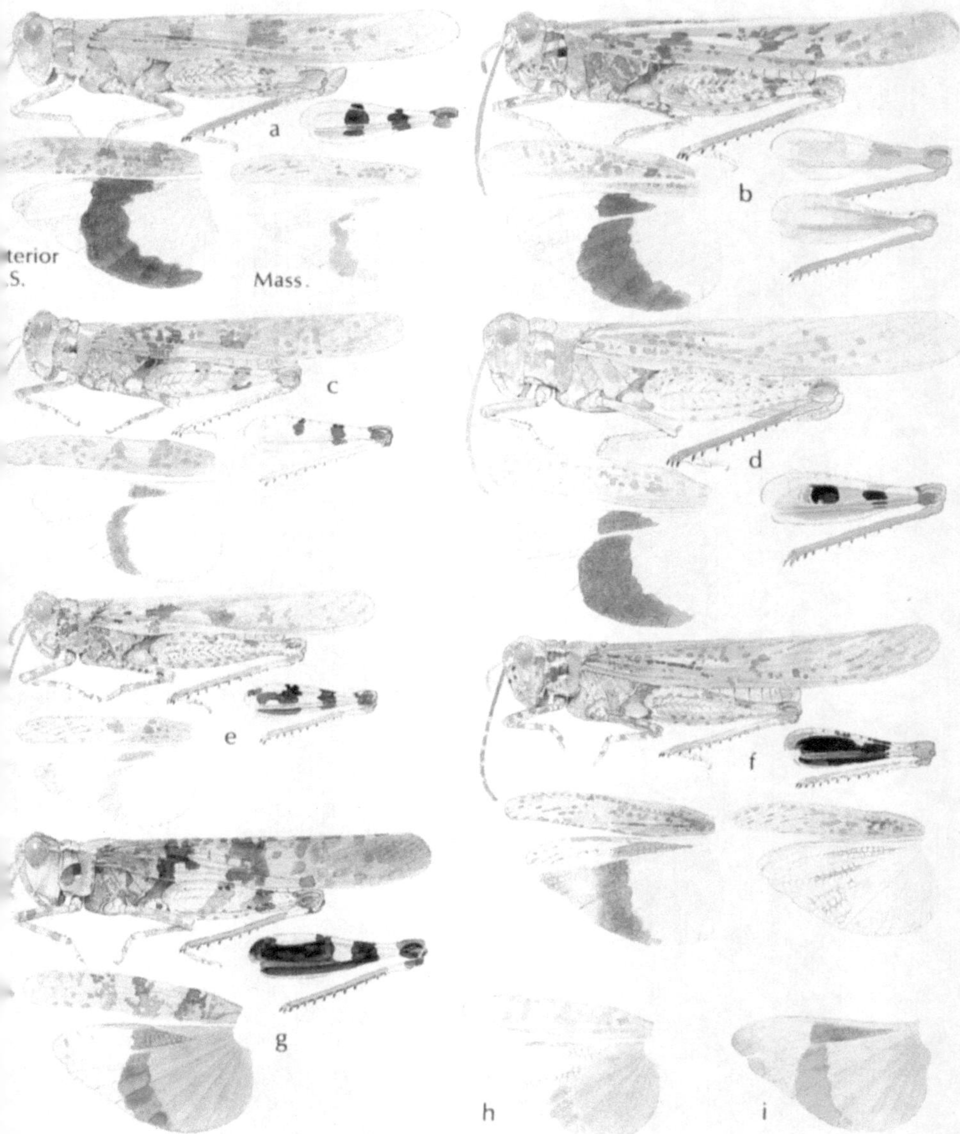

a. *Trimerotropis maritima;* b. *Trimerotropis barnumi;* c. *Trimerotropis whitei;* d. *Trimerotropis agrestis;* e. *Trimerotropis arenacea;* f. *Trimerotropis gracilis;* g. *Trimerotropis cyaneipennis;* h. *Trimerotropis leucophaea;* i. *Trimerotropis inyo*

a. *Trimerotropis latifasciata;* b. *Trimerotropis melanoptera;* c. *Trimerotropis pistrinaria;* d. *Trimerotropis modesta;* e. *Trimerotropis californica;* f. *Trimerotropis tolteca;* g. *Trimerotropis titusi;* h. *Trimerotropis pacifica;* i. *Trimerotropis topanga*

PLATE 17

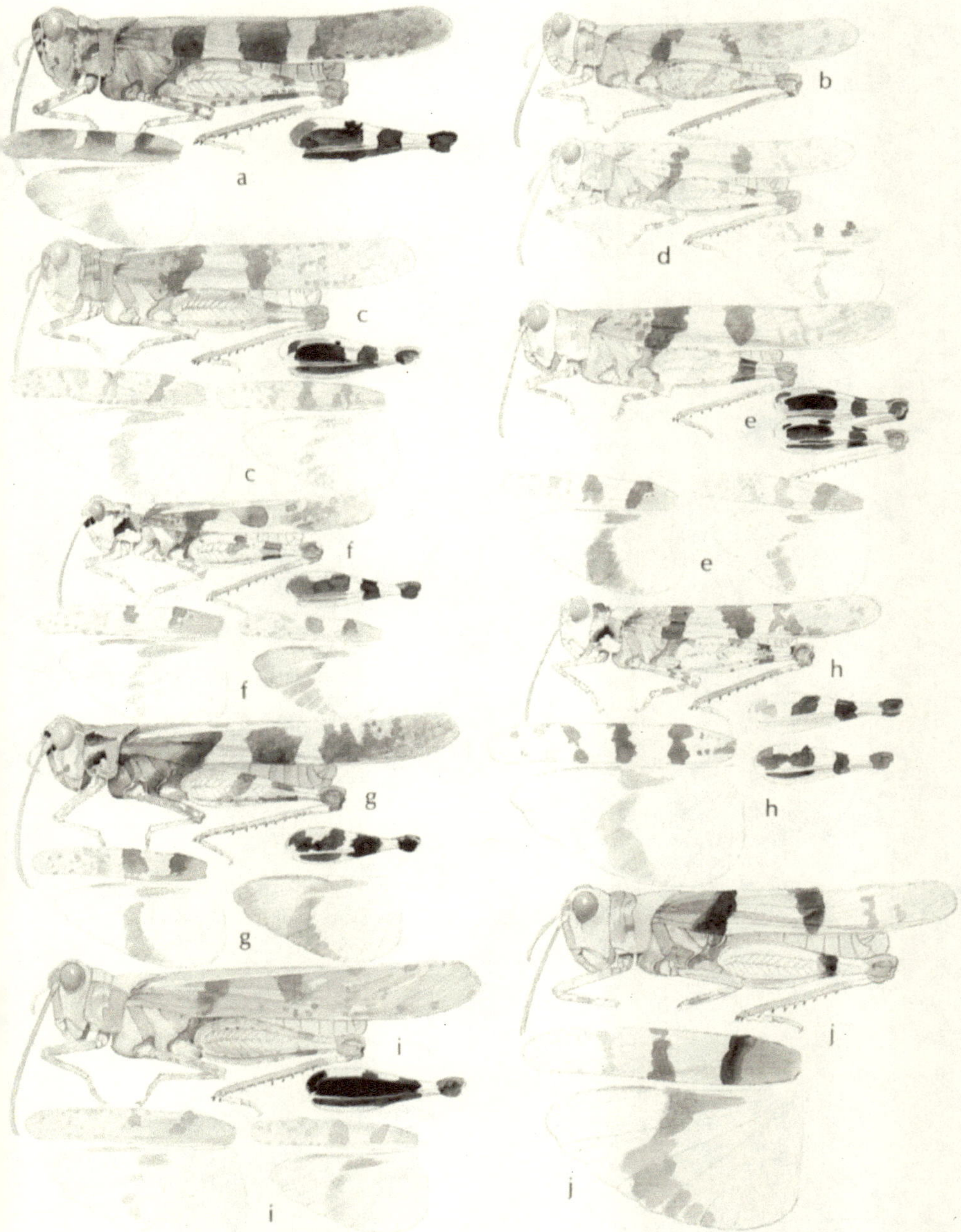

a. *Trimerotropis fontana;* b. *Trimerotropis occidentiloides;* c. *Trimerotropis occidentalis;* d. *Trimerotropis occulens;* e. *Trimerotropis inconspicua;* f. *Trimerotropis koebelei;* g. *Trimerotropis cincta;* h. *Trimerotropis albescens;* i. *Trimerotropis thalassica;* j. *Trimerotropis bifasciata*

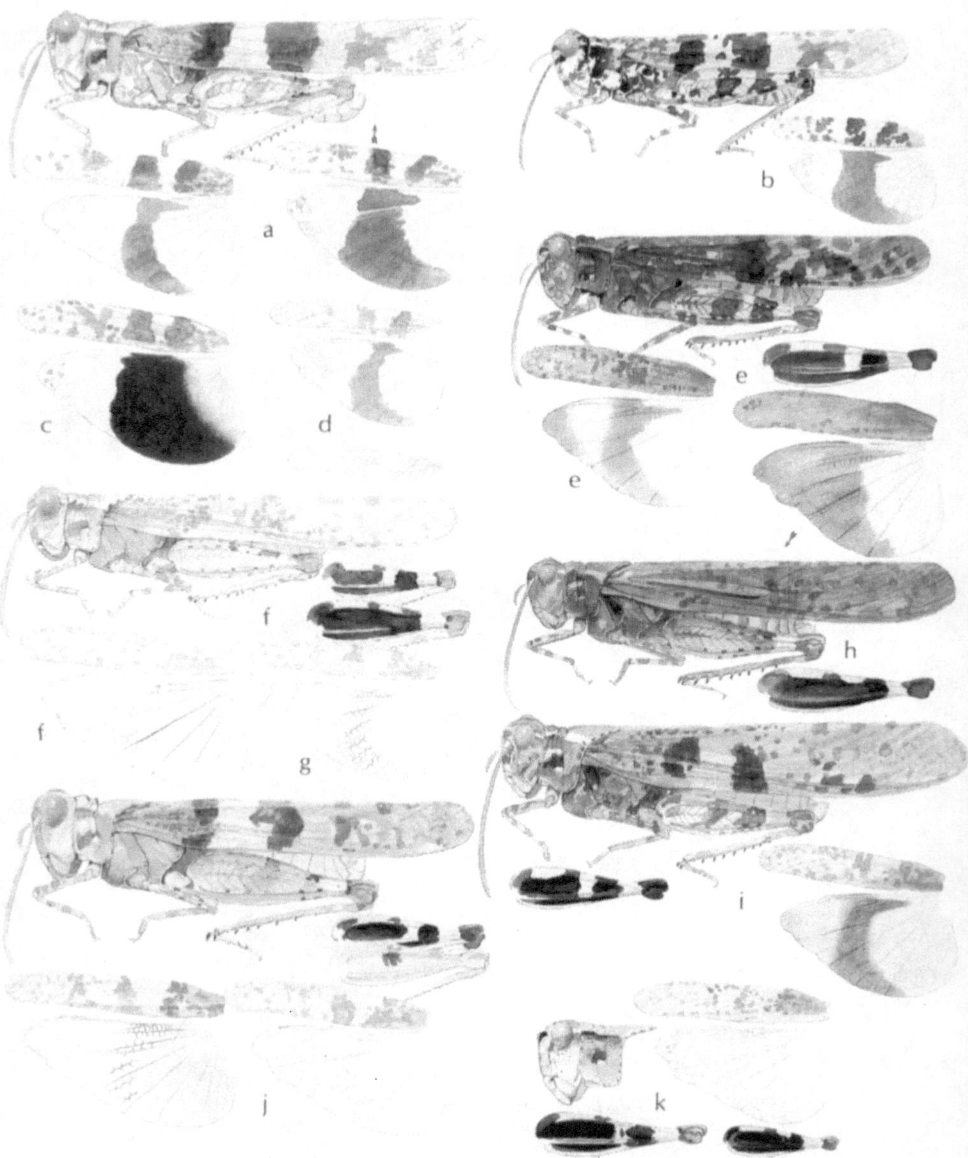

a. *Trimerotropis pallidipennis;* b. *Trimerotropis saxatilis;* c. *Trimerotropis schaefferi;* d. *Trimerotropis huroniana;* e. *Trimerotropis verruculata verruculata;* f. *Trimerotropis sparsa;* g. *Trimerotropis salina;* h. *Trimerotropis verruculata suffusa;* i. *Trimerotropis fratercula;* j. *Trimerotropis pseudofasciata;* k. *Trimerotropis lauta*

PLATE 19

a. *Circotettix undulatus;* b. *Circotettix shastanus;* c. *Circotettix rabula;* d. *Circotettix maculatus;* e. *Circotettix crotalum;* f. *Circotettix stenometopus;* g. *Circotettix carlinianus*

PLATE 20

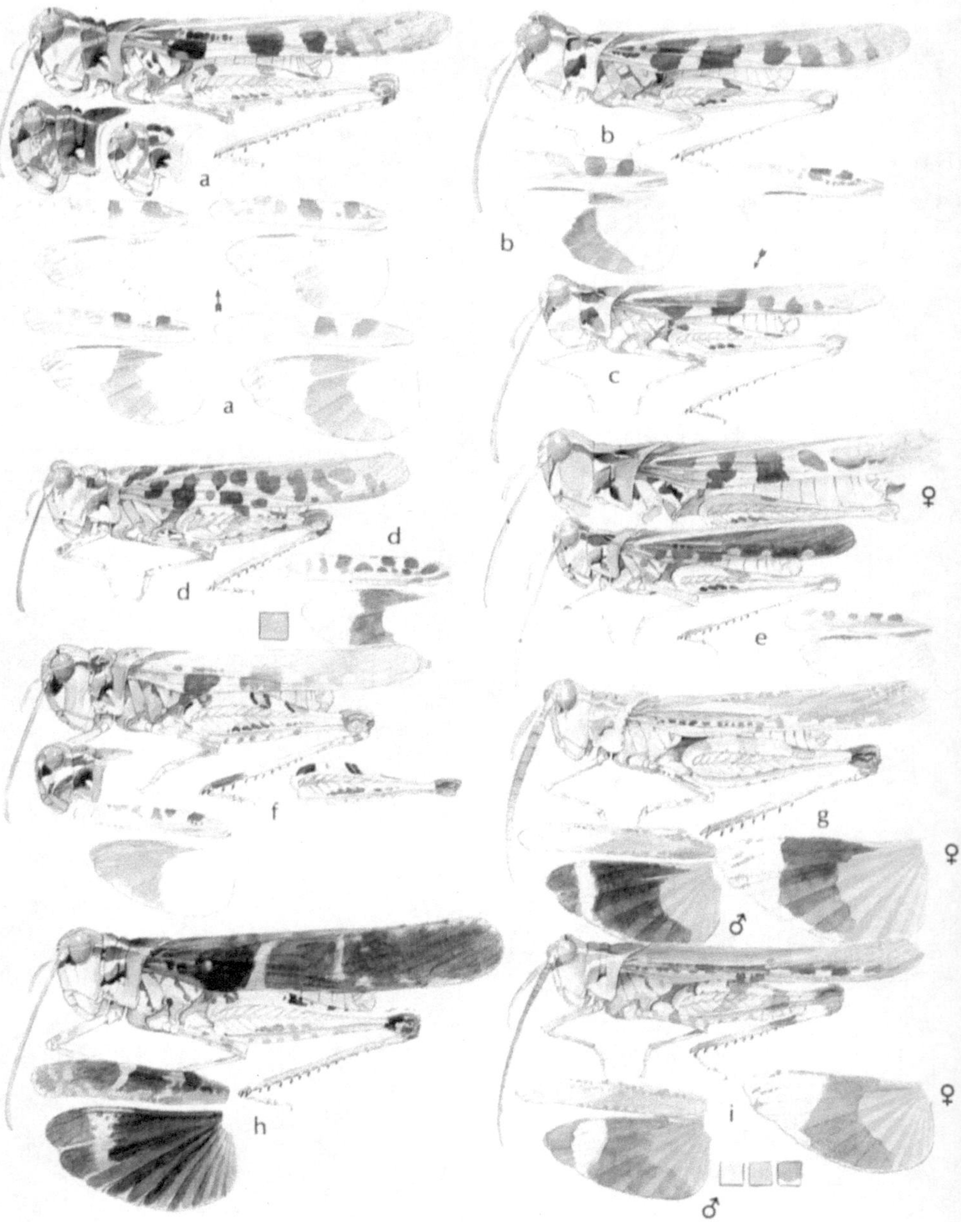

a. *Trachyrhachys kiowa;* b. *Hippopedon capito;* c. *Hippopedon gracilipes;* d. *Trachyrhachys coronata;* e. *Hippopedon saltator;* f. *Trachyrhachys aspera;* g. *Psinidia amplicornis;* h. *Trachyrhachys funeralis;* i. *Psinidia fenestralis*

PLATE 21

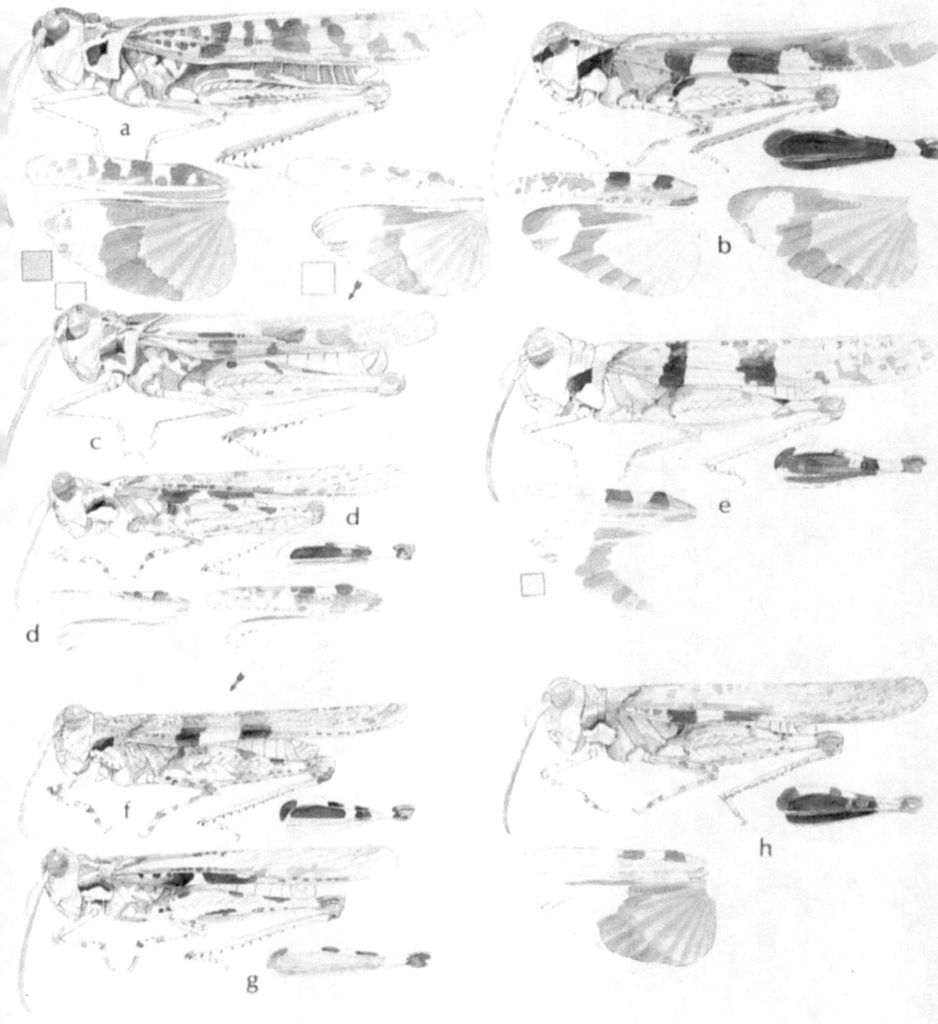

a. *Metator pardalinus;* b. *Mestobregma plattei;* c. *Metator nevadensis;* d. *Trepidulus rosaceus;* e. *Mestobregma impexum;* f. *Trepidulus concinens;* g. *Trepidulus hyalinus;* h. *Mestobregma terricolor*

PLATE 22

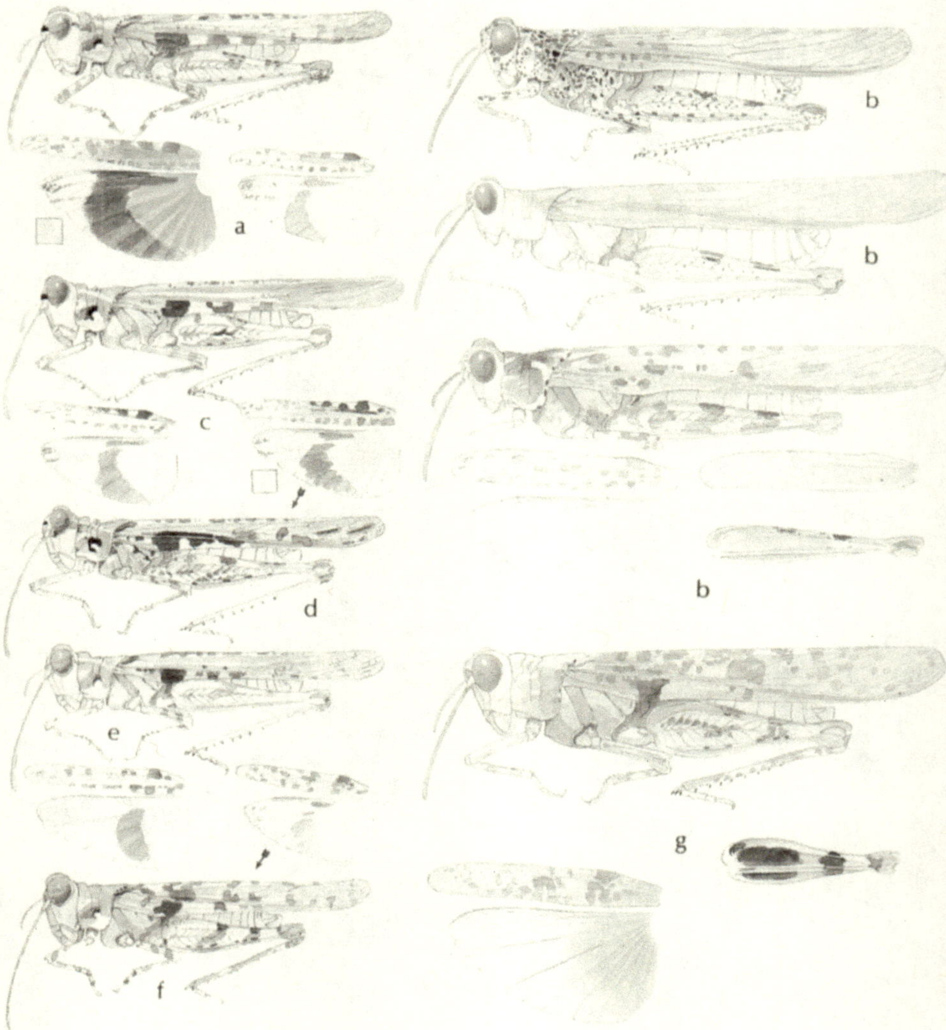

a. *Derotmema haydeni;* b. *Anconia integra;* c. *Derotmema laticinctum;*
d. *Derotmema saussureanum;* e. *Derotmema delicatulum;* f. *Derotmema piute;*
g. *Anconia hebardi*

***Encoptolophus subgracilis* Caudell** Pl. 6

TAXONOMY. Hebard (1935) writes: "Large series now available show that *subgracilis* is at best a poorly defined race of *pallidus*. It is the condition present over most of the very wide range of the species, typical *pallidus* being confined to the areas of extreme aridity from the Panamint and Death Valleys of California and the Amargosa Desert of Nevada south to the Mexican border and differs only in its decidedly more pallid coloration and proportionately larger head and eyes." *E. subgracilis* was subsequently treated as a distinct species, for example by Tinkham (1948). In his Synonymic Checklist Hebard indicated that *subgracilis* was a distinct species, but this opinion was not published. Fig. 31 demonstrates that there is very little overlap in the head and pronotal disk proportions of *pallidus* and *subgracilis* from the regions of the nearest geographic contact. This character, together with differences in the development of the median pronotal carinae, leads me to treat the two forms as separate species.

DISTRIBUTION. Widespread over the southwestern states and through Mexico to Honduras.

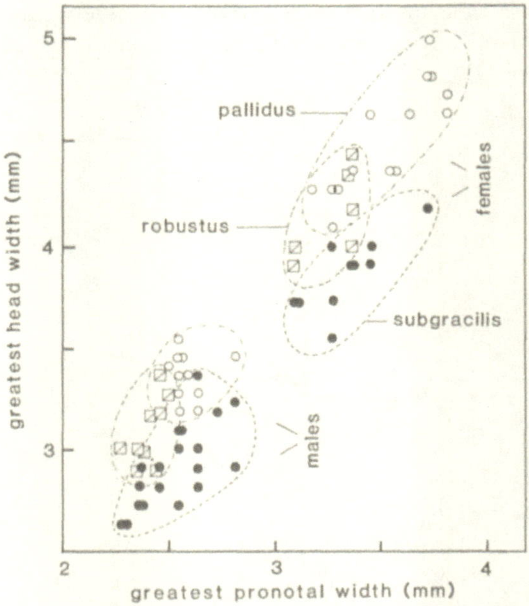

Fig. 31. *Encoptolophus.* Scatter diagram showing separation of *pallidus, robustus,* and *subgracilis.*

RECOGNITION. Figs. 27*c*, 28*b*, *j*, 29*d*, 31. Hindwings largely transparent, faintly greenish yellow in basal part, turning smoky in outer third. Hind tibiae blue to distal two-thirds. Differs from *costalis* and *sordidus* as follows: fastigium distinctly narrower; median pronotal crest quite low; posterior margins of pronotal disk separated by more than 90 degrees (90 degrees or less in *costalis* and *sordidus*); disk of pronotum without a pale X-shaped marking (very rarely such a marking is faintly indicated); metazona without dark markings; dorsal field of forewings almost always without a pale streak along the wing angle east of Arizona, often with such streaks in Arizona. Forewings extend more than one head width beyond ends of hind femora. Abdomen straw-yellow and without black stripes or markings. Rare females are green on the head, pronotum, and hind femora. In specimens from southern Mexico the lateral carinae of the prozona form definite protuberances (almsot small teeth). The smallest males of this species are from Chiapas, Mexico, the largest from Honduras.

Body length to ends of forewings 18–26 mm in males, 24–33 mm in females; Honduras males, 25–29 mm; one female 32 mm.

HABITAT. Grasslands. Ball et al. (1942) report: "A very common species in cultivated fields in the irrigated regions of southern Arizona and in damper areas on the desert and into the desert grassland. Feeds mostly on grasses but causes some damage to crops, especially alfalfa and truck crops."

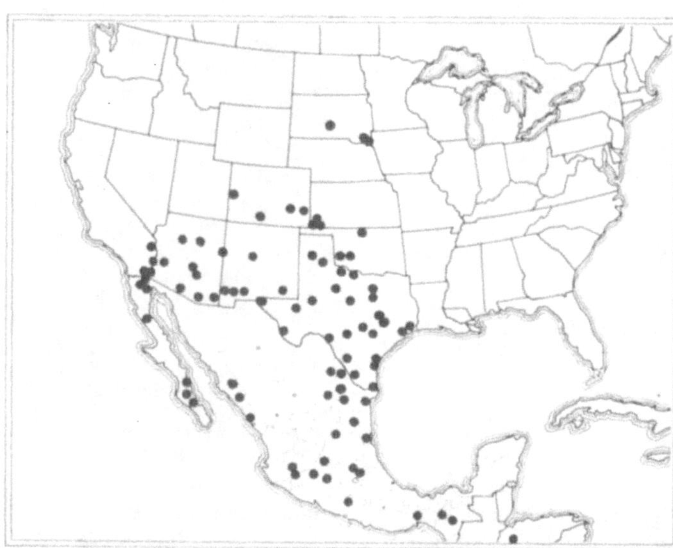

Encoptolophus subgracilis

LIFE CYCLE. In southern Arizona this species overwinters in all stages. Adults have been collected from March to December, but are most common in August (Ball et al. 1942). In Texas adults are most common from August to October (Isely 1937). In central Mexico adults have been collected from May to October.

REFERENCES. Caudell 1903a, Hebard 1935b, Ball et al., 1942, Strohecker, Middlekauff, and Rentz 1968.

***Encoptolophus pallidus* Bruner** Pl. 6

DISTRIBUTION. Deserts of southern California and southern Nevada.

RECOGNITION. Figs. 27*h*, 38*i*, 31. Hindwings entirely transparent, sometimes very faintly yellowish near base. Overall body color sandy or straw-yellow, with dark markings on sides. In very pale individuals the sides of the body are virtually without dark markings. Most similar to *subgracilis,* but differs in having a wider head. In *pallidus* the greatest head width is from 1.21 to 1.44 (mean 1.30) times as wide as the pronotal disk at the rear, in *subgracilis* it is from 1.03 to 1.23 (mean 1.16) times as wide. *E. pallidus* also has a very low median pronotal carina. Body length to end of forewings 20–24 mm, 26–31 mm in females.

HABITAT. Deserts.

LIFE CYCLE. Adults from April to September.

REFERENCES. Bruner 1893a.

***Encoptolophus robustus* Rehn and Hebard** Pl. 6

DISTRIBUTION. Extreme southwestern California and Baja California.

RECOGNITION. Figs. 27*f*, 28*d*, 29*e*, 31. Hindwings largely transparent, but with a faint smokiness at the distal margin. Most similar to *subgracilis,* but differing as follows: median pronotal crest low but sharper; lateral pronotal carinae distinctly defined at anterior and posterior ends, but not continuous through constricted area; forewings barely extend beyond ends of hind femora; head appears to be as broad as pronotum (top view) (distinctly narrower in *subgracilis*). It differs from *californicus* in having interrupted lateral carinae, distinct dorsal femoral dark bands, and no dark pigmentation on the lateral lobes against the lateral carinae. Forewings with or (usually) without a pale streak along the wing angle. Hind tibiae blue in distal two-thirds and with a pale ring in the first quarter. Body length to end of forewings 17–19 mm in males, 20–25 mm in females.

HABITAT. Specimens have been collected from along the Los Angeles River, at Laguna Beach, and on the range at the Santa Margarita River but the precise habitats have not been described in published works.

LIFE CYCLE. Adults from July to November.

REFERENCES. Rehn and Hebard 1910, Strohecker, Middlekauff, and Rentz 1968.

***Encoptolophus californicus* (Bruner)** Pl. 6

DISTRIBUTION. Southern two-thirds of California; see Strohecker, Middlekauff, and Rentz 1968 for localities.

RECOGNITION. Figs. *27b, 28c, 29f.* Hindwings mostly clear but with a faint pale greenish yellow tinge near base and with a scarcely visible smokiness at distal margin. This species differs from *subgracilis* as follows: frontal ridge deeply grooved; fastigium deeply hollowed (V-shaped groove); top of head usually with a low median carinula that extends back to pronotum; lateral pronotal carinae continuous from anterior to posterior margins. Hind femora with a very small marking on upper marginal area near middle. Overall body color of males is brown to dark gray-brown, females occasionally greenish. Both sexes have distinct spots on the lateral field of the forewings. Lateral lobe of pronotum is usually darkly pigmented dorsally. Outer face of hind fe-

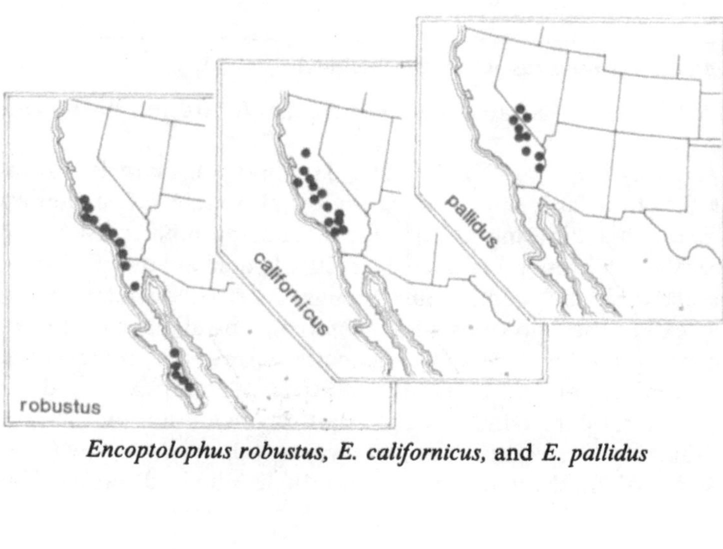

Encoptolophus robustus, E. californicus, and *E. pallidus*

mora banded. Body length to end of forewings 18–20 mm in males, 22.5–28.0 mm in females.

HABITAT. Rehn and Hebard (1910) report: "A female of this species was taken among sand dunes at Alamitos Bay . . . while one male was captured along the Mohave River at Cottonwood."

LIFE CYCLE. Adults have been collected from March to September.

REFERENCES. Bruner 1905, Rehn and Hebard 1910, Strohecker, Middlekauff, and Rentz 1968.

Genus SHOTWELLIA Gurney

DISTRIBUTION. The only species, *S. isleta,* is known from southern New Mexico and from Zacatecas and Durango, Mexico.

RELATIONSHIP. This genus is most similar to *Encoptolophus* species, especially *pallidus* and *subgracilis;* the male aedeagus is similar to those of *Encoptolophus* and *Chortophaga.* In his discussion of possible relationships, Gurney mentions a number of other genera but not *Encoptolophus* or *Chortophaga.*

RECOGNITION. Figs. 27*i,* 28*e, k.* See *S. isleta.*

REFERENCES. Gurney 1940.

Shotwellia isleta Gurney — Pl. 6

DISTRIBUTION. The six known specimens of this species were collected at the following localities: vicinity of Isleta Indian Reservation, Bernalillo County, New Mexico, August 22, 1938 (one male and one female); Gómez Palacio, Durango, Mexico, May 1918 (one female); about 60 km northeast of Zacatecas City, Zacatecas, Mexico, at edge of salt pan, November 13, 1981. The first three specimens are at the National Museum in Washington, the last three are at the Philadelphia Academy of Natural Sciences.

RECOGNITION. Fig. 27*i,* 28*e, k.* Hindwings without strong bands, yellow in basal area. Differs from *Encoptolophus* as follows: lateral field of forewings with one strong dark band and without a dark band above the base of the hind femora; hindwing with an Rs vein; fastigium small relative to eye and without a posterior median carinula; pronotum without a median ridge, but with a raised line near front and rear margins; hind femora without dark bands in medial area; hind tibiae yellowish and somewhat short. Body length to end of forewings 31–33 mm in males.

HABITAT. New Mexico specimens were collected "in the Rio Grande Valley where the Upper Sonoran and Lower Sonoran life zones merge." The Gómez Palacio specimen is from "a small area of short grass in which irrigation has permitted the growing of cotton" (Gurney 1940). I collected three males in Zacatecas on baked mud at the edge of a salt lake.

LIFE CYCLE. Adults have been collected in May, August, and November.

REFERENCES. Gurney 1940.

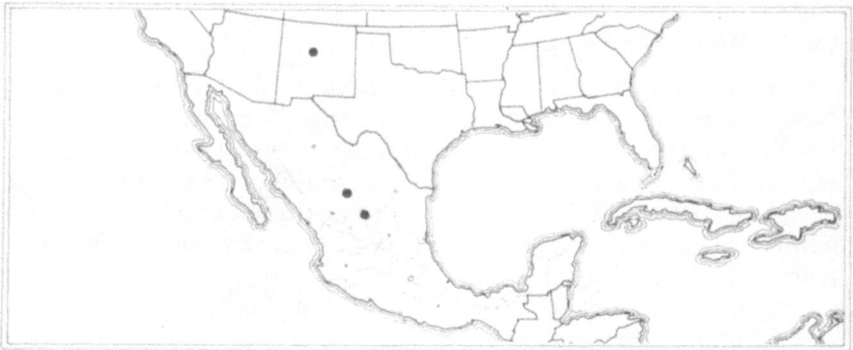

Shotwellia isleta

Hippiscus Genus Group

Hebard (1928) included the following genera in this group, which he called Group Hippisci: *Hippiscus, Pardalophora, Xanthippus, Cratypedes, Sticthippus, Agymnastus,* and *Leprus. Camnula* was omitted but was included in his checklist (Fig. 32). To this group can be added two previously unsuspected members: *Hadrotettix* and *Heliastus.* The former is added without hesitation, the latter tentatively. *Hadrotettix* was previously thought to be related to *Trimerotropis* and *Circotettix,* and *Heliastus* was considered by Hebard (unpublished synonymic checklist) to belong to the Group Heliasti, which also included *Cibolacris* (the latter is now placed in the Gomphocerinae). The two genera were suspected of belonging to the Hippiscus group when the male epiphallus was examined. This feature shows little variation across the subfamily, but the posterior lophi in the Hippiscus group are peculiarly trilobed (Fig. 62). Also most of the genera have two converging pale streaks on the dorsum of the folded wings. *Hadrotettix* and *Heliastus* lack pale streaks but possess trilobed epiphalli, which led me to look for other characteristics shared with the Hippiscus group. *Hadrotettix* is further associated with the Hippiscus group by the following

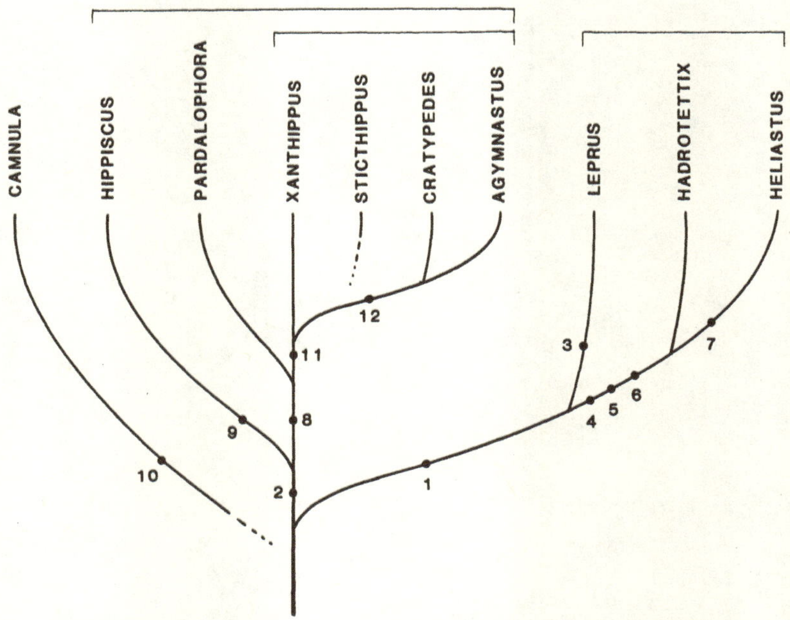

Fig. 32. Hypothetical phylogeny of genera in the Hippiscus genus group (see Appendix 1 for explanation).

characters: hindwing banding similar to *Leprus;* low pronotal median carina and shape of pronotal disk as in *Leprus;* long, black antennae as in some *Xanthippus;* absence of male crepitation in mating behavior, as in *Xanthippus, Hippiscus,* and *Pardalophora.* The association between *Heliastus* and the Hippiscus group is less direct, but the following similarities between *Heliastus aztecus* and the Hippiscus group species are noteworthy: male epiphallus strongly bilobed; antennae long and black; forewings with distinct crossbands (as in *Hadrotettix*); pronotum without a prominent median crest and shaped somewhat as in *Hadrotettix* and *Leprus;* males without a crepitation flight; habitat similar to *Hadrotettix.*

• Identification of Genera

Camnula (prairies and mountain grasslands of United States and Canada; Figs. 33*d,* 34*i;* Pl. 11)

1. Hindwings entirely transparent
2. Lateral pronotal carinae well defined throughout their length

Hippiscus (eastern United States, south to southern Mexico; Figs. 33*a,* 34*a;* Pls. 7, 9)

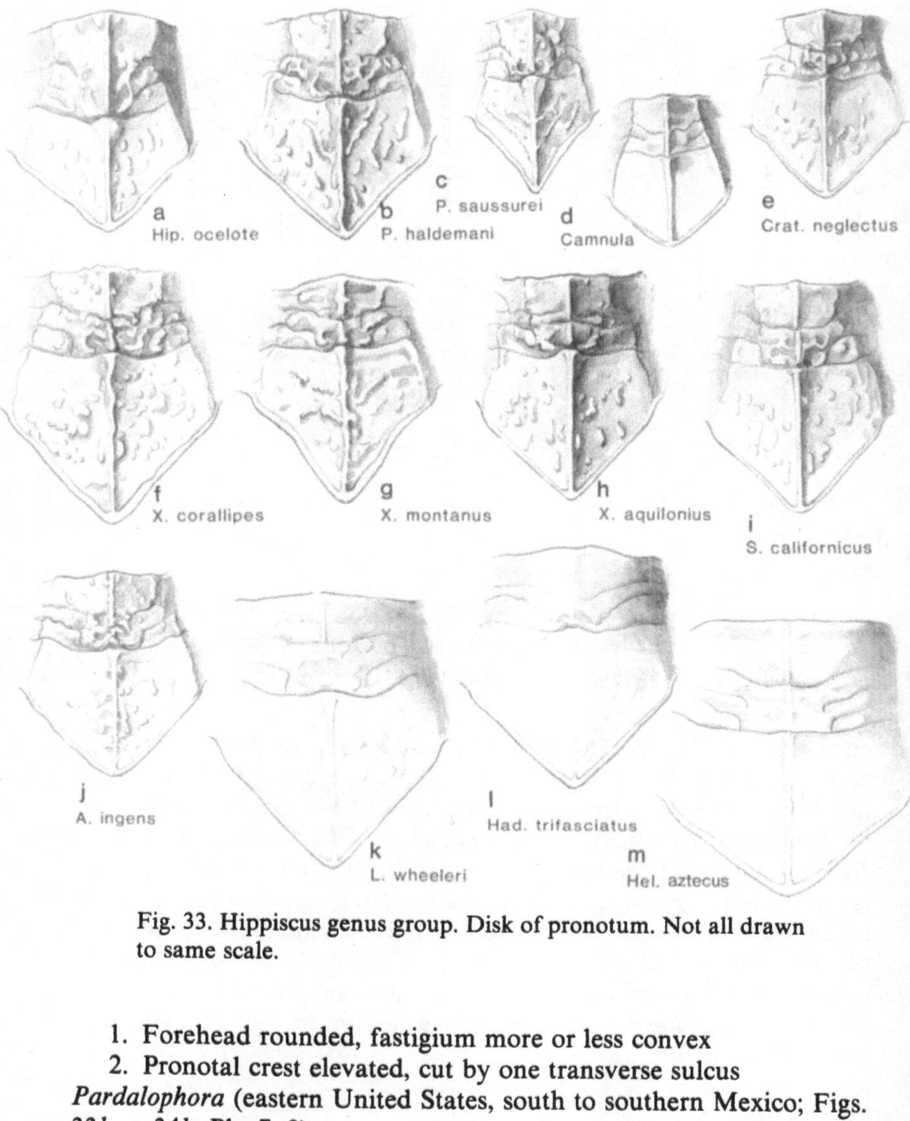

Fig. 33. Hippiscus genus group. Disk of pronotum. Not all drawn to same scale.

1. Forehead rounded, fastigium more or less convex
2. Pronotal crest elevated, cut by one transverse sulcus

Pardalophora (eastern United States, south to southern Mexico; Figs. 33*b, c,* 34*b;* Pls. 7, 9)

1. Forehead rather angulate
2. Pronotal crest as in *Hippiscus,* elevated and cut by one sulcus; principal sulcus does not cut lateral carinae

Xanthippus (prairies and western United States, south to southern Mexico; Figs. 33*f–h,* 34*c, d;* Pls. 8, 9)

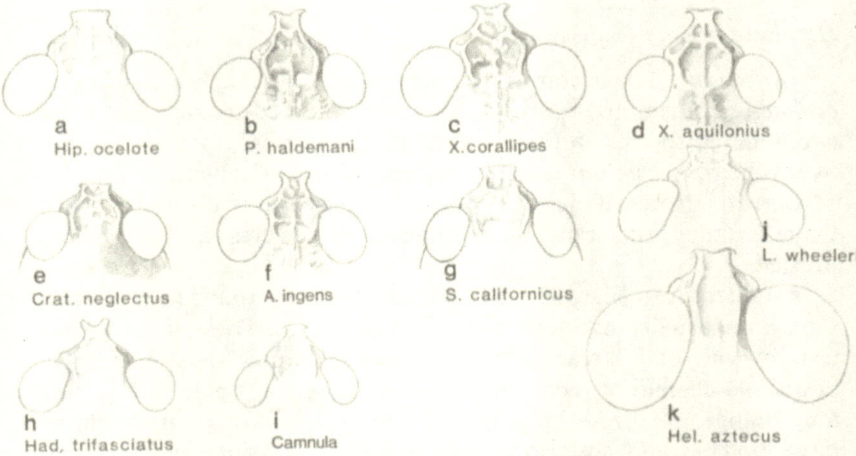

Fig. 34. Hippiscus genus group. Fastigium of head. Not all drawn to same scale.

1. Pronotum roughly sculptured on prozona; median line cut by two sulci; principal sulcus cuts lateral carinae
2. Forewings with many large distinct spots

Cratypedes (northwestern states; Figs. 33*e*, 34*c*; Pls. 8, 9)

1. Pronotum as in *Xanthippus* (see above)
2. Forewings with small scattered spots or uniform on lateral field

Agymnastus (California; Figs. 33*j*, 34*f*; Pls. 7, 9). Superficially similar to *Cratypedes* but:

1. Inner face of hind femora blue-black and with bright orange band at narrowest part
2. Females much larger than males and flightless

Sticthippus (California; Figs. 33*i*, 34*g*; Pls. 7, 9)

1. Fastigium distinctly wider than long
2. Lophi of male epiphallus bilobate (trilobate in above genera)

Genus HIPPISCUS Saussure

DISTRIBUTION. The only species, *H. ocelote*, ranges from the eastern United States to Chiapas, Mexico.

RECOGNITION. See *H. ocelote*.

REFERENCES. See *H. ocelote*.

Hippiscus ocelote **(Saussure)** Pls. 7, 9

TAXONOMY. I have compared the types of *Hippiscus rugosus* and *H. ocelote* and examined specimens from over the entire range of these two nominal species and can find no morphological discontinuity between them. *H. rugosus* is therefore treated as a synonym.

DISTRIBUTION. Widespread over the eastern two-thirds of the United States and ranging southward to Oaxaca and Veracruz, Mexico.

RECOGNITION. Figs. 33*a,* 34*a.* Head is broadly rounded in profile. Fastigium usually has four shallow depressions. Disk of pronotum usually with pale streaks along the lateral carinae. Posterior angle of disk 90 degrees or more. Basal area of hindwings pale yellow, yellow, orange, rosy red, or pale red. Forewings have clearly defined large blotches and posteriorly converging ivory stripes on the dorsal field. Hind femora usually have oblique dark bands on the outer face and four dark bands on the inner face (the proximal two bands are sometimes united into a single band). Hind tibiae are yellowish, sometimes with an indistinct brownish band in the second quarter. Body length to end of forewings 28–40 mm in males, 39–54 mm in females.

BEHAVIOR. I have never seen flight displays or heard crepitation by males. The presence of colored hindwings suggests, however, that the wings are sometimes used as displays, perhaps only when males fly to females. Courting males produce ordinary stridulation and tap the heads of females with their antennae. On occasion such tapping is accompanied by pronounced lateral rocking of the male's head. Similar behavior is observed in *Pardalophora phoenicoptera* and *Xanthippus corallipes.* Femur-tipping and shaking signals are performed in male–male interactions.

HABITAT. In Kansas Smith (1981) reports: "The species was found in communities ranging from native prairie to bare areas and escaped alfalfa. It was common and most frequently found in communities of weedy grasses and forbs and native prairie sites." In Arizona Ball et al. (1942) report: "Feeds on range grasses almost entirely . . . [inhabits] the live oak association in tall grass on rather dry hillsides." In Indiana Blatchley (1920) notes that "it frequents only dry, upland localities, being especially abundant in open woodland pastures, timothy and clover meadows, along roadsides and rail fences." In Virginia, Fox (1917) reports it from "dry, open, untilled grass lands, especially those dominated by *Andropogon* and other coarse grasses, but not infrequent in such succulent grasses as timothy."

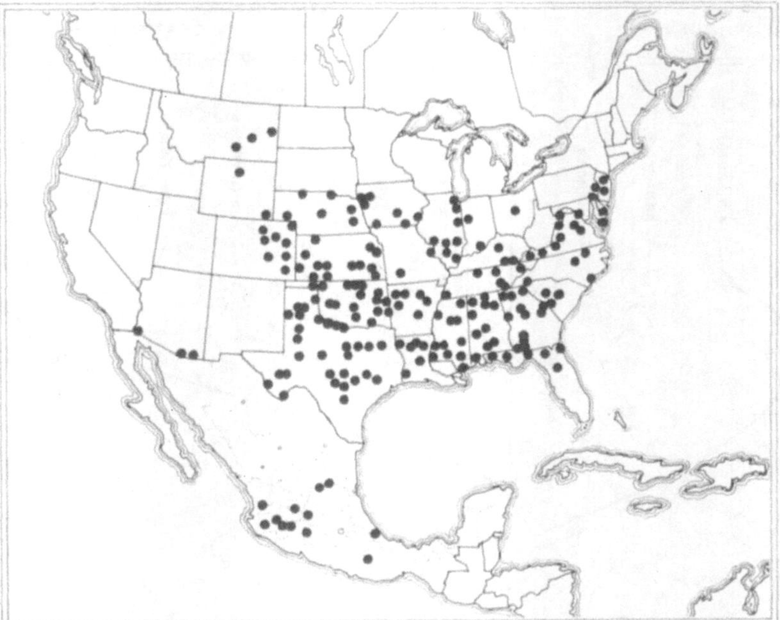

Hippiscus ocelote

LIFE CYCLE. In Arizona eggs and adults overwinter. Eggs overwinter, adults in late summer to late fall in the rest of the United States and Mexico.

REFERENCES. Fox 1914, 1917, Blatchley 1920, Ball et al. 1942, Coppock 1962, Otte 1970, Smith 1981.

Genus PARDALOPHORA Saussure

DISTRIBUTION. From northern Canada through eastern half of United States to northeastern Mexico.

TAXONOMY. This genus is very similar to *Xanthippus;* the only reliable feature separating the two genera is the median pronotal ridge, which is cut twice in *Xanthippus* and once in *Pardalophora.* Whether this genus is monophyletic and distinct from *Xanthippus* is in doubt. In *P. haldemani* some individuals have the pronotal ridge cut by two sulci, making them almost indistinguishable from *Xanthippus corallipes.* In superficial aspect *P. haldemani* appears more similar to *X. corallipes* than to the other *Pardalophora* species.

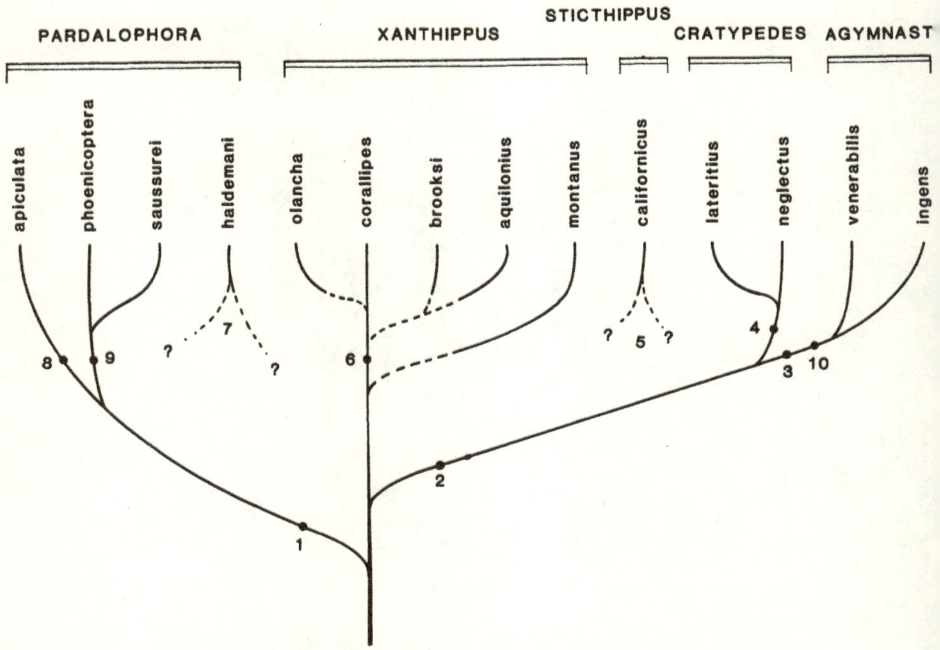

Fig. 35. Hypothetical phylogeny of species in *Pardalophora, Xanthippus, Sticthippus, Cratypedes,* and *Agymnastus* (see Appendix 1 for discussion).

Three species "groups" can be recognized among the four species: *phoenicoptera* and *saussurei* in one group (blue and black inner hind femora); and *apiculata* and *haldemani* each by itself. But *haldemani* may be shown to belong to the *Xanthippus corallipes* group.

RECOGNITION. See Taxonomy, above, and Identification of Genera in the Hippiscus group.

REFERENCES. Saussure 1884.

• Identification of Pardalophora Species

apiculata

1. Forewing area between veins Cu_1 and Cu_2 largely dark
2. Inner face of hind femora yellowish to orange and with two or three black bands (excluding knees); if two, the two basal bands are fused into single broad band
3. Posterior angle of pronotal disk 90° or less

4. Basal area of hindwings rosy red or pale red

haldemani

1. Forewing area between Cu_1 and Cu_2 with several spots
2. Inner face of hind femora unicolorous yellow, orange, or red
3. Posterior angle of pronotal disk usually morte than 90 degrees
4. Basal area of hindwings yellow or reddish

phoenicoptera

1. Forewing area between veins Cu_1 and Cu_2 with several spots
2. Inner face of hind femora bluish in basal half to two-thirds and with two to four black bands, two basal bands sometimes obscured by dark blue color; inner face of knee black
3. Posterior angle of pronotal disk about 90 degrees
4. Basal area of hindwings deep orange to rosy red, rarely yellow

saussurei

1. Forewing area between Cu_1 and Cu_2 with several spots
2. Inner face of hind femora as in *phoenicoptera,* but inner face of knee not black
3. Posterior angle of pronotal disk less than 90 degrees and each margin usually slightly concave
4. Basal area of hindwings yellow or orange-yellow

***Pardalophora apiculata* (Harris)** Pls. 7, 9

DISTRIBUTION. Widespread in central and northeastern United States and Canada, and following borders of the prairies and forest-prairie transition into northwestern Canada.

RECOGNITION. Basal area of hindwings is rosy red or pale red. Forewing area between veins Cu_1 and Cu_2 (directly above the stridulatory vein) is entirely dark (spotted in all other *Pardalophora* species). Posterior angle of pronotal disk is usually 90 degrees or less. Lateral margins of pronotal disk often form rather well-defined lateral ridges continuous from front to back. Dorsal field of forewings has very broad converging pale stripes; in some specimens almost the entire dorsal field is pale. Inner face of hind femora is yellowish with two or three black bands (excluding knees); sometimes the two proximal bands are fused into a single broad band. Inner face of hind knees is not black (black in *phoenicoptera*). Hind tibiae yellowish. Body length to end of forewings 32–42 mm in males, 43–58 mm in females.

BEHAVIOR. Solitary males perform flight displays in which they flash their colored wings, but these are not accompanied by crepitation. Between displays males walk about and occasionally stridulate. I observed one male make four flights, about two feet in height and eight to ten feet in distance. Compared to *Arphia* species, flight displays are rare. Pair formation is often achieved without flight displays when males observe and approach moving females. During courtship males perform O-stridulation and V-stridulation. The former is of two kinds: the first consists of one to four strokes of both femora in unison; the second consists of a rapid series of strokes with one femur, produced in bursts of four to eighteen pulses. Vibratory stridulation is quite similar to that in *Arphia* species. It is produced only by stationary males that have just lost females and by wandering solitary males. In male–male encounters, femur-tipping is the most common signal, but when males touch, the femur-shaking signal is given.

HABITAT. In Canada the species is found in park lands and pine and spruce forest clearings (Brooks 1958). Walker (1895) notes that in Ontario it lives "on light sandy soil covered preferably with rather long grass and generally with other plants, as lupine, scrub oak, blueberries, etc." In New England it lives in "upland pastures, abandoned fields and unimproved lands, often partly overgrown with weeds and bushes" (Morse 1920). In the midwestern states it inhabits roadsides and old fields, occasionally along with *phoenicoptera.*

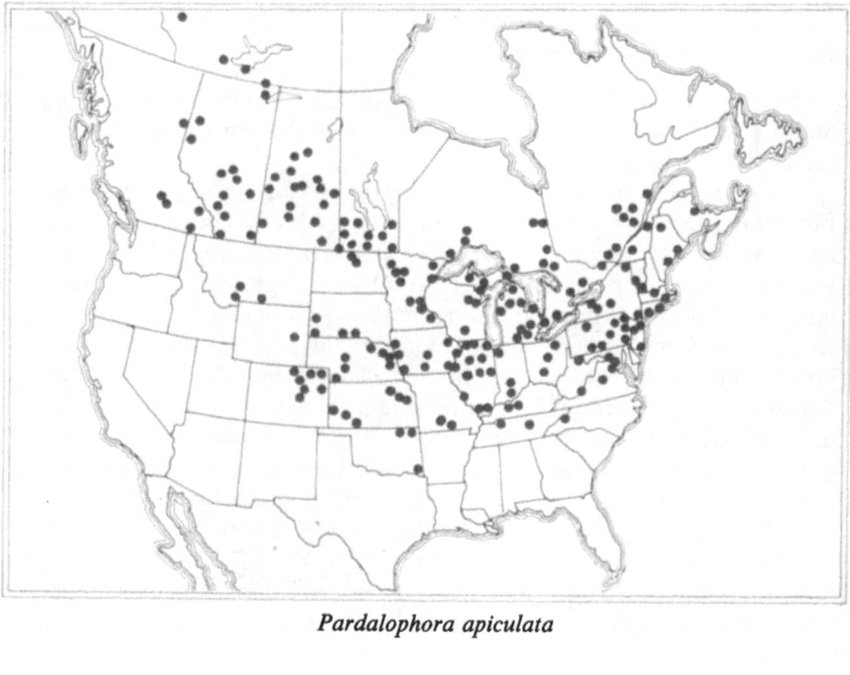

Pardalophora apiculata

LIFE CYCLE. The species overwinters in the fourth nymphal stage in Michigan, and adults are most common in late May and early June. Brooks (1958) claims that it has a two-year life cycle in the western provinces of Canada, overwintering in the egg stage the first winter and in a late nymphal stage during the second winter. In Iowa and South Dakota adults are most abundant from May to July (Hebard 1925b, Froeshner 1954). In eastern Kansas Smith (1981) collected nymphs from July through the following May; nymphs overwintered in the fourth nymphal stage, and adults were found from early April to early June.

REFERENCES. Walker 1898, Morse 1920, Hebard 1925b, Cantrall 1943, Brooks 1958, Otte 1970.

Pardalophora haldemani **(Scudder)** Pls. 7, 9

DISTRIBUTION. Oklahoma to eastern Montana and North Dakota and east through Iowa and Illinois to Michigan.

RECOGNITION. Figs. 33*b*, 34*b*. Basal area of hindwings is yellow or reddish. Forewing area between veins Cu_1 and Cu_2 is spotted (not as in *apiculata*). Differs from all other species in having the inner face of the hind femora unicolorous yellow, orange, or red (except for the crescent of the hind knee, which is dark brown). Posterior angle of pronotal disk is 90 degrees or less. Hind tibiae yellow, orange, or red. Body length to end of forewings 31–38 mm in males, 42–53 mm in females.

BEHAVIOR. Courtship is quite different from that of *apiculata* in that only femur-tipping and femur-shaking signals are given—these signals are performed only in male–male interactions in *apiculata*. During femur-shaking the hind tibiae strike the ground to produce a rapid, audible drumming. In male–male interactions femur-tipping is the main signal, and shaking is rare.

HABITAT. In northern Illinois the species is confined largely to open sandy areas with sparse vegetation (Blatchley 1920). Hebard (1936) reports: "It is one of the few species to reach its greatest abundance in the grasslands from east of the Great Plains to Illinois. In Kansas (Beamer 1917), it inhabits "old, rocky, cactus covered pastures and hillsides [and is] seldom found in tall grass and weeds." In the George Reserve in Livingston County, Michigan, Cantrall (1943) reports that *haldemani* was common prior to 1943, and *apiculata* was scarce. By 1968 *haldemani* had been completely replaced by *apiculata*, which was then very abundant. In eastern Kansas Smith (1981) reports it from native prairie to weedy and bare areas.

LIFE CYCLE. In eastern Kansas Smith collected nymphs from August through the following May. Overwintering is in the second, third, and fourth instars. Smith found adults from May to July. In Michigan (Cantrall 1943), it overwinters in the third and fourth nymphal stages, and adults were most common in late June and early July.

REFERENCES. Beamer 1917, Hebard 1936, Cantrall 1943, Otte 1970, Smith 1981.

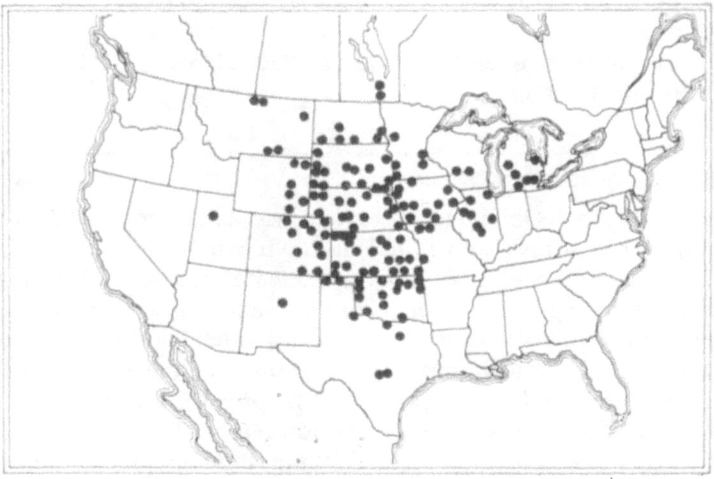

Pardalophora haldemani

Pardalophora phoenicoptera (Burmeister) Pls. 7, 9

DISTRIBUTION. Southeastern United States, north to Maryland, Ohio, and Illinois; west to central Texas and Oklahoma.

RECOGNITION. Basal area of hindwings is deep orange to rosy red. Head, pronotum, and hind femora are often greenish. Inner faces of hind femora are bluish and with two or three black bands (excluding knees); these may be obscured by dark blue coloration. Hind tibiae are yellow to orange. This species is likely to be confused with *saussurei* in the western part of its range, but it differs as follows: posterior angle of pronotal disk about 90 degrees and each posterior margin straight; inner faces of hind knees black. Body length to end of forewings 36–42 mm in males, 45–55 mm in females.

BEHAVIOR. At five locations where I collected this species, I saw no flight displays. O-stridulation was the only courtship signal. In male–male encounters femur-tipping and shaking signals were given.

HABITAT. Exposed sunny hillsides, open woodland meadows, rocky outcroppings, usually places with coarse grasses such as *Andropogon.* In Indiana Blatchley (1920) writes: "[Found in] uplands . . . open grain field, open bare places in the woods, roadsides and timothy meadows." Morse (1907) reports: "In early summer it is one of the commonest species of the genus from Arkansas eastward, frequenting fields and waste grounds, and sometimes entering the more open forests." In the eastern half of Oklahoma it inhabits grasslands and open meadows in woodlands (Coppock 1962). In Texas Isely (1935) refers to it as "an occasional open post oak, sandy soil species." In Kansas (Beamer 1917) it is found only in the eastern counties where "it is a lover of low ground, preferably covered by tall grass, weeds, shrubs and even trees. The dividing line between the habitat of this species and *haldemani* is quite marked. The former was often found occupying the brushy ravines and slopes of a hill, with the latter holding undisputed sway on the flat, barren top."

LIFE CYCLE. Overwinters in late nymphal stages. In southern Indiana adults occur from late June to July. In southeastern states adults occur from early spring to July.

REFERENCES. Morse 1907, Rehn and Hebard 1916, Beamer 1917, Blatchley 1920, Coppock 1962, Otte 1970.

***Pardalophora saussurei* (Scudder)** Pls. 7, 9

DISTRIBUTION. Northeastern Mexico north through central and eastern Texas to Oklahoma and Kansas.

RECOGNITION. Fig. 33*c*. Basal area of hindwing is yellow or orange-yellow. This species resembles *phoenicoptera* in having the inner face of the hind femora bluish and with two or three dark bands that may be obscured by dark blue coloration. It differs from *phoenicoptera* as follows: posterior angle of pronotal disk is usually less than 90 degrees, and each posterior margin is slightly concave; inner face of hind knees is not black, but dark brown on the crescent. *P. saussurei* may be confused with *Xanthippus corallipes;* it differs as follows: median pronotal carina is cut by one sulcus, and posterior angle of pronotal disk is much less than 90 degrees. Body length to end of forewings 39–50 mm in males, 53–63 mm in females.

BEHAVIOR. Flight displays were not observed in a relatively dense population in northeastern Texas. *Spharagemon cristatum* was also present and performed numerous displays.

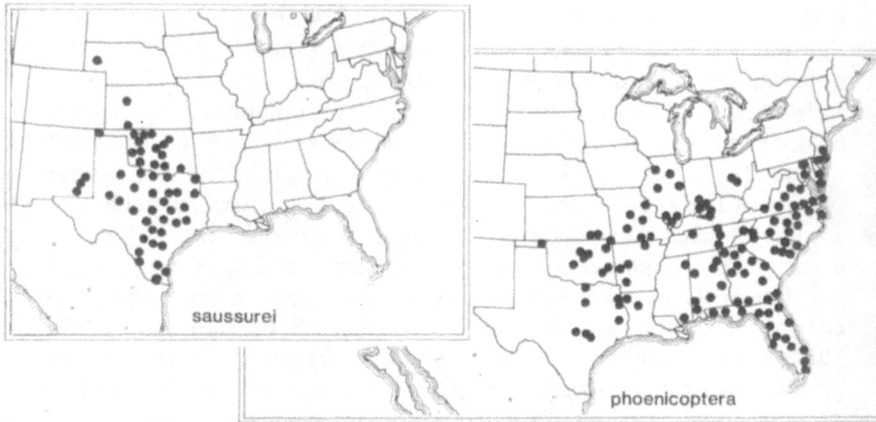

Pardalophora saussurei and *P. phoenicoptera*

HABITAT. At three localities in eastern Texas I found dense populations by the roadside and adjacent areas in semi-open and open vegetation with scattered bare spots of ground. Isely (1937) pictures a typical habitat with medium-density vegetation and with grass two or more feet in height.

LIFE CYCLE. In northeastern Texas the species overwinters in a late nymphal stage, and adults are most common in May and June (Isely 1937), but adults have been collected as late as September.

REFERENCES. Isely 1937, Otte 1970.

Genus XANTHIPPUS Saussure

DISTRIBUTION. Alaska to central Mexico extending east to Minnesota and Manitoba.

RECOGNITION. Figs. 33*f–h,* 34*c, d. Xanthippus* can be distinguished from *Cratypedes, Agymnastus,* and *Sticthippus* by the more distinct spotting on the forewings. These spots are numerous, rather large, and usually well-defined in *Xanthippus* (Pls. 8, 9). In four genera the pronotal disk is rather broken up in the prozonal region; usually the median pronotal ridge is low and indistinct between the two transverse sulci. In *Pardalophora* and *Hippiscus* the ridge is higher and is cut by one sulcus. However, specimens of *P. haldemani* sometimes have the crest cut twice and so may be confused with *X. corallipes* unless wing and leg color are examined. For other characters see identification of genera in the Hippiscus group.

REFERENCES. See references under each species.

▪ Key to Xanthippus Species

CALIFORNIA AND OREGON

1. Inner face of hind femora mostly reddish: **corallipes**
   Inner face of hind femora mostly dark blue: **olancha**

CANADA AND ALASKA

1. Top front of head distinctly horizontal: **aquilonius**
   Top front of head sloping downward: **2**
2. Forewings with distinct large spots. Hindwings rosy red. Hind tibiae deep orange to red at least on inner face. British Columbia, Alberta, and south: **corallipes**
   Forewings with smaller, less clearly defined spots. Hindwings pale yellow. Hind tibiae mostly yellowish, sometimes slightly orange. Yukon Territory and Alaska: **brooksi**

WESTERN UNITED STATES (EXCEPT CALIFORNIA)

1. Forewings with clearly defined dark spots: **corallipes**
   Forewings with poorly defined dark spots: **montanus**

***Xanthippus corallipes* (Haldeman)** Pls. 8, 9

TAXONOMY. This widespread species shows much geographic and altitudinal variation across its range and has troubled taxonomists for a long time (Hebard 1929: 337). It has been named eighteen times, eleven times by Scudder. The species ranges in altitude from low, hot grasslands to cool mountain meadows up to 11,000 feet or more. Specimens from mountain populations are consistently smaller than lowland ones, sometimes half the length (Alexander 1941). Mountain populations are also darker in body color, and the forewings of females often do not reach the end of the abdomen. Variation in hindwing coloration, color of the inner hind femora, and forewing spotting also has contributed to the recognition of many subspecies. Hebard studied a larger series of specimens from across the range of the species and discovered intermediate conditions, suggesting that most forms were the result of adaptation to local climatic conditions. I have not used the subspecific names here. Intermediate conditions between the various named forms are common and cannot be named with confidence. Hebard's specimen labels also indicate uncertainty as to the subspecies assignment of certain specimens.

On the subject of subspecies, Strohecker (1952) writes:

> Numerous names have been applied to specimens of this species, originally described from Utah material. Many of these names have been relegated to synonymy, but persisting in racial status are *pantherinus,*

latefasciatus, leprosus, cupidus, and *altivolus.* Rather than showing geographic correlations, these "races" occur in certain environments over much of the range of the species. *X. corallipes buckelli* . . . however, is a definite geographic race of the extreme northwestern part of the range of the species. Two seasons of collecting in the vicinity of Las Vegas and New Mexico yielded interesting lots of *corallipes.* At Montezuma, in the Gallinas Canyon, I took small, dark specimens, the males about an inch in overall length, the females about one and one-eighth inches but with tegmina not shorter than the body. According to the descriptions these specimens are referable to *cupidus* or *altivolus.* Not over eight miles from Montezuma, on the plains northeast of Las Vegas, are found individuals best referred to the race *leprosus,* while at La Liendre, twenty miles southeast of Las Vegas, occur individuals of similar size but lighter coloration and among these some specimens showing the elytral pattern and dark inner face of the hind femur characteristic of the race *pantherinus.* While the three localities cited are close together, they undoubtedly have different moisture characteristics; the plains vegetation is wholly grass, including grama, while at La Liendre grass is scarce and yucca abundant. Quite certainly the floor of Gallinas Canyon, in addition to greater moisture, has lower average temperatures. Subject to air drainage from high mountains and protected from direct sunlight for much of the day by the canyon's steep walls, the water of Gallinas Creek remains frozen long after ice has disappeared elsewhere.

It appears from examination of these and other specimens from New Mexico, Arizona, Utah, and Nevada that about all gradations of size, elytral maculation and wing length may be encountered within rather small areas widely scattered in the range of the species. While genetic bases for these differences are not ruled out, it seems probable, in light of such experiments as Faure's that *X. corallipes* is a species of great adaptability and plasticity and the races mentioned above are ecophenes rather than genetic units.

Xanthippus affrictus Scudder is placed as a synonym of *X. corallipes* on the basis of the lectotype. I do this with some uncertainty, for the female specimen has smaller spots than typical Arizona *corallipes.* Other members of the type series were from the upper Missouri River, Dakota, Nevada, Colorado, and Utah. Hebard (1929: 337) states: "[The lectotype] represents a species closely related to *Xanthippus montanus* in color and markings, but is structurally distinct."

DISTRIBUTION. Central Mexico to British Columbia; California east to Texas, Oklahoma, and the Dakotas.

RECOGNITION. Figs. 33*f,* 34*c.* This species varies considerably across its range. Mountain populations are consistently darker and smaller than lowland populations. Pl. 8 shows the range of variation known in the species. Forewings have clearly defined dark spots and always have posteriorly converging pale stripes on the dorsal field.

Inner faces of hind femora are usually bright orange, bright red, or deep red, but in the Great Plains and the grasslands of New Mexico, West Texas, Arizona, and Mexico they may be dark blue. Hind tibiae are usually orange or red; in some individuals they are red on the inner face and yellow on the outer face.

VARIATION. (*a*) Great Plains, low-altitude grasslands of West Texas, New Mexico, and Arizona (*corallipes corallipes* and *corallipes pantherinus*). Body length to end of forewings 35–48 mm in males, 45–65 mm in females. Inner face of hind femora either red or deep blue. Basal area of hindwings always yellow. Specimens with blue inner femora can be confused with *Pardalophora saussurei* and *P. phoenicoptera,* but in *X. corallipes* the median pronotal crest is cut by two sulci (Fig. 34*f*), and the pronotal disk lacks the acutely angulate posterior margin seen in *P. saussurei* (Fig. 34*c*). Specimens with red inner femora can be confused with *P. haldemani,* but the latter have the median pronotal crest cut by one sulcus and are often orange or yellow on the inner femoral face. *X. corallipes* can also be confused with *X. montanus,* but in the latter species the dark spots on the forewings are more irregular in shape and size.

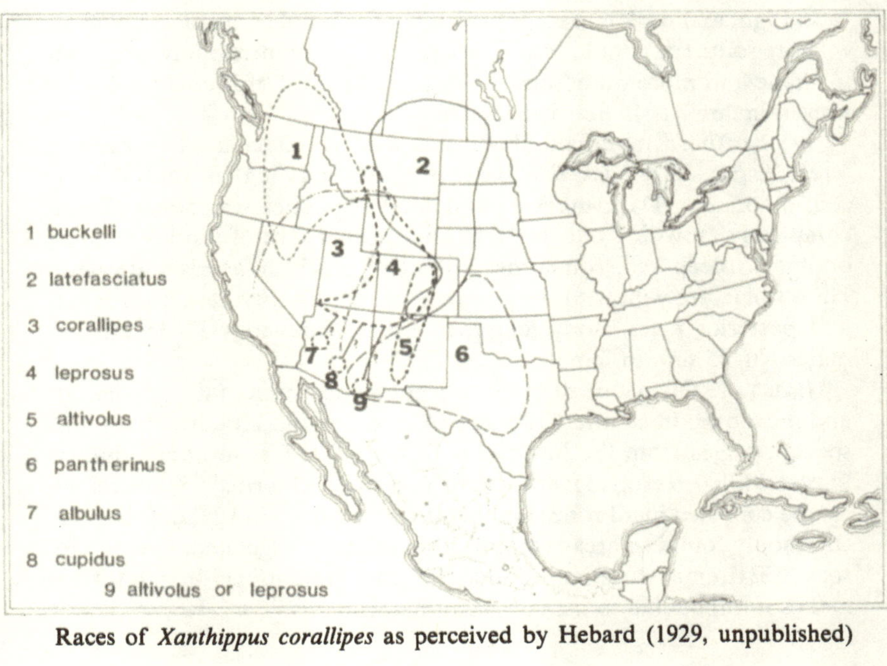

Races of *Xanthippus corallipes* as perceived by Hebard (1929, unpublished)

(*b*) Mountains of Colorado, New Mexico, and Arizona (*X. corallipes altivolus* and *X. c. leprosus*). Mountain populations (above 6,000 feet) are much smaller in size: body length to end of forewings 25–33 mm in males; to end of abdomen 26–36 mm in females. Basal area of hindwings is pale greenish yellow. In females, forewings are relatively shorter than in males and often do not reach the end of the abdomen. Body color is also darker, and dark spots often coalesce, leaving only traces of pale areas normally surrounding dark markings. Inner face of the hind femora is orange to red but often shows traces of two dark stripes on the upper side.

(*c*) Utah. Basal area of hindwings is yellow or pink. Body markings are often dark and coalesce to crowd out the pale areas. Mountain populations are similar to those of Colorado mountains. Low-altitude populations are similar to Great Plains populations. Hind tibiae are yellow to deep orange, inner faces of hind femora orange to orange-red. Female forewings in mountain populations do not reach the end of the abdomen.

(*d*) Nebraska and Dakotas (*X. corallipes latefasciatus*). From Oklahoma through Kansas and Nebraska there is a gradual transition in body shape and coloration. Basal area of hindwings is yellow. Inner face of hind femora is always red to deep red. Hind tibiae are almost always bright red on inner face and often yellow or orange on outer face. Light and dark markings on forewings do not contrast as much as in more southern populations. Dark markings are often more irregular in shape and more numerous. Body length to end of forewings 30–36 mm in males, 41–46 mm in females.

(*e*) Northern mountain states and western Canada (*X. corallipes latefasciatus, X. c. buckelli*). Basal area of hindwings is yellow, rosy red, or orange. Dark markings on forewings coalesce, often almost completely crowding out the light areas. Inner face of hind femora is orange to deep red, sometimes with two blackish bands indicated. Hind tibiae are yellow to deep orange or red, usually darker on inner and posterior faces. Body length to end of forewings 32–38 mm in males, 40–45 mm in females.

HABITAT. Open dry prairies and grasslands, mountain grasslands and meadows. In central Colorado (Alexander and Hilliard 1969) this species ranges from the low plains below 6,000 feet to above timber line at 11,000 feet. Scoggan and Brusven (1972) write: "*X. corallipes* can be collected in all zones and plant communities in Idaho. It is most commonly found in areas varying from sparsely vegetated; gravelly to rocky. Extreme variation in floristic and grasshopper composition makes it impossible to list consistent associations."

LIFE CYCLE. In Texas this species overwinters in a late juvenile instar. Adults are most common in May and June (Isely 1937). In central Colorado (Alexander and Hilliard 1969) this species has a two-year life cycle, spending the first winter in the egg stage and the second as juveniles (see also Pickford 1953, Hewitt and Barr 1967).

> Adults appear by May 1 up to 7,000 feet, by the end of May at 8,500 feet and before the end of June at 11,000 feet. Adults persist into mid-July on the plains and as late as mid-August, rarely later at 6,700 feet or higher. *X. c. leprosus* hatches at both 6,700 and 8,500 feet from last of July through August, the higher altitudes apparently having no significant delaying effects. Most juveniles are in the third instar at the onset of winter, and these individuals can withstand freezing (Alexander 1967). Studies during the winter of 1957–1958 a few miles northwest of Boulder indicated that third-instar juveniles predominated until March. In early April, more fourth than third instars were found; fifth-instar juveniles appeared early that month and increased in relative numbers during that month. [At 11,000 feet on "Baldy"] hatching is apparently during the second week of August and perhaps later. (Alexander and Hilliard 1969)

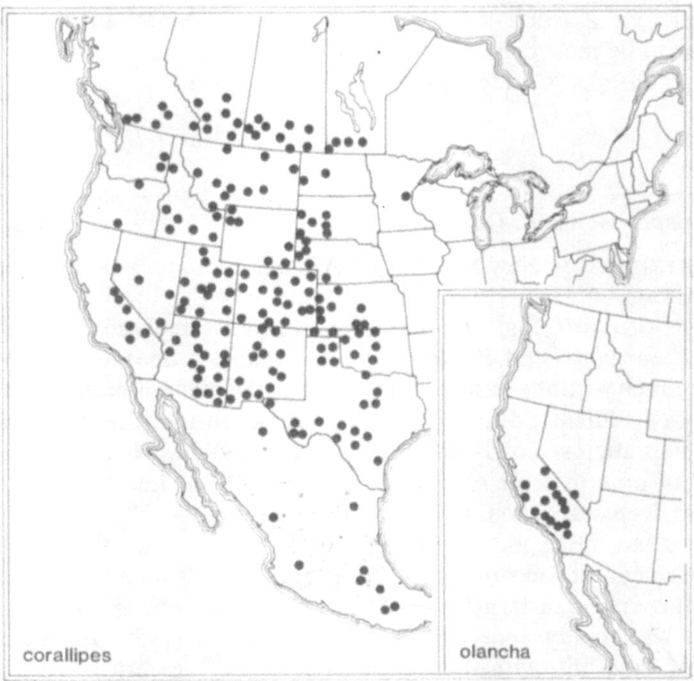

Xanthippus corallipes and *X. olancha*

REFERENCES. Hebard 1929, Alexander 1941, Ball et al. (1942), Pickford 1953, Hewitt and Barr 1967, Alexander 1967, Alexander and Hilliard 1969, Scoggan and Brusven 1972.

Xanthippus olancha (Caudell) Pls. 8, 9

DISTRIBUTION. Mountains of southern California.

RECOGNITION. Basal area of hindwings orange to rosy red. Forewings with strongly contrasting light and dark markings. Overall body color light to dark gray and black. Inner face of hind femora blue through the basal three-quarters, with two black bands and a red band in the narrowest part. Hind tibiae entirely yellow, or yellow on the outer face and orange to red on the inner. Body length to end of forewings 30–36 mm in males, 44–55 mm in females.

HABITAT. Strohecker, Middlekauff, and Rentz (1968) report that the species is "not uncommon in the central and western desert areas of southern California." Their distribution map indicates, however, that the distribution is associated with mountain ranges.

LIFE CYCLE. Adults have been collected from March to August but appear to be most common from May to July.

REFERENCES. Caudell 1921, Strohecker, Middlekauff, and Rentz 1968.

Xanthippus montanus (Thomas) Pls. 8, 9

DISTRIBUTION. New Mexico and Arizona north to Alberta and Saskatchewan.

RECOGNITION. Fig. 33*g*. This species is most likely to be confused with *X. corallipes* and *Pardalophora haldemani*. It does not overlap in range much with the latter species and has the median pronotal carina cut by two sulci. It differs from *X. corallipes* in that the spots on the forewings are less clearly defined. Basal area of hindwings usually yellow, but pink in some specimens from Wyoming. Inner face of hind femora deep orange to red. Hind tibiae occasionally yellow, usually orange or red, often more brilliantly colored on the inner face. Body length to end of forewings 30–40 mm in males, 40–46 mm in females.

HABITAT. In eastern Colorado and Nebraska the species is associated with sandy areas (Hauke 1953). Hebard (1929) reports it in "rolling sand hills with short grass patches and sage brush." In Idaho Hewitt and Barr (1967) note that it inhabits flat dry areas with *Atriplex* and *Artemisia.* I found it on somewhat sandy rolling grassland near Carrizozo, New Mexico, along with *corallipes.*

LIFE CYCLE. Hewitt and Barr (1967) report a two-year life cycle; eggs and nymphs overwinter. In Idaho adults are found as early as April, and by August they have disappeared.

REFERENCES. Hebard 1929, Hauke 1953, Hewitt and Barr 1967.

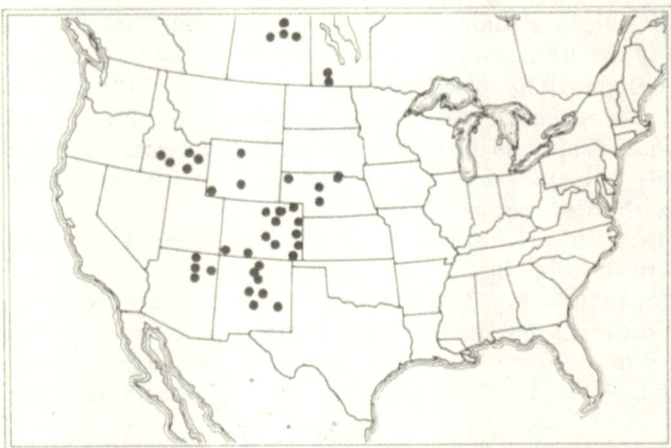

Xanthippus montanus

Xanthippus brooksi Vickery Pls. 8, 9

TAXONOMY. This species was originally described as a subspecies of *corallipes.* I consider it a distinct species for the reasons given below. Intervening populations could reconcile Vickery's and my opinions on the taxonomic status.

DISTRIBUTION. Southern Alaska and Western Yukon Territory.

RECOGNITION. Basal area of hindwings pale yellow. This species differs from *corallipes* as follows: spots on hindwings are more irregular and smaller; inner face of hind femora banded with black in males; dorsal field of the folded forewings has small dark spots; converging pale dorsal wing stripes are often wider and join before the middle of the forewing; hind tibiae mostly yellowish, seldom orange, and never red. Body much smaller. These differences appear to me to be at least as great as between *corallipes* and *montanus.* Body length to end of forewings 24–27 mm in males, 31–38 mm in females.

HABITAT. Not discussed in the literature.

LIFE CYCLE. Adults have been collected in May, June, July, and August.

REFERENCES. Vickery 1967.

***Xanthippus aquilonius* Otte, n. sp.** Pls. 8, 9

TAXONOMY. Recognition of this species is based upon thirty-seven specimens from Vernon, British Columbia; two from Rock Creek, B.C.; and two from Osoyoos, B.C. It is most similar to *corallipes,* from which it differs in forewing spotting and the shape of the forehead (see below). A single male of *corallipes* collected along with a larger series of *aquilonius* supports the idea that these are two separate species. The similarity to *Cratypedes lateritius* is not as close, but the features that distinguish these species are also given below.

DISTRIBUTION. Known only from Vernon, Rock Creek, and Osoyoos, British Columbia.

RECOGNITION. Fig. 34*d.* This species is best compared to *X. corallipes* and *Cratypedes lateritius* from British Columbia. Like those two species it has rosy red hindwings. It differs from one or both species as follows: hindwings with a relatively broader outer dark band; top front

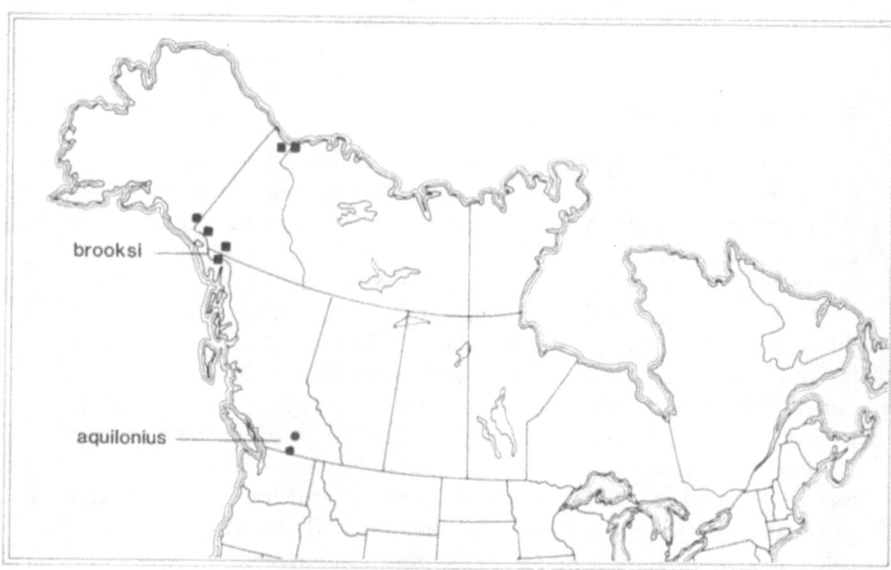

Xanthippus brooksi and *X. aquilonius*

of head more distinctly angulate and fastigium more horizontal; spots on forewings smaller and less clearly defined than in *C. lateritius;* inner face of hind femora with red coloration as in *X. corallipes,* not as in the yellow and black of *C. lateritius* from British Columbia; converging stripes on the dorsal field of the forewings always present (sometimes absent in *C. lateritius*). Hind tibiae yellow to orange. Forewings sometimes strongly spotted, sometimes almost unicolorous. Body length to end of forewings 28–34 mm in males, 39–44 mm in females.

HABITAT. Not known; probably open meadows.

LIFE CYCLE. Adults of this species were collected in May and June.

SPECIMENS. Holotype male, Vernon, B.C., May 31, 1929 (E. R. Buckell), Philadelphia Academy of Natural Sciences. *Paratypes* (all Philadelphia Academy of Natural Sciences): Vernon, B.C., May 26–31, 1929 (E. R. Buckell), seven males, three females; May 9, 1930, five males, four females; May 11, 1931, three males; May 17, 1937, six males; May 8, 1931, one male; May 22, 1930, two females. Rock Creek, B.C., June 25, 1922 (E. R. Buckell), two females. Osoyoos, B.C., June 10, 1910 (W. B. Anderson), two males.

Genus CRATYPEDES Thomas

TAXONOMY. Previously only *C. neglectus* was placed in this genus. Here we also include *lateritius,* formerly under *Xanthippus.* These two species are confusingly similar; only the ventral widening of the lateral lobes can be reliably used to separate them.

RECOGNITION. Figs. 33*e,* 34*e.* This genus and *Agymnastus* probably share a common stem. They differ from *Xanthippus* in lacking numerous distinctly defined spots. *Cratypedes* differs from *Agymnastus* as follows: females of *Cratypedes* resemble the males and are capable of flight; those of *Agymnastus* are fat-bodied, flightless, and much bigger than the males; males of *Agymnastus* have blue and orange on the inner face of the hind femora, while *Cratypedes* is usually banded with black and yellow or orange and yellow.

REFERENCES. Thomas 1876.

• Identification of Cratypedes Species

neglectus
Lateral lobes of pronotum widen ventrally

lateritius
Lateral lobes have parallel sides or narrow ventrally

***Cratypedes neglectus* (Thomas)** Pls. 8, 9

DISTRIBUTION. Widespread through the northwestern mountains of the United States and western Canada.

RECOGNITION. Basal area of hindwings always yellow. Lateral lobes of pronotum wider near the bottom than the top. Rostrum and fastigium distinctly shortened compared to the similar *Cratypedes lateritius* and *Agymnastus ingens.* Inner face of hind tibiae yellow with two black bands (excluding knees), or orange to red with distinct or faint dark bands. Hind tibiae almost always orange, deep orange or red-orange, but sometimes yellow or greenish yellow in the Sierra Nevada of California. Forewings often with a charcteristic light and dark pattern. Dorsal field of folded forewings usually has distinct ivory stripes converging posteriorly, but these are sometimes lacking. Body length to end of forewings 24–32 mm in males, 33–40 mm in females.

HABITAT. Sunny clearings in mountain woodlands and brushy mountain slopes. Hewitt and Barr (1967) note: "In Idaho *neglectus* occurs primarily in mountainous areas where it frequents meadows and grassy hillsides. However, it also can be found in drier areas along with sage brush inhabiting species." In Idaho Scoggan and Brusven (1972) also report: "*C. neglectus* has a diverse range of habitats. It can be collected from all of the major community types except the higher portions of the mountain meadow community. It reaches greatest abundance over the Snake River Plains, particularly where sage brush is minimum. [It] probably would not reach economic numbers, but does inhabit prime range communities." In Utah Henderson (1924) notes that the species is occasionally found in higher dry-farm grain fields.

LIFE CYCLE. In Idaho Hewitt and Barr (1967) note: "This species can be collected throughout the summer months and is especially abundant during the months of June, July and August." Idaho hatching time varies from early May to mid-June, depending on altitude (Scoggan and Brusven 1972). In California adults have been taken from May to August (Strohecker, Middlekauf, and Rentz 1968).

REFERENCES. Henderson 1924, LaRivers 1948, Hewitt and Barr 1967, Strohecker, Middlekauff, and Rentz 1968, Scoggan and Brusven 1972.

***Cratypedes lateritius* (Saussure)** Pls. 8, 9

TAXONOMY. This species is quite variable across its range and has been described four times. Here, for the first time, the names *calthulus*

Saussure, *griseus* Scudder, and *sierra* Rehn are treated synonyms of *lateritius*. I base this synonomy on examination of over 300 specimens from across the northwestern United States and British Columbia. Concerning the nominal species *sierra*, Rehn writes: "Closely allied to *Hippiscus calthulus* Saussure from Nevada and *croceus* Scudder from Idaho. The relationship to the former is not as close as to the latter." I have examined the relevant types and believe *croceus* to be a synonym of *C. neglectus*. Rehn's specimens of *sierra* tend to have a smoother (less tuberculate) pronotal disk than other populations outside of California, but individual variation in this character is common, making it doubtful that more than one species is involved. Also the color of the hindwings, the inner faces of the hind femora, and the tibiae vary in a seemingly haphazard way. Such a lack of concordance also leads to the conclusion that a single species is involved (see Variation).

DISTRIBUTION. Northwestern United States and British Columbia.

RECOGNITION. This species is most easily confused with *C. neglectus* and *Agymnastus* males. Unlike *C. neglectus*, in *lateritius* the lateral pronotal lobes become narrower ventrally, and the fastigium and rostrum (dorsal view) are not so short. In *Agymnastus* the hind femora are largely blue-black and have a red band in the narrowest part, and the females are stout with short wings.

Basal area of hindwings yellow, rosy red, pink, or rarely somewhat orange. Forewings highly speckled or with fewer, highly irregular-shaped darker markings. Folded forewings with or without converging ivory stripes along the wing angle. Males have these stripes more often than females. Female forewings in some Utah mountain populations do not reach the end of the abdomen. Inner face of hind femora quite variable in color; usually with two black bands or traces of them next to the knees, a broad one near the narrowest part; background color varies from orange to red. Hind tibiae yellow to orange. Body length to end of forewings 26–33 mm in males, 33–46 mm in females.

VARIATION. (*a*) British Columbia. Basal area of hindwings varies from rosy red to slightly orange-red. Inner face of hind femora yellow with black bands. (*b*) Washington. Like British Columbia populations, but inner face of hind femur bluish along lower marginal area. (*c*) Idaho. As in British Columbia, some specimens are reddish along inner ridges of hind femora. (*d*) Oregon. Basal area of hindwings yellow or rosy red. Inner face of hind femora yellow with black bands or orange with black bands. Most specimens with rosy red wings have orange or reddish tibiae. Most yellow-winged specimens have converging stripes on the dorsal field. (*e*) Utah. Basal area of hindwings yellow or rosy red. Inner face of hind femora always deep orange or reddish, dark bands distinct to indistinct. Hind tibiae orange to deep orange, rarely yellowish. Forewings of females from mountains (Markagunt Plateau, Duck Lake, and the top of the Parowan Mountains) relatively short, sometimes not reaching end of abdomen. (*f*) California and Nevada. Basal area of hindwings yellow, orange, or rosy red. Inner face of hind femora yellow with two black bands, or orange or reddish; if orange or reddish, the dark bands may be quite indistinct. Hind tibiae yellow, orange, or reddish; orange or red color usually most pronounced on inner or posterior faces.

HABITAT. Open pine woodlands with grassy understory in California. In Idaho Hewitt and Barr (1967) report that it inhabits "gravelly soil and is found where the sagebrush extends northward into the timbered mountainous areas." In Idaho the species inhabits the northeastern portion of the Snake River plains (Hewitt and Barr 1967).

LIFE CYCLE. In Idaho Hewitt and Barr (1967) report: "Adults of this species can be found in May and early June. It has a two-year life cycle with the eggs being laid in the latter part of June and throughout July." In California and Nevada adults have been collected from May to September, in Washington from May to July, in British Columbia only in May, in Idaho in July and August, in Utah from April to July, in Oregon from May to June.

REFERENCES. Hewitt and Barr 1967.

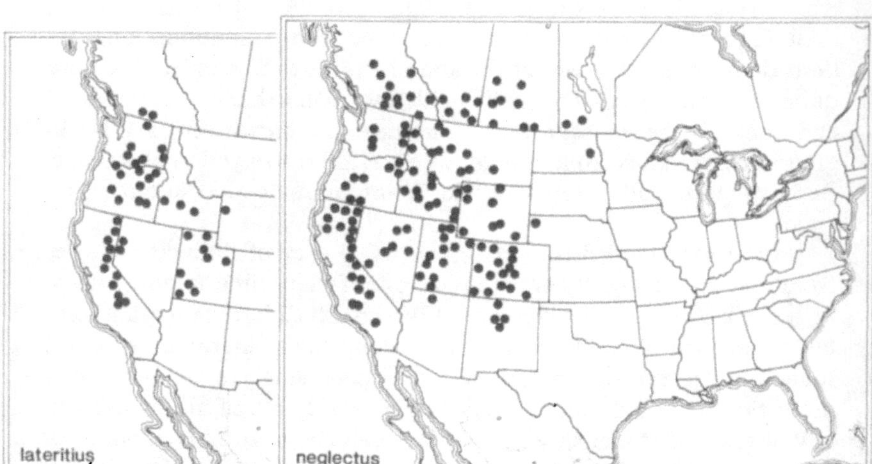

Cratypedes neglectus and *C. lateritius*

Genus AGYMNASTUS Scudder

DISTRIBUTION. California.

RECOGNITION. Figs. 33*a*, 34*f*. This genus is probably most closely related to *Cratypedes*. To distinguish the two, see *Cratypedes* Recognition. This is the only genus of North America Oedipodinae that has flightless females.

REFERENCES. Strohecker, Middlekauff, and Rentz 1968, Rentz 1978.

• Identification of Agymnastus Species

ingens

1. Upper and lower ridges of hind femora narrow suddenly near narrowest point (males)
2. Greatest width of pronotum equals at least half the length of hind femora (males)
3. Mesosternal interspace distinctly wider than metasternal interspace (males)
4. Female forewings do not reach end of abdomen

venerabilis

1. Upper and lower ridges of hind femora narrow gradually (males)
2. Greatest width of pronotum less than half the length of hind femora (males)
3. Mesosternal and metasternal interspaces nearly equal in width (males)
4. Female forewings extend at least to end of abdomen

Agymnastus ingens (Scudder) Pls. 7, 9

DISTRIBUTION. Widespread through hills and mountains in the northern two-thirds of California.

RECOGNITION. Males and females are quite different in size and shape. Males: Basal area of hindwings pale yellow to deep yellow. Overall body color dark brown, but dorsal field of forewings has two converging pale streaks. Inner face of hind femora dark blue and black, with a red band at the narrowest point. Hind tibiae orange to red. *A. ingens* differ from *venerabilis* as follows: upper and lower flanges of hind femora narrow suddenly near narrowest point (gradually in *venerabilis*); greatest width of hind femora at least 0.3 times the greatest length (less than 0.3 in *venerabilis*); greatest width of pronotum (top view) at least half the length of the hind femur (less than half in *venerabilis*); mesosternal interspace distinctly wider than metasternal interspace (nearly equal in width in *venerabilis*). Body length to end of forewings 29–34 mm.

Females: Females of this species are unusual among North American Oedipodinae. They are much wider than males, and the forewings do not reach the end of the abdomen. They appear distinctly pebble-like and escape predators mainly through camouflage. Pronotum is extremely rough. Forewings lack pale lines on dorsal field. Hindwings with a yellow basal area. Inner face of hind femora blue and black except for a red or orange band at the narrowest part. Hind tibiae deep orange. Body length to end of abdomen 33–47 mm.

HABITAT. Hillsides with short grass and much exposed soil. The females resemble pebbles or lumps of soil, are incapable of flight, and are reluctant to move.

LIFE CYCLE. Adults from May to July, most common in May and June.

REFERENCES. Strohecker, Middlekauff, and Rentz 1968.

***Agymnastus venerabilis* Rentz** Pls. 7, 9

DISTRIBUTION. Mountains of eastern California and Nevada.

RECOGNITION. This species is very similar to *ingens* (see *ingens* for discussion of differences). Females of *venerabilis* are much narrower

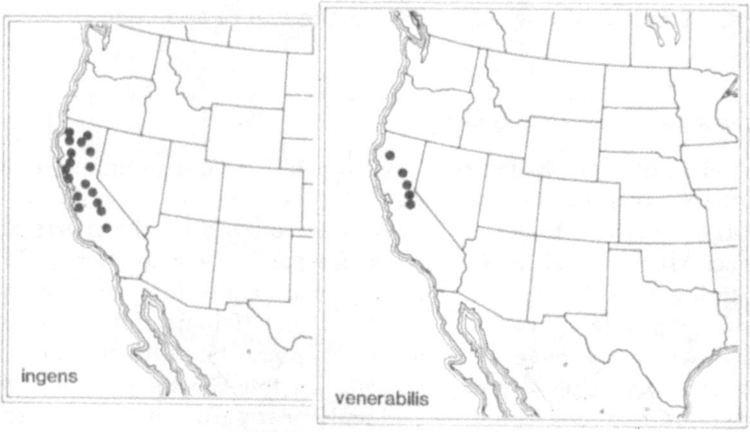

Agymnastus ingens and *A. venerabilis*

and more usual in body form. The forewings extend at least to the end of the abdomen and have distinct pale markings. Inner face of hind femora blue; in females the pale ring at narrowest part is orange to yellowish. Body length to end of forewings 38–40 mm in males, 29–32 mm in females.

HABITAT. Rentz (1978) found *venerabilis* in grassy meadows among giant granite boulders along the Stanislaus River, California.

LIFE CYCLE. Adults have been collected in May, June, July, and August.

REFERENCES. Rentz 1978.

Genus STICTHIPPUS Scudder

DISTRIBUTION. California.

RECOGNITION. Figs. 33*i*, 34*g*. Fastigium wider than long. Male epiphallus not distinctly trilobate, and pronotum not so rough as in *Cratypedes* and *Xanthippus.* See also *S. californicus* Recognition.

REFERENCES. Strohecker, Middlekauff, and Rentz 1968.

Sticthippus californicus (Scudder) Pls. 7, 9

TAXONOMY. *S. californicus* was initially described under *Hippiscus,* then transferred to reside alone in *Sticthippus.*

DISTRIBUTION. Widespread through California in grassy foothills.

RECOGNITION. Figs. 33*i*, 34*g*. Overall body color pale brown to straw-yellow. Forewings distinctly spotted. Basal area of hindwings pale yellow. Dorsal field of forewings always with two pale converging stripes. Upper face of hind femora with four dark bands (including knees). Inner face of hind femora with black knee, yellow to orange background, and usually two dark bands that are darkest dorsally. This species differs from *Xanthippus corallipes* and *olancha* in that the pronotum is smoother and the fastigium is wider than long. Body length to end of forewings 28–32 mm in males, 31–39 mm in females.

HABITAT. Strohecker, Middlekauff, and Rentz (1968) write: "This species is typically an inhabitant of dry grassland areas of canyons and foothill regions, being closely associated with *Oedaleonotus enigma* over much of its range. It is sometimes of economic importance, especially in the western foothill area of the Central Valley from Merced south to Mendota in Fresno County, in the Cuyama Valley of Santa Barbara County and in the Panoche Valley of San Benito County."

LIFE CYCLE. According to Strohecker, Middlekauff, and Rentz (1968): "It appears early in the spring, somewhat earlier than do other economic species, hatching commencing in early March. Adults predominate by May, when egg laying starts. Adults disappear in July. A series of dry years is thought to be conducive to outbreaks."

REFERENCES. Strohecker, Middlekauff, and Rentz 1968.

Sticthippus californicus

Genus LEPRUS Saussure

DISTRIBUTION. Southwestern United States to central Mexico.

RECOGNITION. Figs. 33*k*, 34*l*. Forewings with characteristic broad crossbands and with two very thin converging streaks on dorsum. Pronotum without a median crest and not strongly sculptured; median line cut by two sulci. Head short and broadly rounded in profile. Fastigium largely convex and with a faint median carinula.

REFERENCES. Scudder 1897. See references under each species.

• Identification of Leprus Species

intermedius

1. Basal area of hindwings blue; turquoise in southern California and Baja California; largely black in parts of Nevada.
2. Anterior end of pronotal disk with definite shoulders
3. Lateral lobe wider near bottom than near middle
4. Hind tibiae blue
5. Dark band on hindwings relatively broad

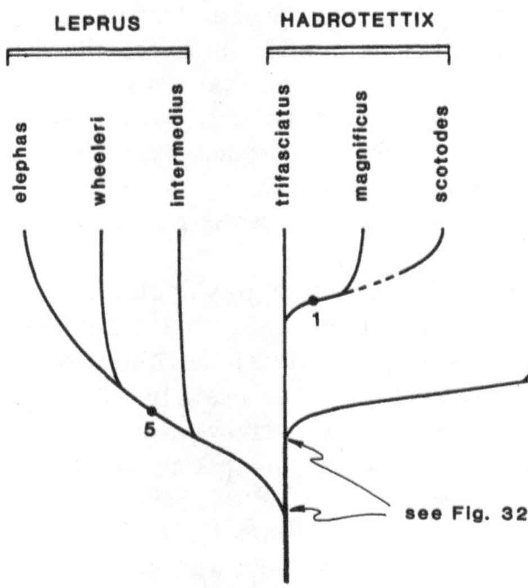

Fig. 36. Hypothetical phylogeny of *Leprus, Hadrotettix,* and *Heliastus* species (see Appendix 1 for explanation).

wheeleri

1. Basal area of hindwings usually yellow-green, occasionally blue
2. Anterior end of pronotal disk rounds gradually into lateral lobes
3. Lateral lobe narrower near bottom than near middle
4. Hind tibiae pale blue to yellow-gray
5. Dark band on hindwings relatively narrow

elephas

1. Basal area of hindwings yellowish
2. Anterior end of pronotal disk usually with definite shoulders
3. Lateral lobe narrower near bottom than near middle
4. Hind tibiae pale blue to yellow-gray
5. Dark band on hindwings relatively narrow

Leprus intermedius Saussure Pl. 10

TAXONOMY. *L. cyaneus* Cockerell is here treated as a synonym of *intermedius.* Since the name *cyaneus* has been so widely used, this name change is unfortunate but unavoidable. Bruner's *interior* is also a synonym. Of this species Bruner (1905: 160) states: "The present insect comes closest to *cyaneus* Cockerell but differs from it in the much more slender hind femora, the relatively smoother head and pronotum, and the shorter and more slender antennae, in addition to the characters [given in the key]." The hind femora are somewhat narrower on the average in specimens from the Salt Lake City area, but all intermediate conditions exist. Other characters mentioned by Hebard fall within the range of variation seen in *intermedius.*

DISTRIBUTION. Widespread over the southwestern United States and northwestern Mexico.

RECOGNITION. Basal area of hindwings blue-gray to blue over most of the range, becoming turquoise in southern California and Baja California. In parts of Nevada (Spring Mountain) the hindwings are largely black but retain blue coloration at the base. In *wheeleri* the hindwings are occasionally blue but usually yellow-green. In *elephas* they are yellowish. In addition to wing color this species can be distinguished from the other two species as follows: anterior end of pronotal disk has definite shoulders (*elephas* is sometimes similar; in *wheeleri* the disk rounds off onto the lateral lobes without a shoulder); lateral lobe is wider near the bottom than near the middle (narrower near the bottom than near the middle in *wheeleri* and *elephas*); hind tibiae blue (pale blue to yellow-gray in *wheeleri* and *elephas*); dark band on hindwings broader; abdomen often has much blue on ventral side (*wheeleri* has blue only in blue-winged specimens). Body length to end of forewings 24–40 mm in males, 35–49 mm in females.

VARIATION. The smallest specimens of this species are from southern coastal California (Santa Barbara and San Luis Obispo areas). The largest ones are from west Texas, New Mexico, and Arizona. The hindwings are often greenish-blue or turquoise in coastal California and Baja California and pale to dark sky-blue over most of the range. The outer black band on the hindwings varies in width and intensity; it is broadest in specimens from Spring Mountain, Nevada, where the black area covers most of the hindwings and the transition from black to blue is gradual. In specimens from northern Baja California the black band is both broad and intense.

HABITAT. In west Texas Tinkham (1948) writes: "[*intermedius*] is found in the mountains whereas *wheeleri* is a species of the plains; [*intermedius*] is seldom found below 5,000 feet elevation, being commonest at about 6,000 feet." In Arizona Ball et al. (1942) note: "A fairly common early season species on rocky hillsides and ridges, also found to some extent in bare areas in the grassland. Lower and Upper Sonoran zones." Strohecker, Middlekauff, and Rentz (1968) note that in California the species inhabits desert wastelands, rocky ridges, and hillsides where vegetation is sparse. Rehn and Hebard (1906, 1909) collected the species on bare, volcanic mountain sides: "[so] steep was the hillside that it was bare of everything but scattered grasses and an occasional sage-brush. When at rest the insects were wonderfully protected by their coloration." In southern California Rehn and Hebard (1909) found it "on rocky hillsides and in rolling grain fields. On Santa Catalina Island it was uncommon in dry places, among rocks and scant grasses in a canyon bed . . . On Echo Mountain the species was numerous at 3,100 feet elevation." Weissman collected the species in cultivated areas in northern Baja California.

LIFE CYCLE. Adults from June to August in California, April to November in northern Baja California, March to September in Arizona, June to October in west Texas, July to September in Nevada and Colorado.

REFERENCES. Ball et al. 1942, Tinkham 1948.

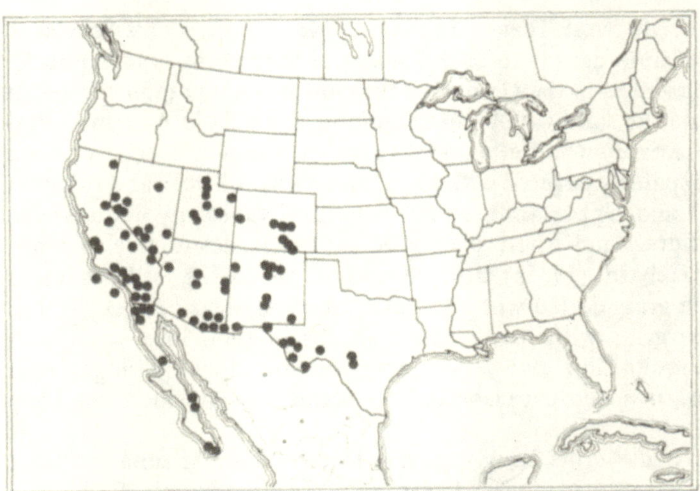

Leprus intermedius

Leprus wheeleri (Thomas) Pl. 10

TAXONOMY. Hebard's *Leprus robustus* is here treated as a synonym of *wheeleri.* Concerning *robustus* Hebard (1935a: 122) writes: "Close relationship is shown to *wheeleri* (Thomas), a more regularly and boldly striped insect with wing disk clear light yellow, but large series of that insect before us from eastern New Mexico and western Texas show no intergradation." Intermediates in hindwing color between the two nominal species, *wheeleri* and *robustus,* from west Texas, New Mexico, Arizona, and Sonora are represented in the Philadelphia Academy of Natural Sciences collection. The boldness of striping varies sufficiently to suggest that only one species is represented.

DISTRIBUTION. Texas to Arizona and northern Mexico.

RECOGNITION. Figs. 33*k*, 34*j*. This species differs from *intermedius* as follows: hindwings usually yellow or greenish yellow, occasionally blue; if blue, the narrower outer dark band is diagnostic. Lateral lobes become continuously narrower ventrally (slightly narrower in the middle section in *intermedius*). Anterior end of pronotum does not have a slight shoulder separating the lateral lobes from the disk. Hind tibiae usually yellowish to pale blue-gray, but blue in blue-winged specimens. In *elephas* the hindwings are pale yellow and the pronotum has a slight shoulder separating the disk from the lateral lobes. Body length to end of forewings 35–45 mm in males, 45–57 mm in females.

HABITAT. In west Texas Tinkham (1948) writes: "This species is much more abundant than *intermedius* and in the grasslands up to 5,000 feet elevation in the Chinati Mountains it is a common species. It occurred on the sandy adobe flat adjoining the Cibolo Arroyo at Presidio." In Arizona (Ball et al. 1942) it is "a common species on bare rocky soil sparsely covered with grass, and rocky hillsides and ridges in the Lower and Upper Sonoran zones [and is] especially abundant in the desert grassland where it may be extremely destructive to range grasses." Hebard (1935a) also reports it on rolling hilltops areas of gramma grass dotted with mesquite, exposed areas of rocky soil in dry yellow grass with many desert shrubs present, in a wash with much Mesquite and low yellow grass on low gravelly ridges with fine green grass and cat-claw and on plains and hillsides and steep slopes.

LIFE CYCLE. Adults from August to December in Arizona and from August to November in west Texas.

REFERENCES. Hebard 1935a, Ball et al. 1942, Tinkham 1948.

Leprus elephas **Saussure** Pl. 10

DISTRIBUTION. Central to northeastern Mexico.

RECOGNITION. Basal area of hindwings pale yellow, sometimes with a very faint tinge of green. It may be confused with *wheeleri,* but usually possesses a very slight shoulder at the junction of the lateral lobes with the pronotal disk (in *wheeleri* they round into one another imperceptibly). Hind tibiae yellow, yellow-green, or blue-gray.

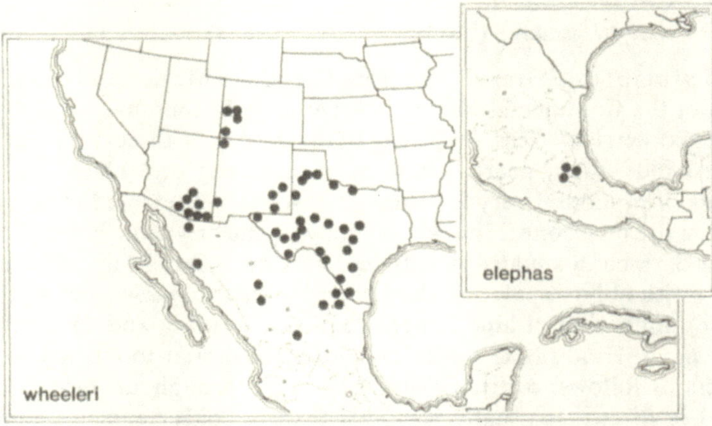

Leprus wheeleri and *L. elephas*

Smaller than *wheeleri:* body length to end of forewings 27–35 mm in males, 40–47 mm in females.

HABITAT. Dry grasslands.

LIFE CYCLE. Adult specimens in the Philadelphia Academy of Natural Sciences were collected in August and September.

REFERENCES. Saussure 1861, Bruner 1905.

Genus HADROTETTIX Scudder

TAXONOMY. The genus includes three species, of which *H. scotodes* is new. *Trifasciatus* was previously the only member of this genus, and *magnificus,* previously in *Trimerotropis,* is assigned to the genus for the first time. The first clue that the three belonged together came from the unique banding and coloration of the hind femora. Other shared characters are given under Recognition.

DISTRIBUTION. South central United States, west to Arizona and south to Durango.

RECOGNITION. Figs. 33*l,* 34*h.* Outer face of hind femora with an oblique black band, inner face with a single pale band. Antennae long and black. Forewings with three crossbands. Hindwings without prominent dark spur directed basally from the black band. Median line of pronotum without a crest. Hind tibiae red. Fastigium shallow, without median carinula.

REFERENCES. See under each species.

***Hadrotettix trifasciatus* (Say)** Pl. 10

DISTRIBUTION. Great Plains from Canada to Mexico and extending west of the Continental divide in Arizona and Montana.

RECOGNITION. Figs. 33*l*, 34*h*. Forewings with three very distinct dark bands, often nearly black. Antennae long and black. Median carina of pronotum very low, almost obsolete, represented by a raised line on the metazona. Hindwings yellow-white or yellow in basal part, outer dark band situated wholly in distal third. Outer face of hind femora with oblique pale and dark bands (similar to *magnificus* and *scotodes*). Inner face of hind femora dark blue to black and with a distal pale band. Hind tibiae red or orange. Most similar to *magnificus* but differs as follows: rostrum slightly concave throughout; second dark band of forewings usually narrower than the pale band preceding it (wider in *magnificus*); inner basal area of hind femora bluish, then turning black (brown or orange in *magnificus*); first dark crossband of forewings with a clearly defined anterior edge (poorly defined in *magnificus*). Body length to end of forewings 26–37 mm in males, 36–50 mm in females.

HABITAT. Grasslands with open patches of shallow, gravelly soil. In

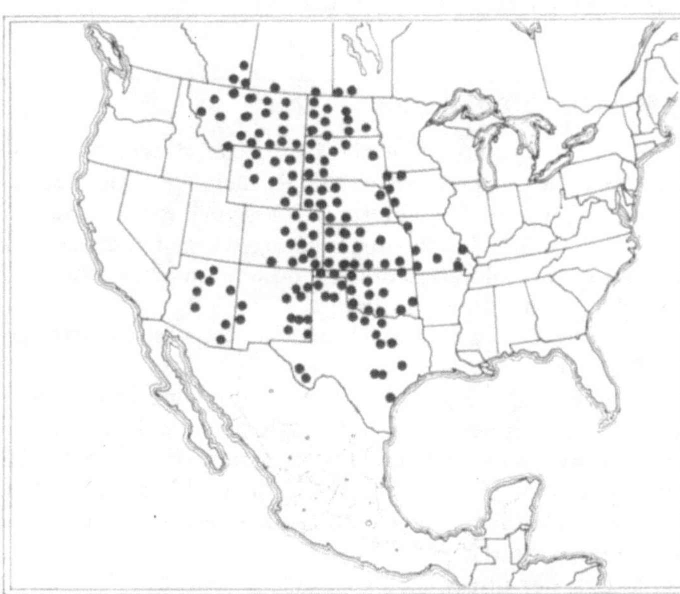

Hadrotettix trifasciatus

Texas Tinkham (1948) writes: "This species is common in the Big Bend area, dwelling as it does in its typical habitat of sparse grass on rocky soil. In North Dakota the gravel pits carry its distribution into the central part of the state, yet in the surrounding territory outside the pits no specimens can be found."

LIFE CYCLE. Eggs overwinter. Adults from June to September.

REFERENCES. Ball et al. 1942, Tinkham 1948.

***Hadrotettix magnificus* (Rehn)** Pl. 10

DISTRIBUTION. West Texas to southeastern Arizona and north to Kansas, south to San Luis Potosi.

RECOGNITION. This is the largest species in the genus. Forewings with three dark (relatively broad) and two pale crossbands. Dark central band of hindwings is wholly in the distal half. Basal area of hindwings whitish with a very slight yellowish green cast. Outer face of hind femora has slanting pale band at the narrowest part. Inner face of hind femora largely black but with a broad pale band at the knee and becoming reddish brown at the base. Hind tibiae orange to reddish. Antennae black. Median pronotal carina very low. Body length to end of forewings 40–49 mm in males, 44–53 mm in females.

HABITAT. Open grasslands. Tinkham 1948 calls it "a member of the Southwestern Grasslands fauna of the Upper Sonoran zone." In Arizona Ball et al. (1942) write: "A rare species found on patches of bare soil in areas of fall grass in the desert grasslands in the Upper Sonoran Zone."

LIFE CYCLE. Eggs overwinter. Adults from July to October.

REFERENCES. Rehn 1907, Ball et al. 1942, Tinkham 1948.

***Hadrotettix scotodes* Otte, n. sp.** Pl. 10

DISTRIBUTION. Central Mexico from Durango to Hidalgo.

RECOGNITION. Overall body color dark reddish brown to blackish. Outer face of hind femora has a characteristic oblique pale band. Inner face of hind femora brown or reddish at the base, becoming black in the second third, and with a yellow to orange pale band at the narrowest part. Hind tibiae orange, often with a blackish band separating the basal third from the distal two-thirds. Forewing crossbands are indistinct or obsolete. Pronotal crest absent, sometimes marked only by a raised line. Body length to end of forewings 28–32 mm in males, 35–38 mm in females.

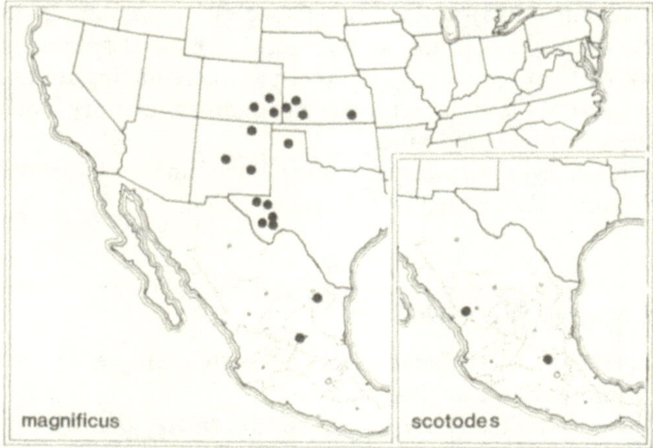

Hadrotettix magnificus and *H. scotodes*

HABITAT. No literature records in Durango; probably open, lightly wooded, mountain slopes.

LIFE CYCLE. Adults in late summer.

SPECIMENS. The holotype male is from Highway 40, 137 mi east of the junction with Highway 15, at 7,000 feet, Durango, Mexico. It is deposited in the Philadelphia Academy of Natural Sciences collection. Paratypes are from the same locality as the holotype (eleven males, fourteen females) and from south of Zacualtipan, Hidalgo, Mexico (one male) and deposited in the Philadelphia Academy collection.

Genus HELIASTUS Saussure

RELATIONSHIPS. In the literature *Heliastus* usually appears at the end of any treatment of oedipodine genera. Because it lacks an intercalary vein, Hebard (unpublished synonymic checklist) placed it with *Cibolacris* under the Group Heliasti. But *Cibolacris* was subsequently shown to be a member of the Gomphocerinae (see Volume 1), leaving *Heliastus* alone in the group. The first clue that it belongs in the Hippiscus group and is related to *Hadrotettix* came from a comparison of epiphalli. A closer look at *Heliastus,* especially *benjamini,* revealed other similarities with *Hadrotettix:* absence of pronotal sculpturing; smooth fastigium; forewings with distinct crossbands (obsolete in sand-inhabiting species); long dark antennae; hind femora with one dark band on outer face.

Heliastus and *Hadrotettix* probably share an exclusive common ancestry; since diverging, *Heliastus* has lost its stridulatory vein, traces of which are still visible in *Heliastus infrequens* and *H. rubellus.* Two species, *sumichrasti* and *subroseus,* no longer resemble *Hadrotettix,* but their relationship to *benjamini* and *aztecus* is not in doubt.

RECOGNITION. Figs. 34*m,* 35*k.* These are small insects. Forewings lack an intercalary vein. Primary lobe of the bilobed epiphallus is quite enlarged. Fastigium narrow. Hind tibiae banded black, ivory, and red.

▪ Key to Heliastus Species

1. Lower posterior margin of lateral lobes with distinct lower projection. Lateral lobes distinctly flare outward (viewed from above). Frontal ridge with deep, narrow V-shaped groove: **2**
   Lower posterior margin of lateral lobes angulate or rounded, usually without distinct lower projection. Lateral lobes do not distinctly flare out. Frontal ridge shallowly grooved or with a U-shaped groove: **3**
2. Basal area of hindwings pink with a distinct central gray band. Coastal dunes of Texas and possibly northern Tamaulipas: **subroseus**
   Basal area of hindwings yellow, yellow-orange, orange, or rosy red; smoky or clear distal to colored area and not forming distinct central dark band. Mexico and Central America, usually more abundant along coastal areas: **sumichrasti**
3. Lower posterior margin of lateral lobes broadly rounded; on forewings, first crossband (above base of hind femur) very distinct across dorsum: **4**
   Without above combination of characters. If lower posterior margin is rounded, then first forewing crossband is not distinct: **aztecus**

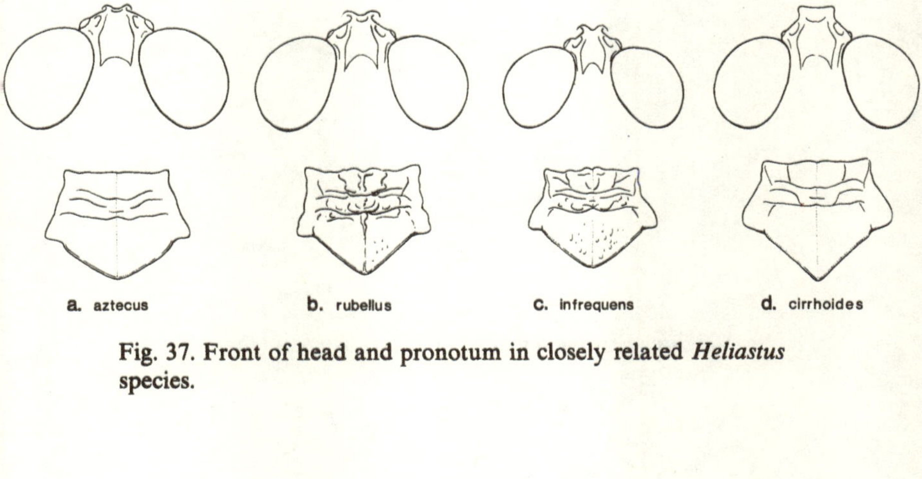

Fig. 37. Front of head and pronotum in closely related *Heliastus* species.

4. Hindwings blue in basal area. Southern Michoacán: **infrequens**
   Hindwings yellow or rosy red in basal area: **5**
5. Hindwings yellow in basal area. Frontal ridge wide and flat (Fig. 37*d*). Highlands of Oaxaca: **cirrhoides**
   Hindwings rosy red. Frontal ridge narrow and grooved (Fig. 37*a*, *b*): **6**
6. Pronotum rough, prozona with definite shoulders. Mexican highlands of Michoacán and Guerrero: **rubellus**
   Pronotum smooth, prozona without shoulders. West Texas to southern Arizona, southwest to Sinaloa: **benjamini**

***Heliastus benjamini* Caudell** Pl. 11

DISTRIBUTION. Western Texas to southern Arizona, south to Chihuahua, Sonora, and Sinaloa, Mexico.

RECOGNITION. Very similar to *aztecus* but differs as follows: forewing crossbands very distinct across dorsum; lower posterior margin of pronotal lateral lobes broadly rounded (angulate or with a definite projection in *aztecus*); hindwings without a darkened area at periphery. Head width (through eyes) 1.08–1.21 times as wide as posterior width of pronotal disk (0.93–1.08 in *aztecus*). Body length to end of forewings 18–23 mm in males; 25–33 mm in females.

HABITAT. Common on gravelly ground with sparse vegetation, especially in dry river washes.

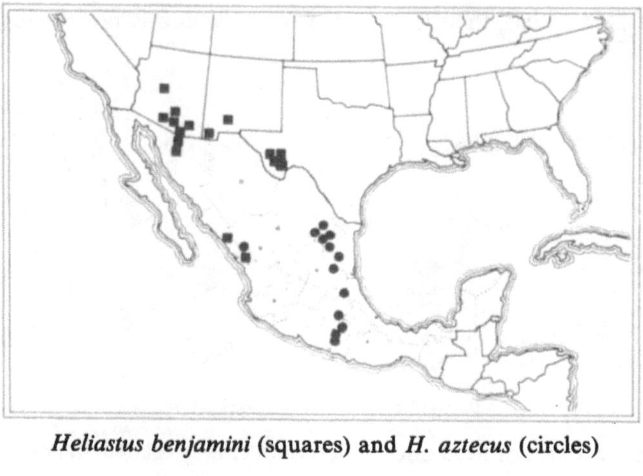

Heliastus benjamini (squares) and *H. aztecus* (circles)

BEHAVIOR. Males seem to perform no solitary flight displays. Males were highly sensitive to the movements of other individuals in the field; they approached moving individuals by flying or walking. As males approach females they perform two signals: in the first they silently raise and lower the hind femora, in the second they shake the hind legs vigorously such that the hind tibiae drum against the ground. Only silent femur-raising motions were seen in male–male interactions.

LIFE CYCLE. Adults July to November.

REFERENCES. Caudell 1905, Ball et al. 1942, Otte 1970.

***Heliastus aztecus* (Saussure)** Pl. 11

DISTRIBUTION. Central Mexico.

RECOGNITION. Figs. 34*m*, 35*k*. Differs from *sumichrasti* as follows: head width (through eyes) 0.67–0.77 times the greatest pronotal width (through lateral lobes) 0.60–0.68 times in *sumichrasti*). Lateral lobes angulate, with or without a lower posterior projection. Differs from *benjamini* as follows: forewing crossbands not very distinct on dorsum; head width (through eyes) 0.93 to 1.08 times as wide as posterior width of pronotal disk (1.08–1.21 times in *benjamini*). Basal area of hindwings deep orange to rosy red. Hind tibiae banded with black and ivory in proximal half and reddish in distal half. Inner face of hind femur has three black bands (including knee). Body length to end of forewings 20–27 mm in males, 27–34 mm in females.

HABITAT. Gravelly ground and dry stony washes.

BEHAVIOR. Reproductive behavior very similar to that of *benjamini* (Otte 1970).

LIFE CYCLE. Adults have been collected from May to December.

REFERENCES. Otte 1970.

***Heliastus sumichrasti* (Saussure)** Pl. 11

TAXONOMY. *H. costaricensis* Rehn, based on a single female from Turrialba, Costa Rica, is here treated as a synonym of *sumichrasti*. I find this specimen to be essentially indistinguishable from twenty-three specimens of *sumichrasti* from Costa Rica. *H. venezuelae* Saussure and *guatemalae* Saussure are similarly indistinguishable from *sumichrasti* and are also treated as synonyms.

DISTRIBUTION. Southeastern Texas, most of Mexico and Central America south to Venezuela and Colombia.

RECOGNITION. This species is highly variable in patterning. It usually lacks distinct crossbands on the forewings, but occasional specimens have dark spots aggregated to form indistinct bands. Head width (through eyes) 0.60–0.68 times greatest pronotal width (through lateral lobes) (0.67–0.77 in *aztecus*). Frontal ridge forms a narrow V-shaped groove (a wider V-shaped groove in *aztecus*). Hindwing color is variable, from yellow-orange to deep orange and rosy red in basal area, often becoming smoky brown distally, but never with a strong dark band. Differs from *aztecus* in always having a prominent projection on the lateral lobes. Differs from the very similar *subroseus* in color pattern of hindwings (dark band in *subroseus* is more clearly defined, and basal area is pink); pronotum distinctly wider when viewed from above; central dark band on inner face of hind femora distinctly broader. Body length to end of forewings 22–28 mm in males, 27–37 mm in females.

HABITAT. Open, bare patches of ground, sandy river beds, and ocean beaches.

BEHAVIOR. I have never seen solitary flight displays in this species. Very likely the colored hindwings are seen by other members of the species only during approach flights. Males shake their femora vigorously while courting females. In this movement the femora strike the forewings; unlike *aztecus*, in *sumichrasti* the tibiae do not drum against the ground. In male–male interactions the femora move up and down silently and are shaken as in courtship.

LIFE CYCLE. Adults from July to December in Mexico, all year round in central and northern South America.

REFERENCES. Bruner 1905, Otte 1970.

Heliastus sumichrasti

Heliastus subroseus Caudell Pl. 11

TAXONOMY. *H. sumichrasti subroseus* Caudell (1907) was described from specimens taken on sand dunes along the Texas Coast. Hebard (1973: 368) writes in a footnote: "[*subroseus*] was described as a variety of *sumichrasti*, but I find it to be a distinctive species peculiar to the sandy strand areas of the Gulf coast." Below I indicate how the two species differ.

DISTRIBUTION. Presently known only from the Texas coastal dunes from Brownsville to Galveston. This distribution coincides remarkably with that of *Trimerotropis schaefferi.* Probably the species extends along the coastal dunes to Tamaulipas, Mexico. Only *sumichrasti* was found along beaches in the vicinity of Tampico.

RECOGNITION. Pale gray to pale yellowish brown, closely matching the color of the sand beaches. Generally resembles *sumichrasti* very closely but differs as follows: hindwings pink in basal area and with a more clearly defined dark band; pronotum narrower when viewed from above; central dark band on inner face of hind femora wider. Hind tibiae banded with black, ivory, and red, much as in *sumichrasti* and *aztecus.* Body length to end of forewings 25–28 mm in males, 32–36 mm in females.

HABITAT. Sand dunes with sparse vegetation in back of the beach along the Texas coast.

LIFE CYCLE. Adults from May at least to November.

REFERENCES. Caudell 1904b, Hebard 1937.

Heliastus rubellus Otte, n. sp.

DISTRIBUTION. Known only from two localities in Michoacán and Guerrero.

RECOGNITION. Fig. 37. Very similar to *infrequens,* but with orange instead of blue hindwings. Superficially similar to *aztecus,* but pronotum strongly sculptured and with definite shoulders on prozona; hindwings extend less than one head length beyond hind femora, and inner face of hind femora as in *infrequens* (Pl. 11). Differs from *cirrhoides* in shape of frontal ridge, wing color, and shape of pronotum. Body length to end of forewings 18.5–21.5 mm in two males, 24–27 mm in females.

HABITAT. Not known; if the habitat is similar to that of related species, then it probably inhabits eroded or sparsely vegetated ground.

LIFE CYCLE. Adults collected between July 10 and September 24.

SPECIMENS. Holotype male, one paratype male, and nine paratype females from 9.4 to 17.5 mi south of Tzitzio, Michoacán, Mexico, September 23–24, 1959 (Cantrall and Cohn), University of Michigan Museum of Zoology and Philadelphia Academy of Natural Sciences. One female, 16 road mi east northeast of Taxco, Guerrero, Mexico, 4,000 feet, September 16, 1959 (Cantrall and Cohn), University of Michigan Museum of Zoology.

Heliastus infrequens Otte, n. sp. Pl. 11

DISTRIBUTION. Known only from three localities in the Tepalcatepec drainage system in Michoacán, Mexico.

RECOGNITION. Differs from all other *Heliastus* as follows: hindwings blue in basal area and becoming smoky gray distally; inner face of hind femora red-orange and unbanded along lower inner marginal area, with a broad basal dark band (*rubellus* is similar); hind tibiae reddish or deep orange and with or without a black ring in proximal third. Body length to end of forewings 18.5–22.0 mm in males, 26–30 mm in females.

HABITAT. I collected this species on overgrazed hillsides on gravelly soil among oak trees in the oak belt of the coastal mountain ranges of Michoacán.

LIFE CYCLE. Adults have been collected between June 16 and November 8.

SPECIMENS. Holotype male and two paratype females from 26 miles north of La Mira on road between Playa Azul and Uruapan, Michoacán, November 8, 1981 (Otte), Philadelphia Academy of Natural Sciences. One male and female from Apatzingan, Michoacán, 400 meters, July 20, 1947 (Hubbell), University of Michigan Museum of Zoology. Two females from Aguililla, 2,800 feet, June 15, 1958 (Wellman and Duellman) University of Michigan Museum of Zoology.

Heliastus cirrhoides Otte, n. sp.

DISTRIBUTION. Known only from one male collected at Huajuapan, Oaxaca, Mexico.

RECOGNITION. Fig. 37. Differs from its presumed closest relatives as follows: frontal ridge wide and largely flat (grooved in related species): hindwings yellowish in basal area (blue or rosy red in related species); lower inner marginal area of hind femora yellowish and with a distal black band (similar to *aztecus*); pronotum flattened and pointed posteriorly (more rounded in other species); prozona with definite shoulders (no shoulders in *aztecus*); hind tibiae entirely deep orange (banded in *aztecus* and some *infrequens*). Body length to end of hindwings of only male is 21 mm.

HABITAT. Probably sparsely vegetated gravelly ground in mountains.

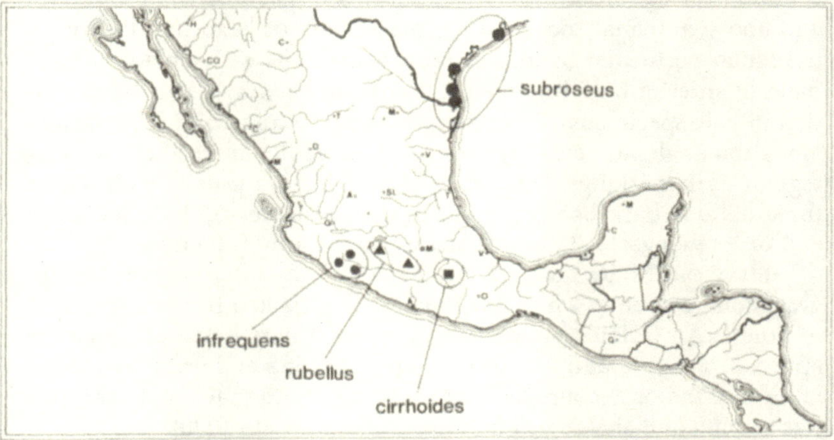

Heliastus infrequens, H. subroseus, H. rubellus, and *H. cirrhoides*

LIFE CYCLE. Only male collected September 13.

SPECIMENS. Holotype male, Huajuapan, Oaxaca, Mexico, 1,597 meters, September 13, 1948 (H. O. Wagner), University of Michigan Museum of Zoology.

Genus CAMNULA Stål

DISTRIBUTION. See *C. pellucida.*

*RELATIONSHIP. While the genus, with one species, seems to be a member of the Hippiscus group, it does not appear to have any close relatives within the group. Hebard did not include it in his G*roup Hippisci in his 1928 paper, but it is included in his unpublished Synonymic Checklist.

RECOGNITION. Figs. 33*d,* 34*i.* Differs from other Hippiscus group genera as follows: hindwings transparent and unbanded; lateral pronotal carinae well developed; aedeagus elongate.

REFERENCES. See *C. pellucida.*

Camnula pellucida **(Scudder)** Pl. 11

DISTRIBUTION. Wide-ranging across the western third of the United States and eastward across southern Canada and northern United States.

RECOGNITION. Figs. 33*d*, 34*i*. Hindwings transparent. Forewings mottled. Lateral margins of pronotal disk clearly defined, forming a continuous ridge from front to back. Area of head beneath fastigium and above antennal sockets often black. Side of head usually with a triangular postocular band. Lateral pronotal lobe with a black vertical band in anterior half, bordered in front and back by a pale band (absent in pale specimens). Forewings have straw-colored or ivory stripes along the angle and converging near the middle; lateral field with irregular dark markings. Outer face of hind femora usually with one to three oblique dark bands (absent in pale specimens). Body length to end of forewings 20–25 mm in males, 25–31 mm in females.

BEHAVIOR. Males lack a flight display. They often form small aggregations, evidently when some males are attracted by the movement of others. An isolated male walking across a bare patch of ground is apt to attract several other males. Femur-tipping and ordinary stridulation are the only courtship signals of males. In male–male encounters tipping and shaking signals are common (Otte 1970).

HABITAT. Grassy pastures, usually on high ground. In Arizona it is usually found above 7,500 feet in mountain meadows, but it may extend to lower elevations in some places. Ball et al. (1942) note: "A very important range grasshopper in limited areas. The nymphs may migrate in swarms for which reason the species is often called the 'warrior grasshopper.' The adults may take flight in swarms but do not migrate for great distances. The species may move into cultivated areas and cause very serious damage to vegetables and other crops, particularly grain. Range grasses are the most important food, but barren or overgrazed land is preferred for egg deposition." In Idaho the species is found almost exclusively in densely vegetated grasslands. Scoggan and Brusven (1972) write:

> It occurs in the fescue-wheatgrass, wheatgrass-bluegrass and mountain meadow communities. It thrives in rolling areas with loam soil and moderate precipitation. In these areas permanent oviposition beds are found, usually on or near hilltops or knolls. When first instars emerge, numbers in and near these beds may reach 1,000 or more per square yard in outbreak years; numbers up to 100 per square yard are common. This species, contrary to most band-wings, dwells mostly on vegetation rather than on the ground. It is a serious economic pest by virtue of its high biotic potential and preferences for grasses.

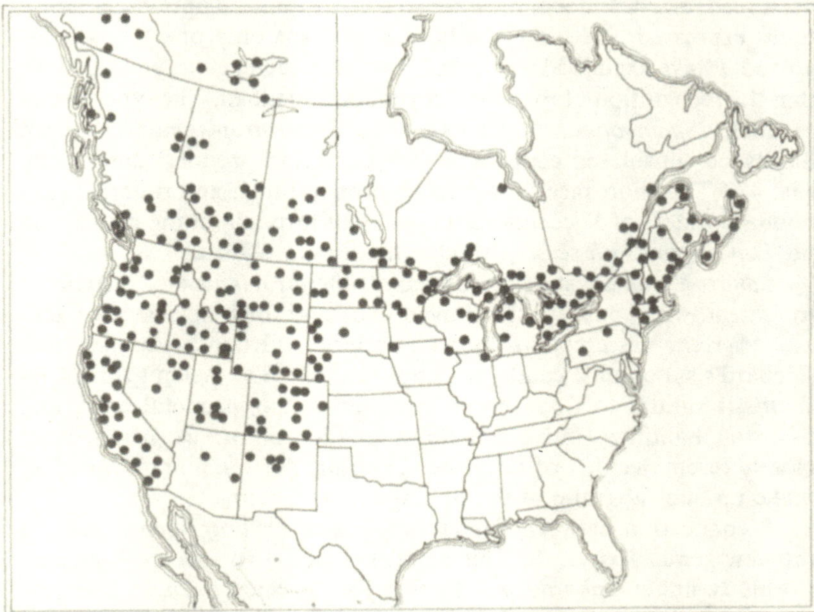

Camnula pellucida

LIFE CYCLE. Eggs overwinter. Adults from early June to September in Arizona, and from late July to September in Idaho and New England.

REFERENCES. Blatchley 1920, Ball et al. (1942), Mulkern et al. 1964, Otte 1970, Scoggan and Brusven 1972.

Sphingonotus Genus Group

Along with the large Old World genera *Sphingonotus* and *Mioscirtus,* the following North American genera are included in this genus group: *Conozoa, Microtes, Trimerotropis, Spharagemon, Dissosteira,* and a new genus, *Eximacris.* Other Old World genera should be included, perhaps, but they have not been studied. The relationships implied by this grouping may come as a surprise to taxonomists who are familiar with these genera. The associations among them are admittedly somewhat tenuous and are estabished in part through connections between individual species belonging to different genera; see Fig. 38.

The relationship between *Sphingonotus* and *Trimerotropis* has long been recognized. Indeed, they are so similar that the opinion is sometimes expressed, at least verbally, that perhaps only one genus is involved. I have examined numerous species in each genus, and I believe that the recognition of two distinct genera is justified. The type species of each, *Sphingonotus caerulans* and *Trimerotropis maritima,* can hardly be considered congeneric (see head and pronotal differences, Fig. 48). The short face, low median pronotal line, and rounded pronotal shoulders of *S. caerulans* is consistent throughout the genus, and no *Trimerotropis* possesses this combination of features.

Conozoa was always placed near *Trimerotropis* in any systematic treatment, with some species moved back and forth between the genera. *Microtes* was previously placed under the Group Spharagemones (Hebard's synonymic catalogue, unpublished). The Asian genus *Mioscirtus* is similar to *Conozoa* in fastigium shape, pronotal crest, and forewing banding; and the male epiphallus of *M. wagneri* (Evers) closely resembles that of *Conozoa. Conozoa, Microtes,* and *Mioscirtus* make up one subgroup of the Sphingonotus Group.

A connection between *Trimerotropis* and *Spharagemon* is seen in the new genus *Eximacris.* This genus contains two species, *superbum* (formerly under *Spharagemon*) and a new species, *phenax. E. super-*

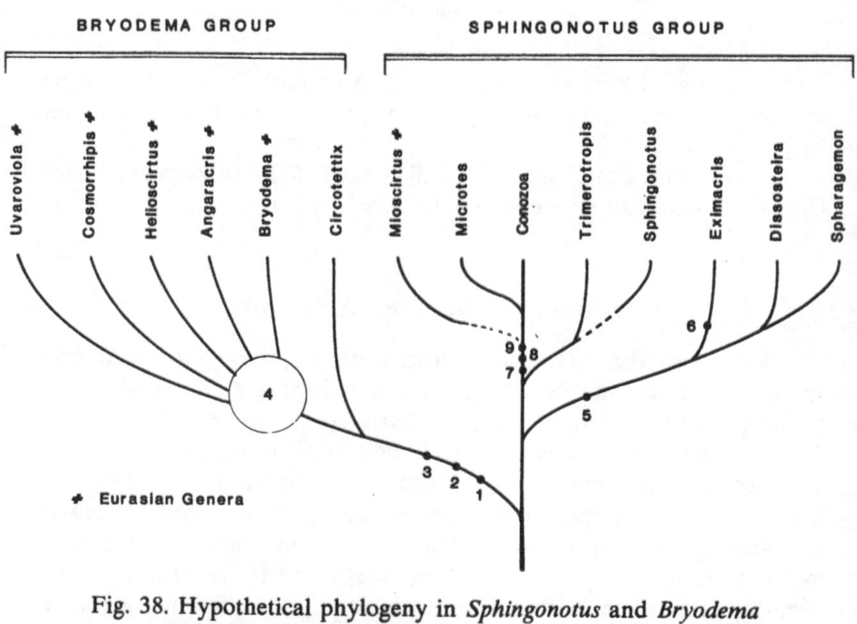

Fig. 38. Hypothetical phylogeny in *Sphingonotus* and *Bryodema* groups (see Appendix 1 for explanation).

bum has a high median pronotal crest with one transverse sulcus, much like *Dissosteira* and several *Spharagemon* species, while *E. phenax* has a low crest cut by two sulci, therefore resembling *Trimerotropis.* These species are placed in a new genus because they share several highly distinctive color patterns and do not fit easily into *Spharagemon, Dissosteira,* or *Trimerotropis.*

Other links between *Trimerotropis* and *Spharagemon* are seen in several species. *Spharagemon equale* has the pronotum sometimes cut by two sulci (*Trimerotropis* character) and sometimes by one (*Spharagemon* character). Specimens with the former condition were named *Trimerotropis humile,* but are now considered to be conspecific with *Spharagemon equale. Trimerotropis campestris,* now in *Spharagemon,* is very similar to *S. equale* except that it usually has the pronotal crest cut twice. *Spharagemon bunites* also has the median crest cut twice but otherwise resembles *Spharagemon campestris* and *S. equale* closely.

RECOGNITION. No single feature or set of features can be used to define this group. The clustering within the Sphingonotus Group is done on the basis of pairwise comparisons of species and genera such as those discussed above. The group is worldwide in distribution but poorly represented in South America.

Genus SPHARAGEMON Scudder

DISTRIBUTION. Widespread through the eastern three-quarters of the United States, Canada, and northern Mexico.

TAXONOMY. In this work several changes are made in the traditional classification of the genus (Fig. 38). First, *S. superbum* and a new species related to it are placed in the new genus *Eximacris* (see that genus for discussion).

Second, *Trimerotropis campestris* is also placed in the genus. It had been placed in *Trimerotropis* because the pronotal crest is cut by two sulci, but emphasis on this character led to difficulties in other genera as well (see *Conozoa, Microtes,* and *Eximacris*) and created difficulties within species. *Trimerotropis humile* now appears to consist of certain specimens of *Spharagemon equale* in which the pronotal carina is cut twice. Intraspecific variation in the number of carinal cuts is also seen in *Conozoa carinata, C. clementina, Microtes occidentalis, Spharagemon saxatile,* and *Pardalophora haldemani.*

The third change is that the genus *Scirtetica* is here treated for the first time as a synonym of *Spharagemon.* The former was not adequately defined by Saussure in relation to *Spharagemon,* and the distinguishing characteristics provided by subsequent authors do not ade-

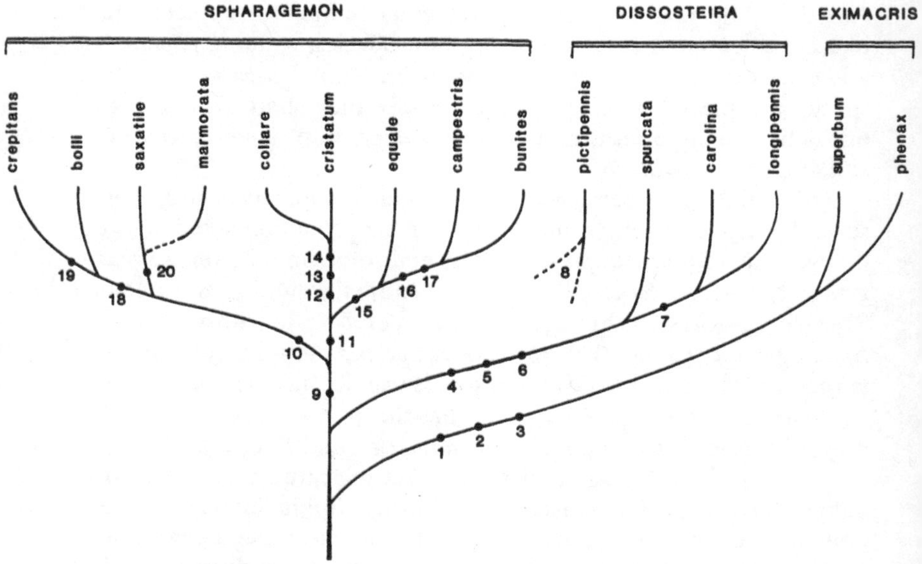

Fig. 39. Hypothetical phylogeny in *Spharagemon, Dissosteira,* and *Eximacris* (see Appendix 1 for explanation).

quately separate them. Bruner 1905: 116 separates *Spharagemon* and *Scirtetica* in a final couplet, stating that in *Spharagemon* the forewings are "rather broad, the intercalary vein is more or less flexuous, the pronotal crest is anteriorly elevated, profoundly intersected, arcuate on the metazona as seen laterally." Each of these characters may apply to one or two species, but none applies to all. Blatchley (1920) also separated the genera in a final couplet of his key (p. 247), using Bruner's characters.

RECOGNITION. Fig. 40. Usually separated from *Trimerotropis* and *Conozoa* by the median pronotal crest, which is cut once, except in *campestris, bunites,* and some *equale.* Unlike *Dissosteira* and *Eximacris,* in *Spharagemon* the hindwings are always yellowish in color (tending toward orange in *marmorata*).

REFERENCES. See under each species.

▪ Key to Spharagemon Species

1. Pronotum with median pronotal carina forming high ridge (at least as high as one-half the eye height) (Figs. 40*i*, 41*a, c, d*): **2**

   Pronotum with low ridge (less than half the eye height) (Figs. 40*h*, 41*b*): **5**

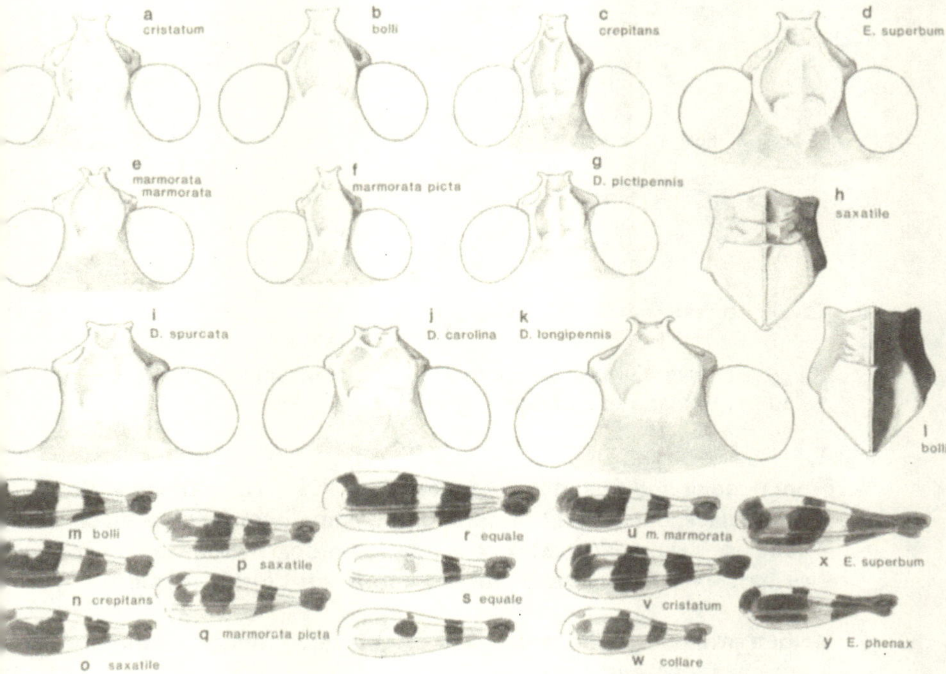

Fig. 40. *Spharagemon, Dissosteira* and *Eximacris.* Fastigium, pronotum, and inner face of hind femur.

2. Hind tibiae entirely red or deep orange, without pale or black bands. Body finely speckled: **3**
   Hind tibiae red or deep orange and with black or pale bands. Body more uniformly brown, often wit broad bands on forewings: **4**
3. Posterior margin of pronotal disk not as strongly acute-angulate (Fig. 41*d*). Male body length to end of forewings almost always less than 29 mm; female body length 27–37 mm. Pronotal crest not high (Fig. 41*d*). Northern United States and southern Canada: **collare**
   Posterior margin of pronotal disk strongly acute-angulate (Fig. 41*c*). Male body length to end of forewings 20–39 mm; females 34–45 mm. Pronotal crest very high (Fig. 41*c*). Southern United States, north to Kansas and Virginia, south to northeastern Mexico: **cristatum**
4. Frontal ridge and fastigium narrow (Fig. 40*c*). Hind tibiae with very broad black band. Florida and extreme southern Georgia: **crepitans**

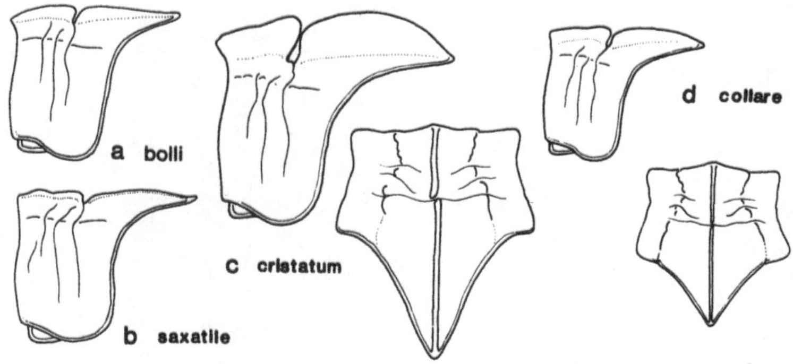

Fig. 41. *Spharagemon.* Comparison of pronotal profiles in *bolli* and *saxatile,* and in *cristatum* and *collare.*

Frontal ridge and fastigium broader (Figs. 40*b*). Hind tibiae with short black band in all eastern states, sometimes lacking black band in Texas, New Mexico, and Colorado. Widespread over eastern North America, but absent from Florida peninsula: **bolli**

5. Median pronotal carina usually cut by two sulci. Inhabiting mountain meadows and clearings: **6**
   Median pronotal carina usually cut by one sulcus: **7**
6. Hind tibiae blue. Sierra Nevada range: **bunites**
   Hind tibiae orange or red. Rocky Mountains: **campestris**
7. Inner face of hind femora orange with two or three black bands (Fig. 40*r, s, t*). Great Plains and Rocky Mountain states: **equale**
   Inner face of hind femora not orange, not banded as in Fig. 40*r, s, t:* **8**
8. Outer face of hind femora with dark bands broader than pale bands. Hind tibiae never blackish at distal end. Pale bands crossing forewings. Ridgetops in Appalachian Mountains: **saxatile**
   Outer face of hind femora with dark bands as wide as light bands. In north, hind tibiae blackish at distal end. Pale bands do not cross forewings. Open, sandy, low woodland: **marmorata**

***Spharagemon collare* (Scudder)** Pls. 12, 13

DISTRIBUTION. Southern and western Canada, northern and western United States. Absent from extreme west. Largely replaced by *cristatum* in Kansas and eastern Colorado and along the Atlantic coast

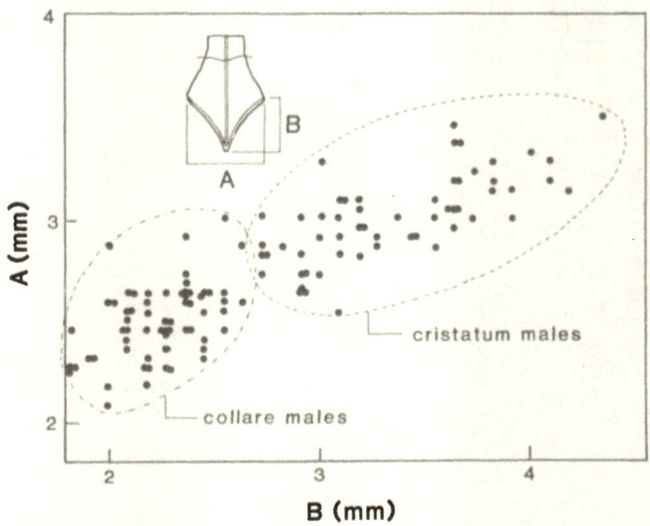

Fig. 42. *Spharagemon.* Scatter diagram showing separation of *cristatum* and *collare.*

between New Jersey and Virginia. Specimens from Virginia seem to be intermediate between *collare* and *cristatum.* Populations from this region should be examined for courtship behavior.

RECOGNITION. Figs. 40*w*, 41*d*, 42, 43. Overall body color sandy to dark gray and strongly speckled. Forewings without dark crossbands, but occasionally the spots coalesce to form three or four indistinct bands. Lower margin of hind femora has a fine fringe of hair. Pronotal crest is high. Hind tibiae entirely red or deep orange. Inner face of hind femora yellowish and with four dark bands; the most proximal one is sometimes indistinct. This species is very similar to *cristatum.* The two species are largely allopatric and so can be recognized on the basis of their distributions. In areas of possible overlap *collare* can usually be distinguished from *cristatum* as follows: (1) male body length to end of forewings almost always less than 29 mm (more than 29 mm in *cristatum*); (2) pronotum not as pointed; and (3) median pronotal crest not as high. Regressions of pronotal dimensions and body length can also be used to separate the two species (Figs. 42, 43). Body length to end of forewings 23–31 mm in males; 27–37 mm in females.

BEHAVIOR. Although males have the ability to crepitate, solitary males seldom do so. Frightened males crepitate as they fly away and

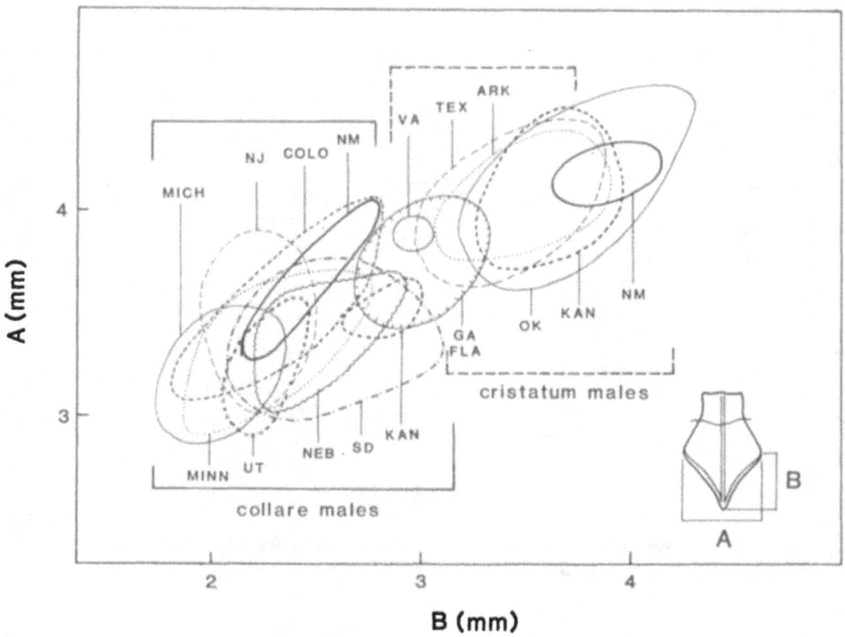

Fig. 43. *Spharagemon.* Separation of *cristatum* and *collare* by state, based on data from Fig. 42.

possibly males flying toward a female do also. In pair formation males watch for moving females, then walk or hop toward them. Males may stand motionless on open ground for long periods, then rapidly approach a moving object, either a male or female of their own species, or an individual of another species.

Unlike *cristatum* males, *collare* males stridulate only as they approach females; they press the hind femora against the forewings on both the upstroke and the downstroke. Courtship in males from Alamosa, Colorado, and from southeastern Michigan was essentially identical (Otte 1970).

HABITAT. This species is most often found in open, sunny, sandy places such as old dunes, blowouts, beaches, sandy cultivated fields, and sandy river margins. In Nebraska (Hebard 1938) it is common in wheat stubble, pastures, and rangeland. Scoggan and Brusven (1972) write:

> [*S. collare* in Idaho] occurs most commonly in early and middle successional stages of disturbed areas, including sand dunes and blowouts. It also

frequents roadsides, gullies and tilled soil areas in the sagebrush grassland community when soil is fine and loose. Preferred habitats include the grasses *Oryzopsis hymenoides, Sporobolus cryptandrus,* and seeded *Agropyron cristatum.* Nymphs are most commonly found in open areas, and when disturbed they do not seek vegetation. Under favorable conditions nymphs may number 10 per square yard and are potentially damaging, particularly in areas tilled and seeded to crested wheatgrass (*Agropyron cristatum*). Except under these conditions *S. collare* seldom reaches damaging numbers and inhabits only poorer range areas. It is mainly a grass and sedge feeder (Mulkern et al. 1969).

LIFE CYCLE. Egg-overwintering; adults from July to late September.

REFERENCES. Morse 1895, Blatchley 1920, Hebard 1934, Cantrall 1943, Hauke 1953, Froeschner 1954, Otte 1970.

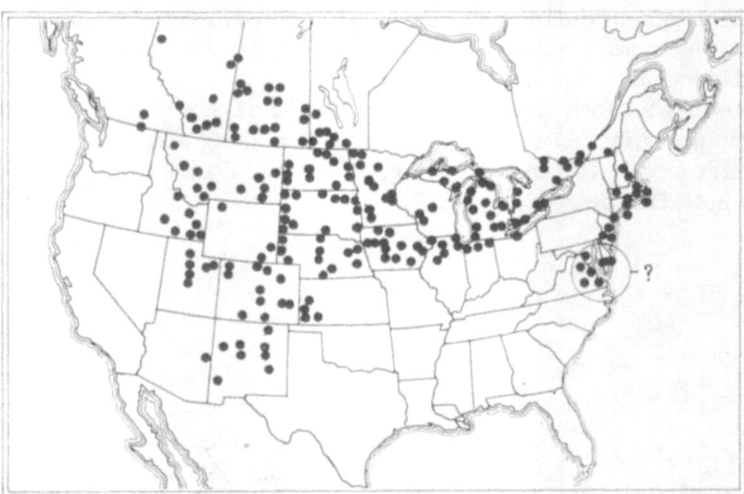

Spharagemon collare

Spharagemon cristatum (Scudder) Pls. 12, 13

TAXONOMY. Hebard (1932, 1937) first considered *cristatum* to be a synonym of *collare.* When courtship behavior was studied, *cristatum* proved to have very different signals. Courtship encounters of *collare* males from Alamosa, Colorado, were essentially identical to those of Michigan males and very different from those of *cristatum* from northeast Texas and northeastern Mexico (Otte 1970). On the basis of a morphological analysis (Figs. 42, 43), the probable contact zones between these two species are believed to be in Kansas and somewhere between Virginia and New Jersey.

DISTRIBUTION. Eastern Mexico north to Kansas and from eastern New Mexico to Virginia. Replaced by *collare* in Kansas and Colorado and in an area between Virginia and New Jersey.

RECOGNITION. Figs. 40*a*, *v*, 41*c*, 42, 43. Very similar to *collare.* Overall body color sandy and strongly speckled. Hind femora with four black bands on inner face; the two proximal bands are rarely joined. Hind tibiae deep orange to red. *S. cristatum* differs from *collare* in the shape of the pronotal disk and in body size. Body length to end of forewings 29–39 mm in males, 34–45 mm in females.

BEHAVIOR. Reproductive behavior in *cristatum* differs from that in *collare* in two important respects: (1) solitary males commonly perform crepitation flights; and (2) courting males produce both femur-shaking and stridulation signals as they approach females. Courtship in northeastern Texas and in Ciudad Valles, San Luis Potosí, Mexico, was identical.

HABITAT. Open sandy grasslands and fields, sandy river margins. Often occurring in overgrown old fields and along roadsides.

LIFE CYCLE. Egg-overwintering; in Texas adults appear in May and last until October.

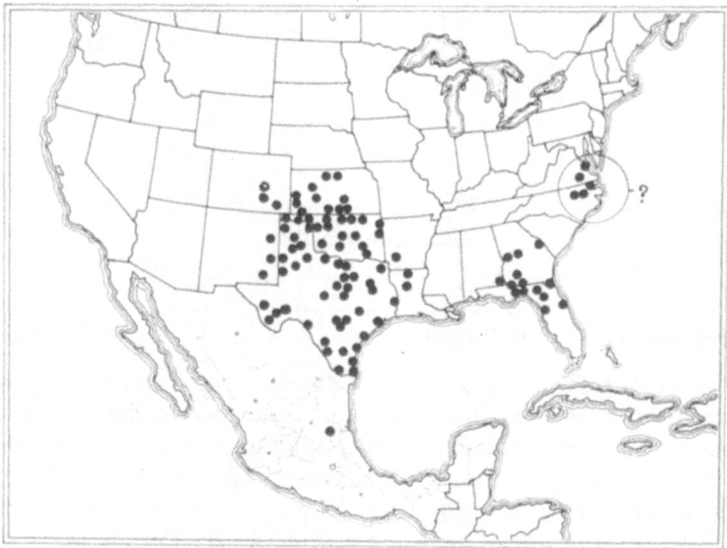

Spharagemon cristatum

REFERENCES. Scudder 1875, Morse 1895, Isely 1937, 1938, Otte 1970, Smith 1981.

Spharagemon bolli Scudder Pls. 12, 13

DISTRIBUTION. This species occurs across the greater part of the eastern two-thirds of the United States and southern Canada; it extends westward to the front ranges of the Rocky Mountains in New Mexico and Colorado and to the Davis and Chinati mountains in west Texas.

RECOGNITION. Figs. 40*b, i, m,* 41*a.* Overall body color medium wood-brown to dark brown. Forewings with two or three darker crossbands. In paler individuals these are largely obsolete. Hind tibiae banded with ivory, black, and red; in far-western populations the black band becomes faint and disappears. The black and ivory bands are usually nearly equal in length. Median carina of pronotum forms a sharp crest (not as high as in *cristatum,* similar to that in *crepitans,* and much higher than in *saxatile* and *equale*). Frons more produced than in *saxatile* and distinctly broader than in *crepitans.* Coloration of some western specimens is similar to that of *cristatum,* but the latter has a strongly acute-angulate pronotal margin, and the banding on the inner hind femora is different.

BEHAVIOR. Crepitation flights by males are common. Between flights males walk about searching for females. Like *cristatum,* this species has incorporated femur-shaking into its courtship behavior. Courting males always shake their hind legs vigorously, drumming the ground with their tibiae at the same time. Shaking is often followed by one to six strokes of stridulation. Femur-tipping almost always precedes femur-shaking in a courtship sequence. Tipping and shaking also occur in male–male encounters.

HABITAT. Often found in open, sunny, woodlands, woods margins, and open woods back of the Great Lakes beaches. Blatchley (1920) writes: "It frequents paths and bare places in dry upland woodland pastures, especially those which include the slopes of high hills, roadsides and stubble fields." Morse (1907:37) notes: "It is one of the few sylvan geophilous locusts, its distribution coinciding with that of dry woodlands. In northern states it frequents old, sandy fields as well as the more open woods and forest paths and margins." Tinkham (1948) writes: "In the Davis Mountains [west Texas] this species was found only at considerable elevations in an oak-sylvan environment where its buff brown coloration matched perfectly with the fallen oak leaves that it was resting upon. On the table-like top of Blue Mountain in the Chinati Range the species was semi-sylvan and confined more to the grassy areas between the trees." Coppock (1962) notes that it inhabits upland woodland pastures in Oklahoma. In the foothills of the Rocky Mountains in Colorado and New Mexico I found this species in small clearings in scrub-oak woodland. In the Appalachians of Pennsylvania I found it living with *saxatile* on open sunny ridges.

LIFE CYCLE. Egg-overwintering. In northern states adults begin to appear in late June and early July and disappear by late August or September. In southern states adults have been collected mainly in July, August, and September.

REFERENCES. Morse 1895, 1907, Fox 1914, Blatchley 1920, Cantrall 1943, Tinkham 1948, Coppock 1962, Otte 1970, Vickery and Kevan 1983.

***Spharagemon crepitans* (Saussure)** Pls. 12, 13

DISTRIBUTION. Florida and southern Georgia. This species is largely allopatric with *bolli,* but their distributions overlap in western Florida.

RECOGNITION. Fig. 40*c, n.* This species is quite similar to *bolli.* Both are dead-leaf brown to dark brown and have sharp, prominent pronotal crests. But *crepitans* has a much narrower frontal ridge, a broader dark band on the hindwings, and a narrower fastigium. On the hind tibiae the black band is much longer (in some females the black band is located mainly between the tibial spines), and either the distal ends of the tibiae are rust-colored (instead of red) or the red band is not longer than the black band.

BEHAVIOR. It is not known if solitary males make crepitation flights. In two males I collected, I observed courtship that was somewhat similar to that of *bolli,* but the males never stridulated after femur-shaking (Otte 1970).

HABITAT. Rehn and Hebard (1914) report: "[In southern Florida and the Florida keys] the only situation in which this species was found was on the ground under bushes in high jungle growth ... where a few scattered individuals and one small colony were encountered. The species was very difficult to capture as the brush was too heavy to use a net and the insects flew up hurriedly. Their flight was observed to be usually low and for short distances." The species has also been collected "in an open field near a stream," "on sandy soil of a live-oak hammock," and at other localities among palmetto scrub.

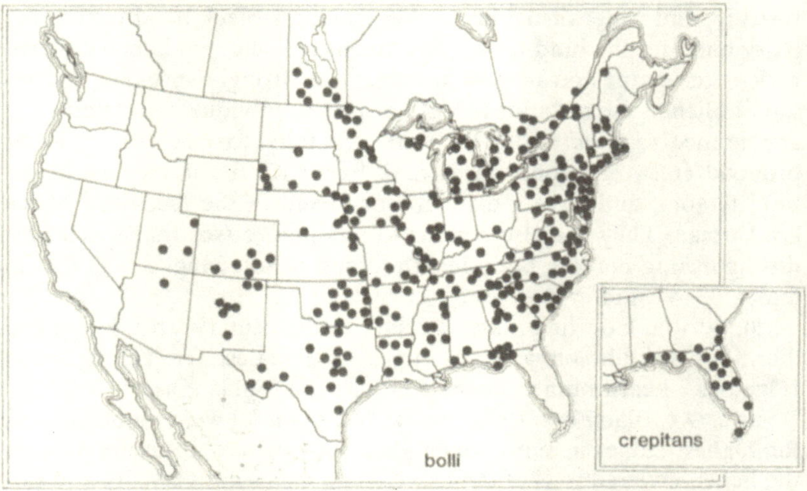

Spharagemon bolli and *S. crepitans*

LIFE CYCLE. Adults have been collected from July through September.

REFERENCES. Rehn and Hebard 1914, Blatchley 1920, Otte 1970.

Spharagemon saxatile **Morse** Pls. 12, 13

TAXONOMY. In 1904 Morse described the taxon *planum* as a variety of *saxatile*. *S. planum* was later treated as a distinct species by Fox (1928) and Hebard (1937: 365). *S. saxatile* was the form found in Connecticut and Massachusetts, while *planum* was found in Pennsylvania, Ohio, and Kentucky. *S. planum* is here considered to be a synonym of *saxatile* for two reasons: (1) the courtship behavior of the two forms is indistinguishable (though few interactions have been recorded so far) and (2) morphological intermediates between the two forms have been collected from Connecticut and New York. The matter deserves further study because of the very different habitats in Pennsylvania and New England.

DISTRIBUTION. Appalachian Mountains from eastern Tennessee and southern Ohio to Massachusetts. Probably extends north to Maine.

RECOGNITION. Fig. 40*h, o, p*. This species is most likely to be confused with *bolli*, but it differs in having a low pronotal crest, shorter rostrum, and wider fastigium, and in lacking a black band next to the ivory band on the hind tibiae. Specimens collected on lichen-covered rock outcroppings on mountain ridges are strongly speckled and resemble lichen encrustations. Most of these individuals are similar in appearance to *collare,* but they are generally darker, have a lower pronotal crest, have only three dark bands on the inner face of the hind femora, and have a distinct ivory band in the first third of the hind tibiae. They are also similar to *marmorata* (see that species for distinguishing marks). Body length to end of forewings 27–32 mm in males, 33–41 mm in females.

BEHAVIOR. Courting males produce both ordinary stridulation and vibratory stridulation as they approach females. On Pennsylvania ridgetops I heard numerous male crepitation flights (Otte 1970).

HABITAT. Blatchley (1920) writes: "[*S. saxatile*] were found only on the highest ledges in sunny open glades . . . [where it] occurs only on the lichen-covered tops of the ledges. The males are very shy and difficult to approach, moving short distances with a quick, almost noiseless flight, in longer ones making a rather loud rattling sound. Their hues accord perfectly with the lichens and gray rocks." Morse (1894) notes: "I have named this species *saxatile* for the reason that it finds life most to its taste in unsettled, somewhat wooded districts of a rocky, often elevated character. Here it finds a congenial home and may be seen during the latter half of the season crawling actively about over the lichened ledges, whose tints harmonize with its own, or flying from one to another, stridulating loudly as it goes." In Pennsylvania it lives on exposed ridgetops, upland old fields, and woods margins (Fox 1914, Blatchley 1920).

LIFE CYCLE. Egg-overwintering. Adults from July to September.

REFERENCES. Morse 1894, 1895, Fox 1914, Blatchley 1920, Otte 1970.

Spharagemon marmorata marmorata (Harris) Pls. 12, 13

TAXONOMY. This species was previously placed in the genus *Scirtetica.* That genus is now treated as a synonym of *Spharagemon.* See generic discussion.

DISTRIBUTION. This subspecies ranges from Minnesota eastward into southern Canada and to New England, and along the east coast southward through New Jersey. It is replaced by *marmorata picta* in Virginia, but intermediate conditions are evident as far south as North Carolina.

RECOGNITION. Fig. 40*e, u.* Basal area of hindwings yellow, with broad outer dark band. Distal section transparent but smoky at tip. Overall body color variable. Many individuals have strongly contrasting black and pale gray or greenish markings, but some are more uniformly brownish. The pale bands on the forewings do not cross the wing. This species is very similar to *saxatile* but differs in the patterning of the forewings and in the banding of the hind tibiae; in the latter the distal end is blackened and there is a distinct black band proximal to the red band. The species may sometimes be confused with *S. collare,* but *collare* never has a black band on the hind tibiae, and it does have a high pronotal crest. Body length to end of forewings 21–26 mm in males, 25–30 mm in females.

HABITAT. In New Jersey Fox (1914) says: "A sand-loving species amost entirely restricted to sylvan situations. Occurs in bare sand or in open scrub or sandy soil." In New England Harris (1862: 179) writes: "One of our prettiest species found in the open places contiguous to or within pitch-pine woods, flying over the scanty grass and reindeer moss which grow in these situations." At Douglas Lake, Michigan, Vestal (1914: 107) found it in open forest among undergrowth of bracken fern and blueberry. He notes: "It appears to be associated with lichen surfaces in rather open situations somewhat after the manner of *Trimerotropis saxatile* McNeill, the markings giving it much of the general appearance of the lichens upon which it rests."

LIFE CYCLE. Egg-overwintering. Adults from July to October.

REFERENCES. Rehn and Hebard 1912b, Vestal 1914, Fox 1914, Blatchley 1920, Vickery and Kevan 1983.

***Spharagemon marmorata picta* (Scudder)** Pls. 12, 13

DISTRIBUTION. Virginia to Florida and westward to central Texas.

RECOGNITION. Fig. 40*f, q.* Basal area of hindwings deep yellow to orange-yellow; black band very broad. Overall body color variable, from sandy brown or pale gray to reddish brown and usually with strong black markings. Upper side of lateral lobes usually black. Pronotal disk usually pale. Lateral field of forewings usually dark reddish brown to black along upper side and pale along lower margin. Dorsal field much lighter than lateral field. Hind tibiae mostly deep orange to red and with a pale ring in the first quarter. In the north there is often a black band next to the pale band. Body length to end of forewings 26–31 mm in males, 27–35 mm in females.

BEHAVIOR. Males perform loud and lengthy crepitation flights. Although I found no distinct aggregations of males, I had the impression that there were cycles in the crepitation activity in a given area of the woods (Otte 1970). Blatchley (1920) made the following observations: "[*S. marmorata picta*] has the most prolonged stridulation [crepitation] of any locust known to me. He zig-zags in his flight almost at right angles, sometimes staying in the air for half a minute, flying all about an acre or more, and finally alighting on the sand within a yard of where he arose."

In addition to femur-tipping and ordinary stridulation, *S. marmorata picta* has incorporated a unique and complicated movement into courtship. It lifts both femora to a vertical position, then jerks the femora forward twice through about 20 degrees, either together or alternately, then lowers one femur before the other, to about the 45-degree position. Then it vibrates one femur (*A*) against the forewing while lowering the femur through 30 degrees or so, then raises it again to the 45-degree position. As *A* is being raised, femur *B* vibrates against the forewing as it is lowered. As *B* is raised, *A* is lowered and vibrates again. This sequence results in three or four bursts of vibratory stridulation. Before making this complicated movement males often produce a two-stroke burst of ordinary stridulation.

HABITAT. Bare sandy places in open woodland, sandy paths and roadsides, and open areas back of the beach.

LIFE CYCLE. In Florida and Georgia this adult is present throughout the year.

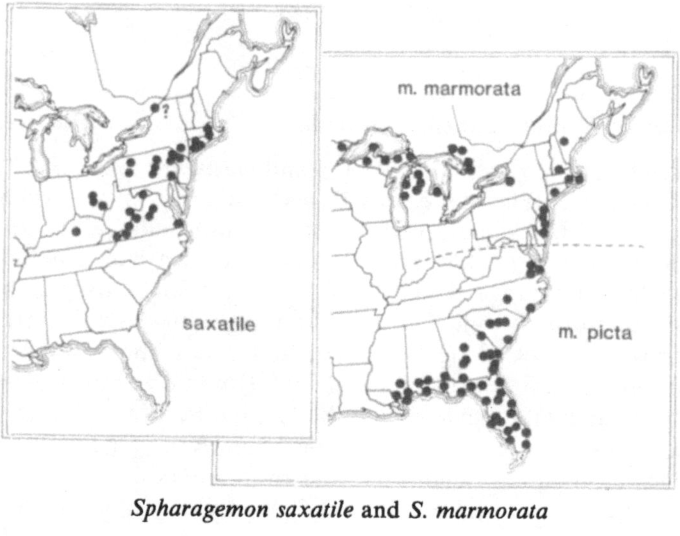

Spharagemon saxatile and *S. marmorata*

REFERENCES. Rehn and Hebard 1912b, 1916, Blatchley 1920, Otte 1970.

***Spharagemon equale* (Say)** Pls. 12, 13

DISTRIBUTION. Widespread in mountain and prairie states from Alberta and Saskatchewan to Texas and northern Mexico.

TAXONOMY. *Spharagemon* (or *Trimerotropis*) *humile* is here treated as a synonym of *equale*. The status of *humile* has always been problematical. It was described by Morse (1895) from two males collected at Garden of the Gods, Colorado. The characters he considered diagnostic cannot be used to distinguish the species from sympatric *equale*. Some individuals have the median pronotal crest cut by two sulci, causing Hebard (1937) to place the species under *Trimerotropis*. As I noted previously, this character presents difficulties in other species as well.

RECOGNITION. Fig. 40*r–t*. *S. equale* is most closely related to *campestris*, which lives in mountain grasslands. It differs from *campestris* as follows: median pronotal crest very low or almost obsolete and usually cut by one sulcus (specimens in which the pronotum is cut twice were previously called *Trimerotropis* or *Spharagemon humile*). *S. equale* overlaps in distribution with *cristatum*, *collare*, and *bolli*. It differs from these three species as follows: forewings with rather distinct bands that often cross the dorsal field; median pronotal carina low, sometimes forming a raised line; inner face of hind tibiae orange, with two distinct dark bands in the distal half and with a smaller, sometimes indistinct band in the proximal half. Hind tibiae are uniformly orange over most of the range. In some specimens from British Columbia, they are yellow. Body length to end of forewings 29–43 mm in males, 36–47 mm in females.

HABITAT. Prairie grasslands with patches of bare ground, rolling grassland in the Rocky Mountain foothills. In Idaho Scoggan and Brusven (1972) report that this species is "associated with climax areas in sagebrush and wheatgrass-bluegrass communities, and is a rare inhabitant of the shadscale community. Vegetation in preferred areas of its range includes a predominance of perennial bunchgrasses . . . some annual grasses and forbs may be present . . . soils in these areas are loam with few rocks. Nymphs are moderately active, and show little preference for either open areas or vegetation when disturbed. Ordinarily it is not an economic pest but contributes to the overall damage of good rangeland, and can be an important component of a damaging complex." It is a more general feeder than *collare,* but prefers grasses or sedges (Mulkern et al. 1969).

LIFE CYCLE. Egg-overwintering. First nymphal instars emerge by mid-June in Idaho. Adults from July into September in northern states and from July into November in Texas.

REFERENCES. Hebard 1937, Brooks 1958, Alexander and Hilliard 1969, Otte 1970, Scoggan and Brusven 1972, Smith 1981.

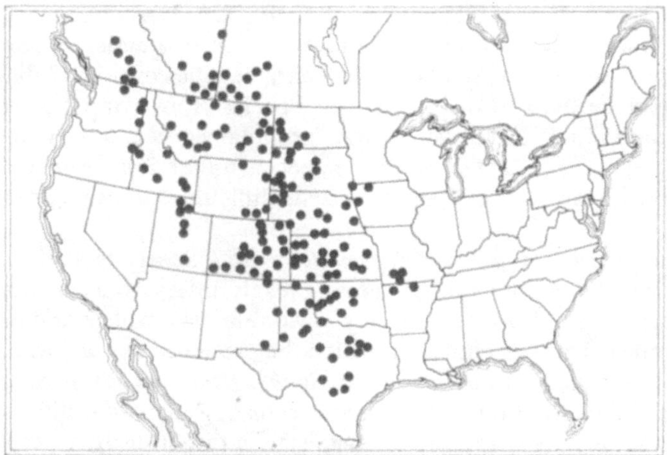

Spharagemon equale

Spharagemon campestris (McNeill) Pls. 12, 13

TAXONOMY. Heretofore *campestris* was always placed under *Trimerotropis* because its pronotal carina is cut twice. But there are two reasons for including it with *Spharagemon:* (1) there is an overall similarity in the color pattern, body shape, and behavior of *campestris* and *equale;* and (2) the pronotal carina is cut once or twice in both species. *S. bunites* from the Sierra Nevada range appears to be a close relative, but in this species the pronotal carina is always cut by two sulci.

Trimerotropis longicornis E. M. Walker from British Columbia is a synonym of *campestris.* Hebard first synonymized the name in 1925; subsequently Vickery (1975) considered it to be a valid species. I have studied the holotype and find it to be largely indistinguishable from *campestris;* the yellow tibiae used to separate it from *campestris* cannot be used, because variation from yellow to red was already known within the species, and similar variation is known in *S. equale.*

DISTRIBUTION. Mountains and hills from western Nebraska and South Dakota westward to Nevada and from New Mexico north to British Columbia and Alberta.

RECOGNITION. Figs. 40, 41. Most closely related to *equale* but separated from that species by habitat and by having a higher pronotal crest, usually cut twice (see *S. equale*). This species may be confused with three *Trimerotropis* species: *latifasciata, pistrinaria,* and *modesta.* It is easy to separate *campestris* from the first two species because it occupies high mountain meadows. *S. campestris* has crossbands on forewings without clearly defined edges (clearly defined in *T. latifasciata* and *pistrinaria*). Pronozona relatively higher than in the other three species. Metazona of pronotum often pale. Outer face of hind femora with at least two dark bands (including the knees). It differs from *modesta* in the banding of the hind femora and hindwings. Inner face of hind femora often orange. Hind tibiae orange or red. Body length to end of forewings 25–30 mm in males, 28–40 mm in females.

HABITAT. Mountain meadows and slopes of the Rocky Mountains and its foothills. Scoggan and Brusven (1972) note: "*campestris* is a high-altitude species in Idaho. It inhabits the mountain meadow and wheatgrass-bluegrass communities, and may extend as low as the transition between sage brush and wheatgrass-bluegrass communities, but rarely below 5,500 feet. Typical habitats are moderately to densely vegetated with *Poa* sp, *Agropyron spicatum, Festuca idahoensis, Carex* sp, and *Artemisia frigida.* Soils range from extremely rocky to a rocky loam. Grasshoppers often associated with *campestris* include *Circotettix undulatus, Bruneria sordida, Melanoplus alpinus, M. infantilis,* and *Camnula pellucida.*" Brooks (1958) reports that in the western Canadian prairies: "*campestris* is a common species of the area, and occurs locally on dry hillside and gravelly spots within the parklands." Hebard (1925) notes that in South Dakota: "This insect not only occurs in the Great Plains but also in hilly or mountainous country, where it is usually limited to open, semiarid, grassy spots. Its distribution in South Dakota is probably very much more local than further west."

BEHAVIOR. Solitary males produce a single sequence of snapping sounds during a flight display. Both sexes crepitate during escape flights. Courting males produce three signals: bursts of stridulation made by vibrating the hind femora against the forewings, silent femur-raising, and ordinary stridulation. In male–male interactions males shake and tip the femora (Otte 1970).

LIFE CYCLE. Adults from July to September. In Idaho Scoggan and Brusven (1972) report: "*campestris* has a broad emergence period. First instars may appear from early June to early July, adults from mid-July to mid-August. In some areas of the state this species may have a 2-year life cycle."

REFERENCES. Hebard 1925c, Ball et al. 1942, Brooks 1958, Otte 1970, Scoggan and Brusven 1972.

***Spharagemon bunites* Otte, n. sp.** Pls. 12, 13

TAXONOMY. This species has the pronotal carina cut by two sulci and so may be confused with *Trimerotropis* species. It is apparently related to *campestris* in the Rocky Mountains.

DISTRIBUTION. This species has been collected in Yosemite National Park and at Sonora Peak and Sonora Pass, Tuolumne County, California.

RECOGNITION. This species is marked with strongly contrasting colors—black and pale brown or gray brown. It differs from the somewhat similar *Trimerotropis fontana* as follows: crossbands on forewings distinct on dorsum; disk of pronotum rounding onto the lateral lobes (usually marked by a distinct ridge in *T. fontana*); metazona relatively pale, prozona black; crest on prozona not nearly so pronounced. Basal area of hindwings yellow, outer dark band distinct, outer third of forewings becoming smoky at tip. Hind tibiae blue but with a pale band next to the black base. Body length to end of forewings 27–29 mm in males, 36 mm in the only female.

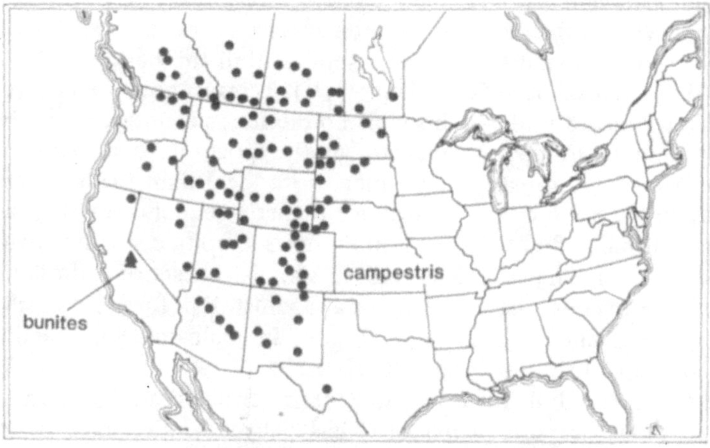

Spharagemon campestris and *S. bunites*

HABITAT. High-altitude mountain slopes in the Yosemite area of California.

LIFE CYCLE. All specimens were collected on August 7 and 8.

SPECIMENS. Holotype male, Glacier Point Road, Yosemite National Park, Mariposa County, California, August 7, 1940, E. R. Helwig, collector. One female and ten males were collected along with the holotype. Fourteen males were collected by Helwig on August 8 at Sentinel Dome, Yosemite. Twenty-nine specimens from the vicinity of Sonora Pass and Sonora Peak, Tuolumne County, are in the California Academy of Sciences collection.

Genus DISSOSTEIRA Scudder

DISTRIBUTION. Entire United States and Southern Canada.

RECOGNITION. Figs. 40*i–k*. Like *Spharagemon, Dissosteira* has a high median pronotal crest deeply cut by one sulcus. Unlike *Spharagemon,* hindwings are never yellow in basal area; forewings lack broad crossbands; hind tibiae are yellowish or bluish gray. Two of the species, *carolina* and *longipennis,* have black wings and so are obviously separate from *Spharagemon.*

REFERENCES. See references under each species.

• Identification of Dissosteira Species

carolina (widespread over United States and southern Canada) Fig. 40*j*

1. Hindwings black with yellowish trim
2. Forewings unicolorous or indistinctly marked
3. Posterior angle of pronotal disk 90 degrees or less

longipennis (prairie states) Fig. 40*k*

1. Hindwings with ivory or transparent rim
2. Forewings usually heavily mottled
3. Posterior angle of pronotal disk 90 degrees or less

spurcata (Utah to California and Washington) Fig. 40*i*

1. Hindwings transparent with smoky outer band
2. Forewings usually with distinct dark marks
3. Posterior angle of pronotal disk more than 90 degrees

pictipennis (Pacific states from Washington to Baja California)

1. Hindwings rosy red or deep pink at base and with distinct outer dark band
2. Forewings with distinct dark marks
3. Posterior angle of pronotal disk about 90 degrees

***Dissosteira carolina* (Linnaeus)** Pl. 13

DISTRIBUTION. This species, variously known as the road duster, the quaker, and the Carolina locust, is distributed over most of the United States and southern Canada. The reason for such an extraordinary range may be that roadsides are such an important habitat. What its precivilization distribution might have been is an intriguing question. The species also lives along sandy beaches and shorelines so its precivilization range may have been similar to that of today. But it has colonized many forested areas via roadsides.

RECOGNITION. Fig. 40*j*. This is one of two large species in which the hindwings are black with a pale outer margin. In *carolina* the black band is relatively broader and the pale outer band more opaque yellow than in *longipennis*. Forewings are not as strongly spotted. Overall body color varies according to the prevailing soil color. All shades of gray, brown, and reddish are represented in collections. Posterior angle of pronotal disk 90 degrees or less (*longipennis* is similar; in *spurcata* the angle is more than 90 degrees). Body length to end of forewings 37–42 mm in males, 42–58 mm in females.

BEHAVIOR. *D. carolina* is perhaps the most noted for its habit of hovering in flight over bare patches of ground (Otte 1970). Males rise almost vertically to a particular height, where they hover. During the first half of each flight the wing-beat rate is about four times that of the second half. The sound produced is not a loud crackling sound like that of *Arphia* species but has a sibilant or lisping quality. Flights commonly last more than fifteen seconds, and at the end of each flight the male drops to the spot from which it began. Hovering flights are less common compared to flights of *Arphia* species; one can wait for half a day sometimes to see a single flight, even if many individuals are present and otherwise active. Although females have not been seen to be attracted to performing males, that is almost certainly the function of the flights, at least in part. Males are often attracted to a hovering male and usually fly to him as he alights. I speculate that they do so on the chance of intercepting females who might also have been attracted. The attracted males mill about, exchanging male identifying signals, then slowly separate.

Sometimes solitary males of *carolina* sit in one spot for many minutes and stridulate, presumably also to attract sexually receptive females. Stridulation consists of alternately raising and lowering the hind femora, pressing and vibrating them against the forewings on the downstroke.

The behavior I observed most frequently in this species was males walking and flying toward other moving individuals. In the literature male aggregations are often mentioned, but they are usually of short duration and develop only because of the males' habit of approaching any moving individual. I speculate that the males cannot distinguish the sexes at a distance and therefore find it advantageous to investigate all moving individuals.

Courtship does not begin until a male is approximately six inches away from a female. Males always tip their femora while approaching, using two kinds of movements: one is silent, and in the other the forewings are struck on the upstroke; both movements consist of a fast upward jerk and a slower lowering. Both sexes tip their femora silently while interacting and while walking about alone. Frequently males switch from silent tipping to acoustical tipping while walking about alone. In male–male encounters both tipping signals were performed.

HABITAT. Open bare roadsides, beaches and shorelines, playgrounds, and along railroad beds. Nymphs usually inhabit thicker vegetation than adults. Blatchley (1920: 272) writes:

> Its numbers appear multiplied because it frequents highways and by-ways of man rather than the pastures and meadows where other grasshoppers are wont to congregate. Moreover, when disturbed, it more often betakes itself to the bare earth than to the green grass. Why this absurd taste asks the person uninitiated in the doings of nature's objects. For the simple reason that the dust of the roadside and the gravel ballast of the railway correspond so closely with the color of its back that its best friends and

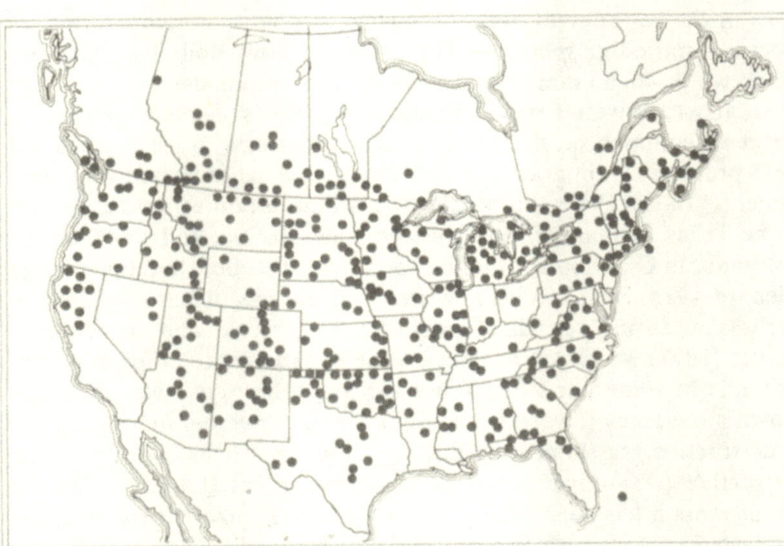

Dissosteira carolina

> worst enemies will overlook it if it will only remain quiet. Yea, even that sharp-eyed connoisseur of grasshopper tid-bits, the turkey gobbler, oftentimes walks right over it, mistaking it for a wayside pebble.

Scoggan and Brusven (1972) note: "*D. carolina* is an indicator of disturbed soil conditions. It is most commonly associated with field margins, roadsides, ditches, blowouts and other disturbed areas with sandy to fine loamy soils. In Idaho it has been collected from all major communities possessing disturbed, fine soils with vegetation consisting of sparse weedy forbs and scattered annual and perennial grasses."

LIFE CYCLE. Egg-overwintering. Adults from June to October in northern states, from May to November in northern Texas, June to November in California, June to September in Arizona.

REFERENCES. Somes 1914, Blatchley 1920, Cantrall 1943, Otte 1970, Scoggan and Brusven 1972, Vickery and Kevan 1983.

Dissosteira longipennis (Thomas) Pl. 13

DISTRIBUTION. Prairies from western Texas to South Dakota.

RECOGNITION. Fig. 40*k*. Like *carolina,* this species has mostly black hindwings with a pale to transparent outer margin. It differs from *carolina* as follows: outer pale band of hindwings not as yellow, more transparent in the distal third, and with larger, more conspicuous dark spots in the transparent area. Forewings strongly mottled (in this respect it is more like *spurcata*). Hind tibiae yellow. Body length to end of forewings 40–50 mm in males, 48–60 mm in females.

HABITAT. In west Texas Tinkham (1948) notes: "This large species, not at all common on the Marfa grasslands, seems to prefer enclosed areas protected from grazing where the grass reaches its natural development." In the 1940s this species was an important economic species in the Texas Panhandle, northeastern New Mexico, and southeastern Colorado. In Oklahoma it inhabits both wheat stubble and open range (Hebard 1938). Hauke (1953) writes: "In the past it has appeared in [Nebraska] during periods of grasshopper outbreaks and migrations." Bruner (1897) wrote: "It has been quite destructive to crops of nearly all kinds in some parts of [Nebraska] and adjoining states, and has shown a tendency toward becoming a leading member among the list of destructive grasshoppers." But Hauke notes: "It has not become a destructive grasshopper in Nebraska as Bruner feared it might. During recent years it has been held in check through control measures at the egg beds in areas south and southwest of Nebraska."

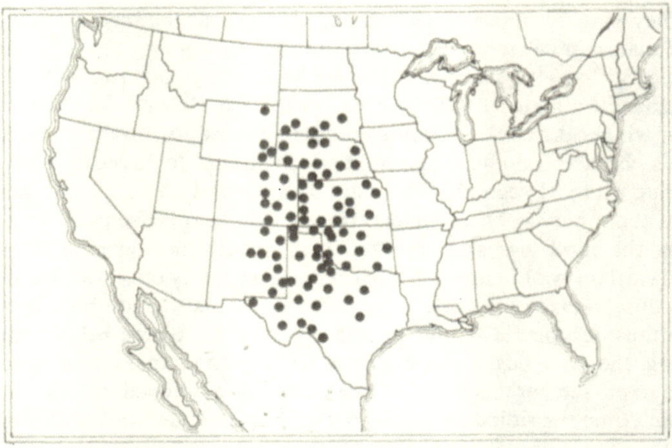

Dissosteira longipennis

LIFE CYCLE. Egg-overwintering. Adults from July to September.

REFERENCES. Bruner 1897, Hebard 1925, Tinkham 1948, Hauke 1953.

Dissosteira spurcata **Saussure** Pl. 13

DISTRIBUTION. California, Nevada, Utah, Oregon, Idaho, and Washington.

RECOGNITION. Fig. 40*i.* Hindwings transparent but with a smoky area near the outer margin. Overall body color quite variable: most individuals are heavily mottled or banded, but some are straw-yellow. In the more contrastingly colored specimens, the disk of the pronotum has an X-shaped ivory marking, the forewings have three to five larger dark markings on the lateral surface, and the hind femora are distinctly banded, especially on the upper face. The hind tibiae are always yellowish. Posterior angle of the pronotal disk is more than 90 degrees (90 degrees or less in *carolina* and *longipennis*). Body length to end of forewings 33–42 mm in males, 37–44 in females.

BEHAVIOR. Kelly and Middlekauff (1961) provide a detailed study on the biology of this species. They report that the males' hovering flight is different from that of *carolina:*

> The hovering flight varies from several inches to several feet above ground level and usually lasts 10 to 30 seconds, occasionally longer. The male begins his flight around 10:00 A.M. after air temperatures have increased,

> and they may be observed performing throughout the daylight hours. A single male can be seen to jump into the air a foot or two above a patch of flattened grass or a bare spot, and there he hovers in one spot with fluttering wings, the flight is noiseless for about 5 to 10 seconds, then with a slower wingbeat and the abdomen pointed almost downward he begins to make a characteristic buzzing sound which lasts 10 to 20 seconds but may continue up to 35 seconds. The male then abruptly drops to the ground, moves about nervously, and makes three or four rasping sounds by sharply raising the hind legs and rubbing them against the tegmina. There is usually a short wait before the flight is repeated. Only once was the nuptial flight observed to culminate in copulation. A male made his flight and while thus occupied a female moved up to the bare area below him. On landing, the male began rasping until sighting the female, whereupon he began to pursue her . . . He circled her, attempting several times to mount, but each time she sidled away. He circled her again and this time was successful.

HABITAT. In Idaho it inhabits dry desert areas, waste places, and hilly country. LaRivers (1924) notes: "[It] is hardly to be considered as having economic importance, though along with many other kinds, each doing its own damage, this one adds its might to the toll taken from the crops each year. It is a desert species and is found frequently on dry farms." Stohecker, Middlekauff, and Rentz (1968) write: "In California *D. spurcata* prefers a well-drained, hot and dry area that contains sparse or spotty vegetation with a rocky or graveled layer over the soil. It appears to be most abundant on deteriorated, overgrazed ranges. While absent from the Indio, Mojave, and Great Basin subdivisions in California, they are quite common in much of the lower foothill belt and valley zones where rainfall averages between 7 and 15 inches per year."

LIFE CYCLE. Egg-overwintering. California adults from mid-June to October. In the San Joaquin Valley of California eggs are deposited from late August to October. Hatching is from April to late May (Kelly and Middlekauff 1961).

REFERENCES. Kelly and Middlekauff 1961, Strohecker, Middlekauff, and Rentz 1968.

Dissosteira pictipennis Bruner — Pl. 13

DISTRIBUTION. Washington, Oregon, California, and Baja California.

RECOGNITION. Basal area of hindwings deep pink to rosy red, surrounded by a dark band that varies in darkness and width, and with a transparent band along the perimeter. Overall body coloration variable: most specimens have three large, dark markings on the lateral field of the forewings and four dark bands on the hind femora, but in specimens the color of yellow-brown dead leaves the dark bands are indistinct. In the more contrastingly marked specimens the pronotal disk has an X-shaped ivory marking. Posterior angle of pronotal disk usually more than 90 degrees. Hind tibiae yellow-gray to blue in the distal three-quarters with an ivory band in the first quarter. Body length to end of forewings 21–29 mm in males, 29–37 mm in females.

BEHAVIOR. Males do not perform a hovering flight like *spurcata* and *carolina;* instead they crepitate during directional flights up to three feet above the ground and distances up to fifteen feet. Crepitation consists of about four bursts of sound per second, each burst consisting of three wing-snapping sounds. The two longest flights I tape-recorded contained ten and fifteen bursts (Otte 1970). Between flights males walk about, occasionally tipping their femora silently. Males have two main courtship signals: in the first they kick out with the hind tibiae, striking the ends of the forewings (this is done three to five times, alternating left and right tibiae); the second is ordinary stridulation. Bursts of ordinary stridulation are produced by three to six synchronous up-and-down movements of the hind femora. Males tip and shake their femora in encounters with other males.

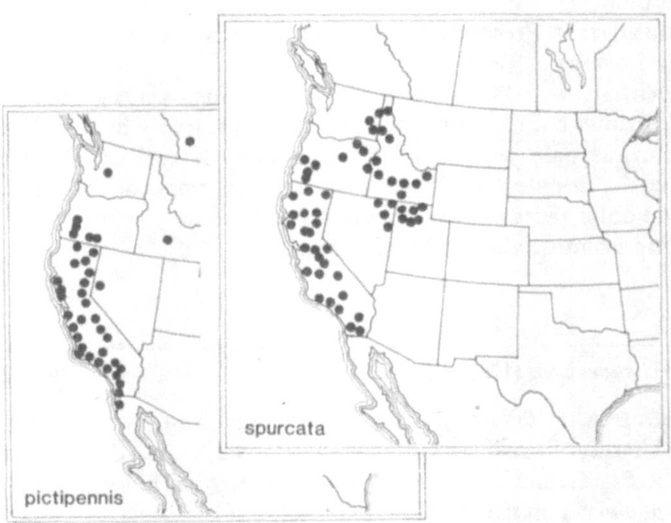

Dissosteira spurcata and *D. pictipennis*

HABITAT. This species lives at higher altitudes than *spurcata,* and the two species seldom associate. In the coastal areas of California it commonly inhabits the oak-grassland association. Strohecker, Middlekauff, and Rentz (1968) report that it is widely distributed in grassy rangelands except in the Indio, Mojave, and Colorado subdivisions.

LIFE CYCLE. Egg-overwintering. Adults from June to October; most common in August and September.

REFERENCES. Kelly and Middlekauff 1961, Strohecker, Middlekauff, and Rentz 1968, Otte 1970.

Genus EXIMACRIS Otte, n. gen.

TAXONOMY. The genus is erected to encompass two species: the type species, *superbum,* described by Hebard in the genus *Spharagemon,* and *phenax,* a new species from eastern Oklahoma. The two species are defined mainly by their unique and nearly identical coloration. The median pronotal crest is very different in the two species and therefore could be used to unite the genus with either *Spharagemon* or *Trimerotropis.* In *superbum* the crest is very high and cut by one sulcus, which would place the species in either *Dissosteira* or *Spharagemon* by the classical method of defining genera. In *phenax* the crest is low and cut by two sulci, placing the species in *Trimerotropis* according to the classical theme.

DISTRIBUTION. Presently known only from two sites, eastern Oklahoma and southern Texas.

RECOGNITION. Figs. 39, 40*d, x, y.* Hindwings smoky brown and without distinct bands, body greenish, forewings rusty brown. Disk or pronotum has pale yellowish markings along lateral carinae and is connected by a pale band across the median ridge of the prozona. Inner and outer faces of hind femora have three dark bands, the distal one taking up more than the distal third.

Eximacris superbum (Hebard) Pls. 12, 13

DISTRIBUTION. This species is known only from two males from Willacy County, Texas. Hebard (1938) believed he had discovered the species at Big Cedar, Oklahoma, but his specimen belongs to a new species, *phenax,* described below.

RECOGNITION. Fig. 40*d, x.* See *Eximacris* Recognition. Similar to *phenax* but with a high pronotal crest. Body length to ends of forewings about 37 mm.

HABITAT. Hebard (1937: 368) writes: "I was greatly surprised when the first male flew a short distance out of a thick growth of low oak shoots near extensive oak thickets in the loose sand region near Katherine. Its dark coloration was in great contrast to this environment bathed in brilliant sunlight."

LIFE CYCLE. The only two males were collected on August 8, 1912.

REFERENCES. Hebard 1937.

Eximacris phenax **Otte, n. sp.** Pls. 12, 13

TAXONOMY. In 1938 Hebard reported finding *superbum* at Big Cedar, Oklahoma. When specimens are viewed from above, one is struck by the great similarity between the Big Cedar specimen and the *superbum* from the southern Texas coast. But in side view one sees strong differences in the size of the pronotal crest, indicating that the two forms belong to different species.

DISTRIBUTION. The only known specimen of this species was collected by Standish-Kaiser on June 24, 1937 at Big Cedar in the Kiamichi Mountains of southeastern Oklahoma.

RECOGNITION. Fig. 40*y*. See *Eximacris* Recognition. Similar to *superbum* but with a low median carina cut by two sulci. Body length to ends of forewings 34 mm.

HABITAT. Not known.

LIFE CYCLE. The only specimen, a male, was collected on June 24, 1937. Its slightly shrunken abdomen and hind femora suggest that it had recently emerged to adulthood.

REFERENCES. Hebard 1938.

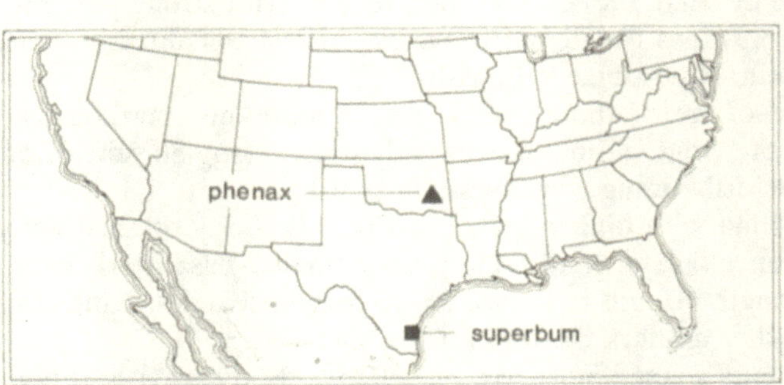

Eximacris superbum and *E. phenax*

Genus CONOZOA Saussure

DISTRIBUTION. Western United States from Washington to Baja California and east to New Mexico.

TAXONOMY. In the recent literature this genus includes only *carinata* and *sulcifrons.* Five other species are here included: *hyalina, texana,* and *rebellis* (moved from *Trimerotropis*), *clementina* (moved from *Scirtetica*), and *nicola* (moved from *Microtes*); see Fig. 44.

RECOGNITION. Figs. 45*c,* 46. Most similar to *Trimerotropis* species but may also be confused with *Lactista.* Differs from *Trimerotropis* by the following combination of characters: dark crossbands on forewings located on lateral surface only, absent on dorsum. Median carina usually well defined on both prozona and metazona. Median pronotal crest cut by one or two sulci; anterior cut often indistinct. Lateral lobes of pronotum have a small tooth in all mainland species; tooth is absent in the two island species.

▪ Key to Conozoa Species

CHANNEL ISLANDS, CALIFORNIA

1. Head and pronotum as in Fig. 46*d.* San Nicolas Island: **nicola**
   Head and pronotum as in Fig. 46*a.* San Clemente Island: **clementina**

MAINLAND

1. Median ridge on metazona well developed (Fig. 46*h*). Median ridge on prozona often scarcely bilobate (Fig. 47*j*). Hind tibiae yellowish: **carinata**
   Median ridge on metazona low (Fig. 46*g*). Median ridge on prozona variable, sometimes strongly bilobate. Hind tibiae reddish, yellowish, or bluish gray: **2**
2. Side of head, thorax, and hind femora with strong pale markings. Forewings usually with narrow pale streak directly beneath Sc vein. Hind tibiae yellowish: **rebellis**
   Side of head, thorax, and hind femora not marked as above. Forewings without narrow pale streak along Sc vein. Hind tibiae reddish, orange, yellowish, or bluish: **3**
3. Prozona with high ridge as in Fig. 47*s.* Body often straw-yellow. Inner face of hind femora with smaller basal dark band. Body length 30 mm in males, 35–40 mm in females. Hindwings with either no dark band or a faint one: **hyalina**
   Prozona shaped more like Fig 47*i, p, g.* Body grayish. Inner face of hind femora with larger basal dark band. Body length less than

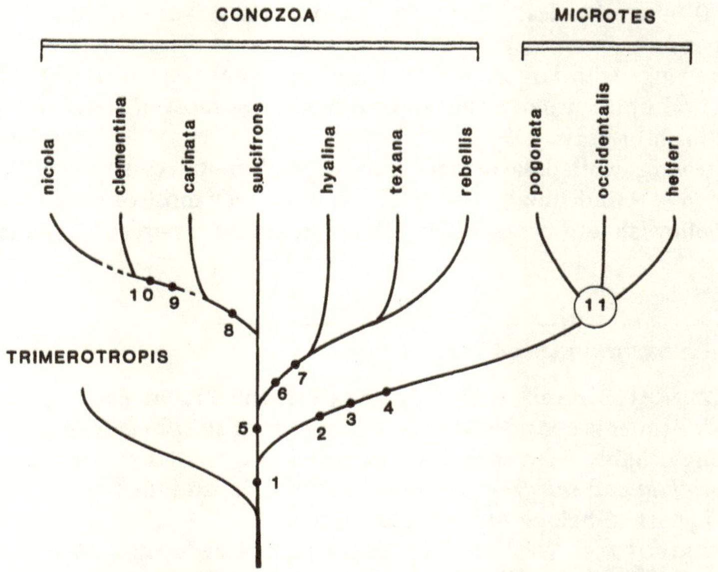

Fig. 44. Hypothetical phylogeny of *Conozoa* and *Microtes* species (see Appendix 1 for explanation).

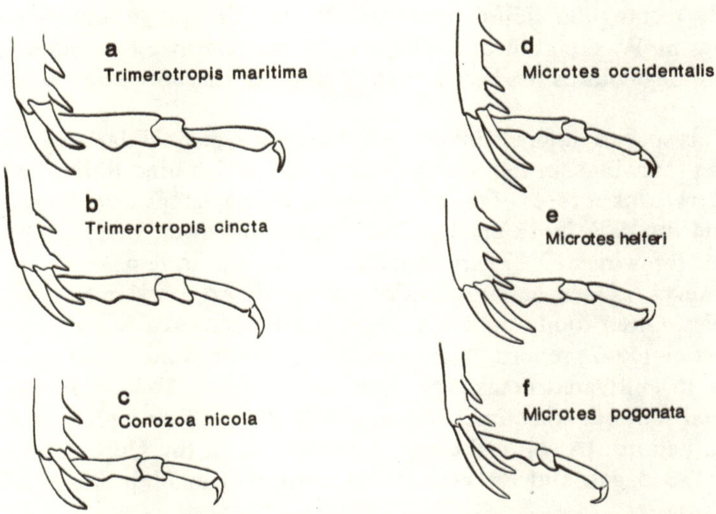

Fig. 45. Comparison of hind tibial spurs in *Trimerotropis, Conozoa,* and *Microtes.*

30 mm in males. Hindwings almost always with distinct dark band: **4**

4. Forewings with larger, darker bands. Pronotal crest as in Fig. 47*i*. Hind tibiae usually reddish or orange over most of range, never bluish or gray: **sulcifrons**

   Forewings with less distinct markings. Pronotal crest as in Fig. 47*n–g*. Hind tibiae usually bluish gray over most of range (but yellowish and occasionally orange in southern part): **texana**

Conozoa sulcifrons **(Scudder)** Pl. 14

TAXONOMY. In this work *Conozoa* (formerly *Trachyrhachys*) *occidentalis* Bruner is considered to be a synonym of *sulcifrons*. The type is indistinguishable from specimens assigned to the latter species. Likewise, *wallula* and *sulcifrons* are indistinguishable, and their type specimens appear to belong to the same species.

DISTRIBUTION. Western New Mexico, Colorado, and Wyoming west to California, and Baja California north to British Columbia.

RECOGNITION. Figs. 46*g*, 47*i*. Basal area of hindwings pale yellow to yellow, and distal area beyond dark band transparent (*carinata* is similar in this respect). Median carina of pronotum forms a low ridge or a raised line on the metazona. Posterior margin of pronotal disk nearly rectangular. Hind tibiae usually red or orange, occasionally yellow in Washington and Oregon. In the southwestern states the reddish leg color is the best way to distinguish this species from *carinata*.

This species differs from *texana* in the more distinct dark markings on the forewings, in the usually orange or reddish hind tibiae, and in the darker inner faces of the hind femora. In specimens from Baja California the dark band on the hindwings is narrower. Body length to end of forewings 19–27 mm in males, 25–37 mm in females.

HABITAT. Open bare ground of deserts, desert grasslands, and dry prairies. Often found on dry stony ground beside roads. In Arizona, Ball et al. (1942) report: "Common in the Lower Sonoran Zone, especially in cultivated areas, low flats, and arroyos. Usually found on alluvial soil, but sometimes in areas of shallow soil and often in very dry situations. [Also] in the sage brush desert of the Upper Sonoran Zone." Scoggan and Brusven (1972) write that in Idaho this species (*wallula*)

> prefers disturbed areas with fine, loose, predominantly sandy soils. It occurs in early and middle stages of dune and blowout succession, along

ditches and roadside areas, and in general where disturbance has kept soils from being stabilized by vegetation. It occurs in two major communities in Idaho, shadscale and sagebrush. Preferred habitats may be sparsely to densely vegetated with annuals, and usually contain sagebrush and rabbitbrush. It is frequently associated with *Trimerotropis pallidipennis, Melanoplus sanguinipes,* and *Oedaleonotus enigma.* By itself it is not a major pest species, reaching highest numbers only in poor unproductive rangeland. It may prefer grasses though it is probably a mixed feeder.

LIFE CYCLE. In Arizona adults from March to November. In northern states adults from mid-July to September.

REFERENCES. Ball et al. 1942, Hewitt and Barr 1967, Scoggan and Brusven 1972.

Conozoa carinata Rehn Pl. 14

DISTRIBUTION. Extreme southwestern United States and Baja California, north to southwestern Utah and east to southwestern New Mexico.

RECOGNITION. Figs. 46*h*, 47*j*. Hindwing coloration as in *sulcifrons.* Differs from *sulcifrons* as follows: median pronotal carina much higher on metazona, posterior margin of pronotal disk acute-angulate. Hind tibiae yellowish. Point at lower posterior angle of lateral lobes usually well developed (poorly developed or absent in *texana*). Body length to end of forewings 18–24 mm in males, 25–35 mm in females.

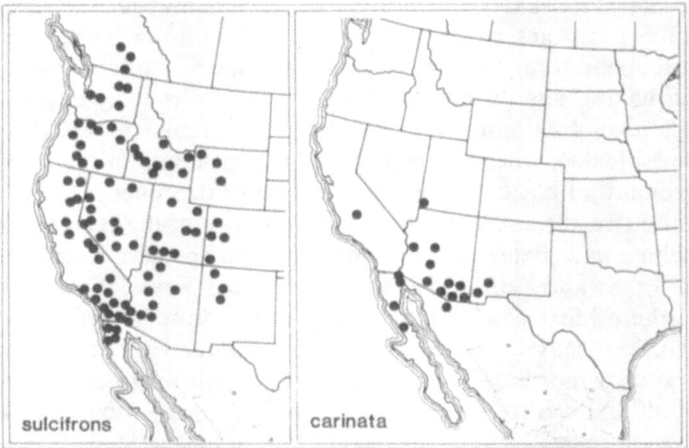

Conozoa sulcifrons and *C. carinata*

HABITAT. Ball et al. (1942) report: "A common grasshopper on the range and in cultivated fields at a higher elevation than *sulcifrons.* Especially common in sandy washes in the desert grassland. Feeds on herbaceous plants such as *Lupinus, Amaranthus,* and *Boerhavia,* as well as grasses."

BEHAVIOR. Males do not stridulate during courtship but shake their hind femora, striking the ground as they do so. Only silent tipping motions were seen in male–male interactions (Otte 1970).

LIFE CYCLE. The species overwinters in the egg stage. In Arizona adults may be found from April to November.

REFERENCES. Rehn 1907a, Ball et al. 1942.

Conozoa texana (Bruner) Pl. 14

TAXONOMY. *Trimerotropis bilobata* and *T. cristata* are synonymized under *Conozoa texana.* This species shows considerable geographic variation in the shape of the pronotum, in body color, and in body size (see Recognition).

DISTRIBUTION. Widespread over western North America from Baja California and northeastern Mexico to British Columbia.

RECOGNITION. Figs. 46*b,* 47*n–q.* Basal area of hindwings pale yellow, sometimes almost colorless; outer dark band variable in darkness. Dark bands on forewings vary in darkness and distinctness. Pronotal crest distinctly higher than metazonal crest. Transition from pronotal disk to lateral lobes less abrupt than in *carinata* and *sulcifrons,* sometimes quite indistinct. Height of prozonal crest quite variable, lowest in pale specimens from Nevada deserts. This species differs from *sulcifrons* in having less distinct dark markings on the forewings and in having blue or gray hind tibiae over most of its range (except in Texas and New Mexico, where orange tibiae are common), and in having a lesser amount of black on the lower margin of the inner femoral face. Color and size variations occur across the range of the species. Specimens with a pale metazona are common in the northern states but absent in Texas and California. Specimens from Texas, New Mexico, and northern Mexico are distinctly larger than those in California and more northern states. The color of the hind tibiae varies from mainly bluish gray in northern states to yellow or yellow-green in parts of Utah and Arizona, to yellow and orange in New Mexico and Texas. Body length to end of forewings 19–28 mm in males, 25–37 mm in females.

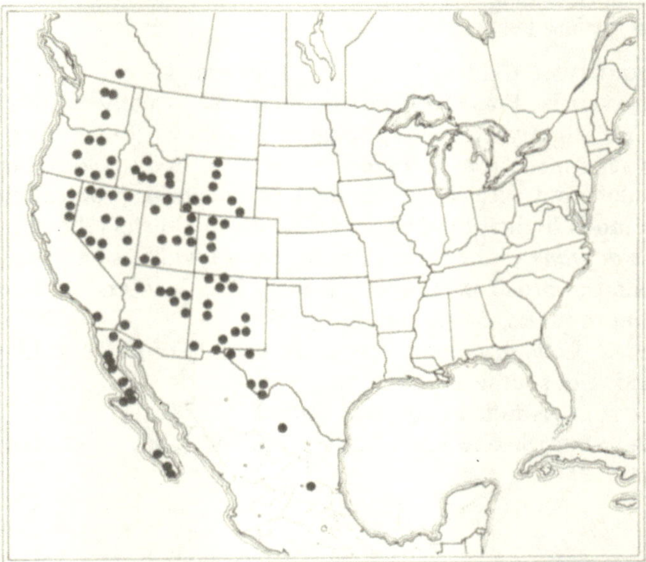

Conozoa texana

HABITAT. Inhabits dry, sandy desert and semidesert areas.

LIFE CYCLE. Adults from July to October in Texas, August to September in northern states, and March to November in southern California and Baja California.

REFERENCES. McNeill 1901, Strohecker, Middlekauff, and Rentz 1968, Otte 1970.

Conozoa rebellis **(Saussure)** Pl. 14

DISTRIBUTION. Northern Baja California, Mojave Desert and Central Valley of California.

RECOGNITION. Fig. 46*r*. Basal area of hindwings pale yellow. Sides of head, pronotum, and second and third thoracic segments contrastingly marked; the pale areas are often white. Forewings usually with a pale streak along and beneath the Sc vein. Hind tibiae yellowish. Body length to end of forewings 22–27 mm in males, 31–37 mm in females.

HABITAT. Low-lying arid regions of the Mojave Desert of California.

LIFE CYCLE. Adults from May to October.

REFERENCES. Saussure 1888, McNeill 1900, 1901, Strohecker, Middlekauff, and Rentz 1968, Otte 1970.

Conozoa hyalina (Scudder) Pl. 14

DISTRIBUTION. Central Valley of California.

RECOGNITION. Fig. 46*s*. Hindwings variable, usually entirely transparent and slightly yellowish, sometimes with an outer dark band. Readily separated from *sulcifrons* as follows: body color often pale straw-yellow, but occasionally as dark as *sulcifrons;* pronotal crest on prozona much higher; hind tibiae yellowish to bluish (yellowish to orange in *sulcifrons*); inner face of hind femora with a small basal dark band (band is broad in *sulcifrons*); body length to end of forewings 32–35 mm in males, 34–40 mm in females.

HABITAT. Low, dry desert flats in the vicinity of Bakersfield, Ceres, and Turlock in southern California.

LIFE CYCLE. Adults in July and August.

REFERENCES. Strohecker, Middlekauff, and Rentz 1968, Rentz and Weissman 1980.

Conozoa nicola (Rentz and Weissman)

TAXONOMY. Rentz and Weissman (1981) placed this species in the genus *Microtes.* They write: "*M. nicola* is placed in *Microtes* on the basis of its robust hirsute appearance, which is no doubt correlated with its arenophilous habits so characteristic of *Microtes,* its buzzing crepitation, and the structure of the head and pronotum." To me *nicola* bears a much closer resemblance to *clementina* and to *Conozoa carinata* than it does to *Microtes.* The paratypes I examined are decidedly slender and the pronotal and head proportions are similar to those of *Conozoa,* not *Microtes. C. nicola* and *clementina* have similar karyotypes: 2N♂ = 23, all telocentric, 2 large, 7 medium, 2 small autosomes; with a secondary constriction.

DISTRIBUTION. San Nicolas Island, southern California.

RECOGNITION. Figs. 46*d,* 47*k.* Hindwings with very pale greenish basal area and an almost obsolete outer dark band. Forewing crossbands vary from indistinct to absent. Differs from *clementina* mainly in having a more roughly textured head and pronotum. Body length to end of forewings 21–23 mm in males, 31–34 mm in females.

HABITAT. Rentz and Weissman (1981) report: "The species occupies stabilized blowouts, coastal and inland sand dunes, and coastal bluff communities. It is the most common grasshopper on San Nicolas Island, is widely distributed, and found in almost every habitat there."

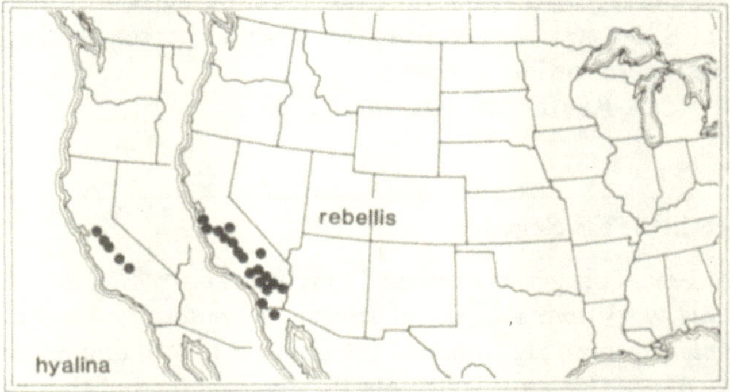

Conozoa rebellis and *C. hyalina*

LIFE CYCLE. Adults from late April to late August but most abundant in July and August.

REFERENCES. Rentz and Weissman 1981.

Conozoa clementina (Rentz and Weissman)

TAXONOMY. Rentz and Weissman (1981) placed this species in the genus *Scirtetica.* They write, however: "This species cannot readily be referred to any presently recognized oedipodine genus in California." *Scirtetica* is here synonymized under *Spharagemon.* But *clementina* is not a *Spharagemon,* and it does fit readily with other *Conozoa* species morphologically, distributionally, and by habitat. The resemblance of *clementina* to *carinata* is especially close; almost the only difference is in the shape of the lateral lobes.

DISTRIBUTION. San Clemente Island, southern California.

RECOGNITION. Figs. 46*a,* 47*l, m.* Basal area of forewings pale greenish (similar to some specimens of *carinata* from Baja California), outer dark band barely indicated. Hind tibiae yellowish. Median pronotal crest well developed on metazona, cut by one distinct sulcus; occasionally a second cut is barely indicated (*carinata* is similar in this respect). Lateral lobes rounded along lower posterior edge. Differs from *nicola* mainly in having a smoother head and pronotum surface. Body length to end of forewings 21–23 mm in males, 29–34 mm in females.

HABITAT. Rentz and Weissman report it from "coastal bluffs, sand dunes, and sparse, rocky grassland."

LIFE CYCLE. Adults were collected from June into August. Rentz and Weissman report that by June 26, 1972, 80 percent of the males captured at Site 1 were crepitating; the other 20 percent and most females were teneral or late instars.

REFERENCES. Rentz and Weissman 1981.

Genus MICROTES Scudder

TAXONOMY. This genus previously included only *Microtes occidentalis*. In this work two additional species, *pogonata* and *helferi* (formerly *Trimerotropis*), are included. Strohecker (1960) and Strohecker, Middlekauff, and Rentz (1968) note the resemblance of these species to *Microtes occidentalis*. In his original description of *T. helferi* Strohecker (1960) says: "Form short and robust for the genus, in general proportions resembling *Microtes*." Strohecker, Middlekauff, and Rentz (1968) note: "[*T. helferi*] is apparently restricted to northern coastal dunes. [It occupies] the same ecological niche as does *Microtes occidentalis* which occurs in central and southern California." Elsewhere they note that *T. pogonata* is "short and stout, in appearance much like *Microtes* and probably more related to that genus than to *Trimerotropis* species, but with median carina of pronotum twice intersected."

In this case, as in others, emphasis on the number of cuts in the median pronotal ridge has probably resulted in an incorrect clustering of species. The resemblance among *occidentalis, helferi,* and *pogonata* appears to be too close to place them in different genera.

Weissman, on the basis of the chromosomes of the three species,

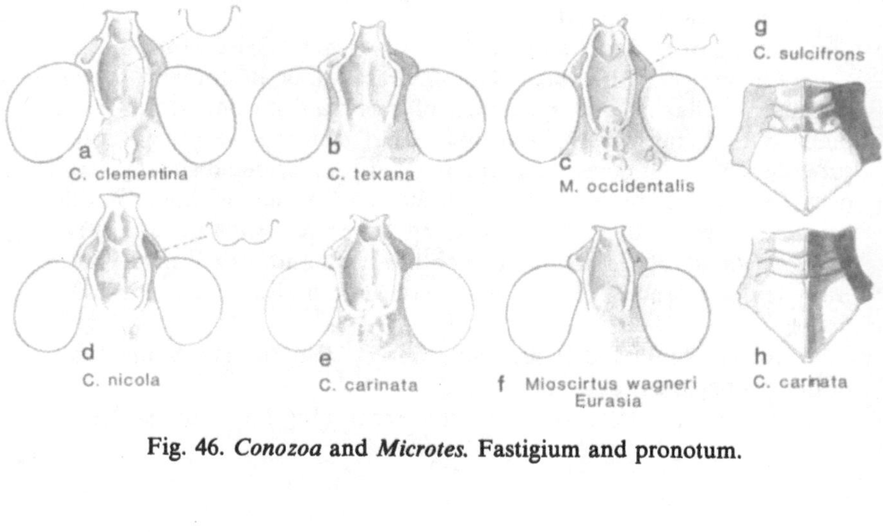

Fig. 46. *Conozoa* and *Microtes*. Fastigium and pronotum.

believes that *helferi* and *pogonata* are not closely related, being from different sections of the genus *Trimerotropis,* and that the morphological resemblance between the two species is due to convergence. Another hypothesis is that metacentric chromosomes in *helferi* evolved independently and are not the result of a shared common ancestry with the so-called section B *Trimerotropis,* but that their peculiar geographic distributions and their morphological resemblance are the result of their descent from a common stem species.

The genus appears to be most closely related to *Conozoa,* which it resembles in the structure of the head and pronotum. The genus is probably derived from a dune-inhabiting member of the *Conozoa* that subsequently split into three species along the California coast.

RECOGNITION. Figs. 45*d–f,* 46*c,* 47*g, h.* Hind tibiae with long apical spurs. Body rather short and hairy and rather strongly speckled. Hindwings very pale green to transparent in basal area. Outer dark band barely indicated. Forewing crossbands do not have sharp margins and do not cross dorsal area. Fastigium elongate and deep, much as in *Conozoa.* Median pronotal crest well developed throughout, cut by one or two sulci.

REFERENCES. Scudder 1900, Strohecker 1960, Strohecker, Middlekauff, and Rentz 1968.

• Identification of Microtes Species

occidentalis (Santa Barbara to Marin County, California) Figs. 45*d,* 46*c,* 47*g, h*

1. Median pronotal crest cut once
2. Hind tibial spurs slightly shorter and stouter
3. Forewings extend beyond ends of hind femora

helferi (northern California) Fig. 45*e*

1. Median pronotal crest cut twice
2. Hind tibial spurs similar to those of *occidentalis*
3. Forewings extend beyond ends of hind femora

pogonata (Santa Barbara to San Luis Obispo, California) Fig. 45*f*

1. Median pronotal crest cut twice
2. Hind tibial spurs especially long and slender
3. Forewings short, scarcely extending beyond ends of hind femora

---

***Microtes occidentalis* (Bruner)** Pl. 14

DISTRIBUTION. Coast of California from Santa Barbara north to Marin County. The type of this species, according to the label, is from the Argus Mountains in the eastern California desert, but it seems highly probable that the specimen was incorrectly labeled.

RECOGNITION. Figs. 45*d*, 46*c*, 47*g*, *h*. Small and wide-bodied. Generally dark gray or gray-brown, but sometimes lighter gray and speckled with dark. Head and pronotum short and sharply sculptured. Forewings usually with two or three dark blotches on lateral field, but not banded across dorsal field. Hindwings pale greenish yellow (sometimes almost clear), outer dark band variable, from moderately distinct to obsolete. Outer face of hind femora with two oblique dark bands or unbanded; inner face mostly black but with two pale bands. Hind tibiae yellowish in pale individuals, grayish or bluish in dark individuals. Body length to end of forewings 17–20 mm in males, 20–26 mm in females.

HABITAT. Strohecker, Middlekauff, and Rentz report that the species occurs commonly on coastal sand dunes.

LIFE CYCLE. Adults from May to September.

REFERENCES. Strohecker, Middlekauff, and Rentz 1968.

Microtes pogonata (Strohecker) Pl. 14

DISTRIBUTION. Coastal California in the vicinity of Santa Barbara and San Luis Obispo, but not in coastal dunes.

RECOGNITION. Fig. 45*f*. A very small, hairy, wide-bodied species quite similar in appearance to *helferi*. Body speckled with dark spots. Unlike *helferi*, there are distinct dark bands visible along the costal margins of the forewings. The forewings are short, scarcely extending beyond the ends of the hind femora. Apical spines of hind tibiae are even longer than in *helferi*. Hind tibiae yellow or banded yellow and brown. Body length to ends of forewings about 16 mm in males, about 23 mm in females.

HABITAT. This species has been collected along the shores of Oso Flaco Lake, San Luis Obispo County, and on the San Marcos Ranch, Santa Barbara County.

LIFE CYCLE. Adults were collected in the first two weeks of July in 1956 and 1959.

REFERENCES. Strohecker 1963, Strohecker, Middlekauff, and Rentz 1968.

Microtes helferi (Strohecker) Pl. 14

DISTRIBUTION. Coastal dunes from northern California to Oregon.

RECOGNITION. Fig. 45*e*. A stout-bodied, hairy species, strongly speckled and without crossbands on forewings. Fastigium and frons with a continuous median groove. Transition from pronotal disk to lateral lobes angulate, distinct throughout. Dorsal face of hind femora usually with three or four dark bands (including knees). Hind tibiae yellowish. This species is most similar to *pogonata* (see *pogonata* for differences). Body length to end of forewings 20–32 mm in males, 29–33 mm in females.

HABITAT. Strohecker (1960) notes: "*helferi* occupies a habitat much similar to that of *T. maritima* . . . it is an occupant of the first line of dunes. Plants with which the insect was associated are *Convolvulus soldanella, Franseria chamissonis,* and *Artemisia psycnocephala.*" The species occupies the same ecological niche as *M. occidentalis* in central and southern California.

LIFE CYCLE. Adults from late June to late August.

REFERENCES. Strohecker 1960, Strohecker, Middlekauff, and Rentz 1968.

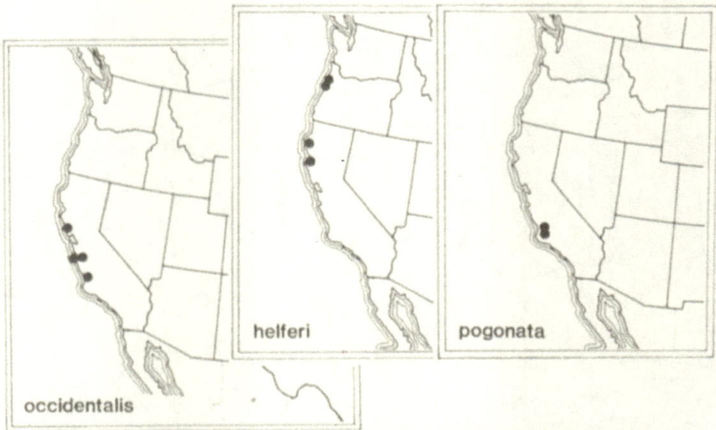

Microtes occidentalis, M. helferi, and *M. pogonata*

Genus SPHINGONOTUS Fieber

TAXONOMY. Three North American species of *Sphingonotus* have been described, all from the Caribbean islands: *Sphingonotus haitensis* (Saussure) from Hispaniola, *S. jamaicensis* Saussure from Jamaica, and *S. cubensis* Saussure from Cuba. The last species was first described as a subspecies of *S. caerulans,* a European species. Rehn and Hebard (1938) synonymized *S. jamaicensis* under *S. haitensis* and considered *cubensis* to be a distinct species. I have examined 170 specimens studied by Rehn and Hebard and others, and I believe a single species, showing some interisland differentiation, is involved. One can recognize two subspecies, one from Cuba and another ranging from Jamaica east through Hispaniola, Puerto Rico, and as far as Nevis.

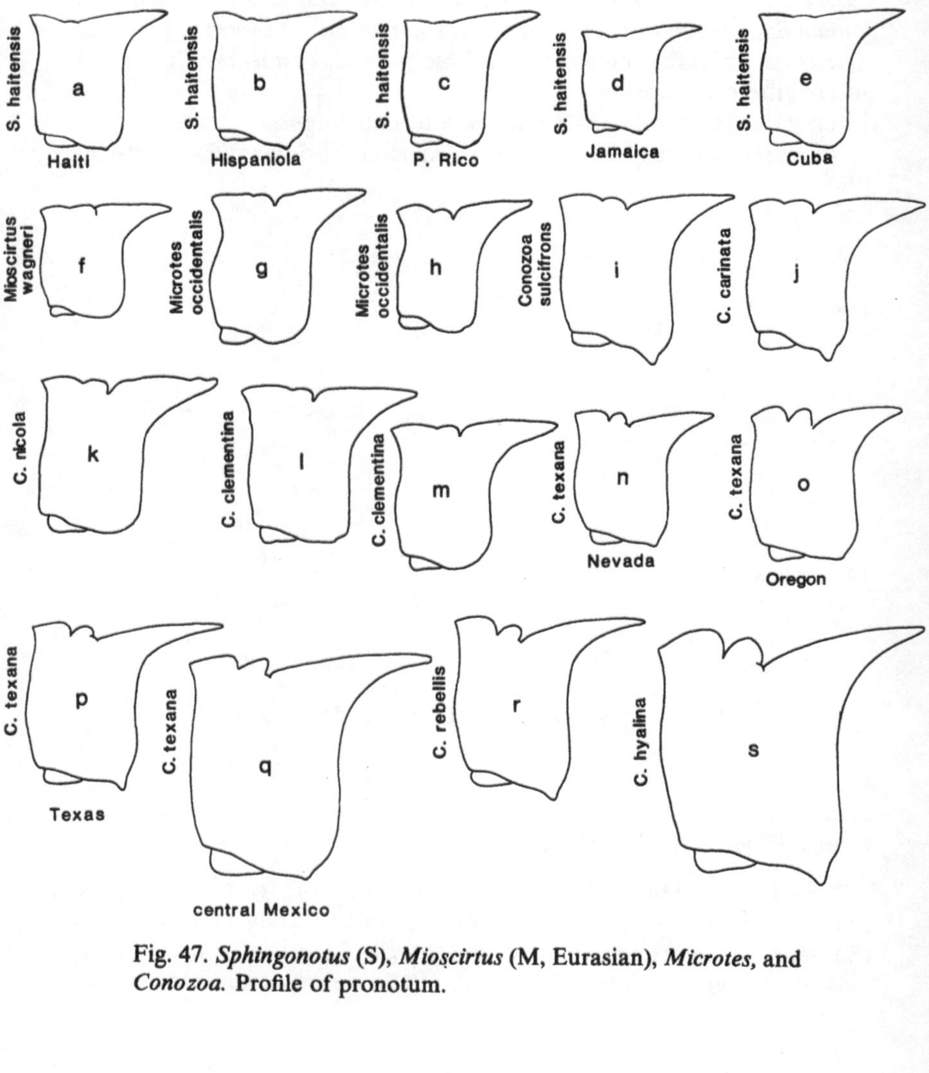

Fig. 47. *Sphingonotus* (S), *Mioscirtus* (M, Eurasian), *Microtes,* and *Conozoa.* Profile of pronotum.

Until appropriate crossing experiments are done, the true taxonomic status will remain in doubt.

S. haitensis is much closer to *S. caerulans* from Europe than to any North American *Trimerotropis*. The species may well have been introduced into the Caribbean by man. Indeed, Saussure considered Cuban specimens to be conspecific with *S. caerulans*. The only other genera of Oedipodinae that have penetrated the Caribbean region are *Psinidia* and *Chortophaga*. In these two genera the relationships between island and North American mainland forms are close, and there is even some doubt that they have speciated (see *Chortophaga cubensis*).

RECOGNITION (Caribbean species only). Fig. 47*a–e*. Basal area of hindwings blue, with or without an outer dark band. Forewings with at least two strong crossbands. Median pronotal carinae very low, not much more than a raised line, and cut by two or three transverse grooves. Hind tibiae yellow-green to blue.

Sphingonotus haitensis haitensis (Saussure) Pl. 14

DISTRIBUTION. Caribbean islands: Jamaica, Navassa, Hispaniola, Puerto Rico, and Nevis.

RECOGNITION. Fig. 47*a–d*. Basal area of forewings light blue. Dark band on hindwings located nearer periphery than in *S. h. cubensis*. Lateral lobe with a more distinctly indicated projection. Hind tibiae more green than blue and with a darker annulus near center. Body length to end of forewings 17–22 mm in males, 23–31 mm in females.

HABITAT. Dry areas with coral, cactus, and sand; beaches, dry sandy stream washes, and bare spots on dry hills.

LIFE CYCLE. Adults throughout the year.

REFERENCES. Rehn and Hebard 1938.

Sphingonotus haitensis cubensis (Saussure) Pl. 14

DISTRIBUTION. Cuba, Bahamas.

RECOGNITION. Fig. 47*e*. Basal area of forewings light blue. Outer dark band varies from absent to strongly defined. Hind tibiae blue. Lower margin of lateral lobe does not have a distinct projection, but sometimes a projection is barely indicated. Body length to end of forewings 21–23 mm in males, 29–32 mm in females.

Sphingonotus haitensis

HABITAT. Sandy pine woods, short dry grasses in low hills, beaches.
LIFE CYCLE. Adults throughout the year.
REFERENCES. Rehn and Hebard 1938.

Genus TRIMEROTROPIS Stål

TAXONOMY. The genus can be broken up into a number of species groups based on overall morphological resemblance. The position of some species remains doubtful (Fig. 48). White (1951) and Weissman and Rentz (1980) divide *Trimerotropis* into two groups: section A species have only acrocentric chromosomes; section B have some metacentric chromosomes. Although members of section B tend to be related, morphological, behavioral, and ecological information does not support the notion that A and B groups are monophyletic. White's scheme is discussed in more detail under *Circotettix*.

DISTRIBUTION. Canada to Chile and Argentina, but with distributional gaps in Central America.

RECOGNITION. Figs. 49, 50, 52. Because this is a large genus with considerable morphological variation among its species, it is not easy to characterize the group without comparing it to other genera individually. The figures can be used to match specimens being determined.

The median ridge of the pronotum almost never forms a sharp ridge and usually is very low on the metazona (in *Conozoa* the ridge is more clearly defined on the metazona). Pronotal crest always cut by two sulci (also true for the Psinidia group, *Spharagemon campestris, S. bunites,* some *S. equale,* and some Hippiscus group species). In all species not living in sandy habitats the forewing bands cross the dorsum (never true in *Conozoa*). Eyes are relatively large, but not huge and protruding as in Psinidia group. *Anconia hebardi* is similar, but it has

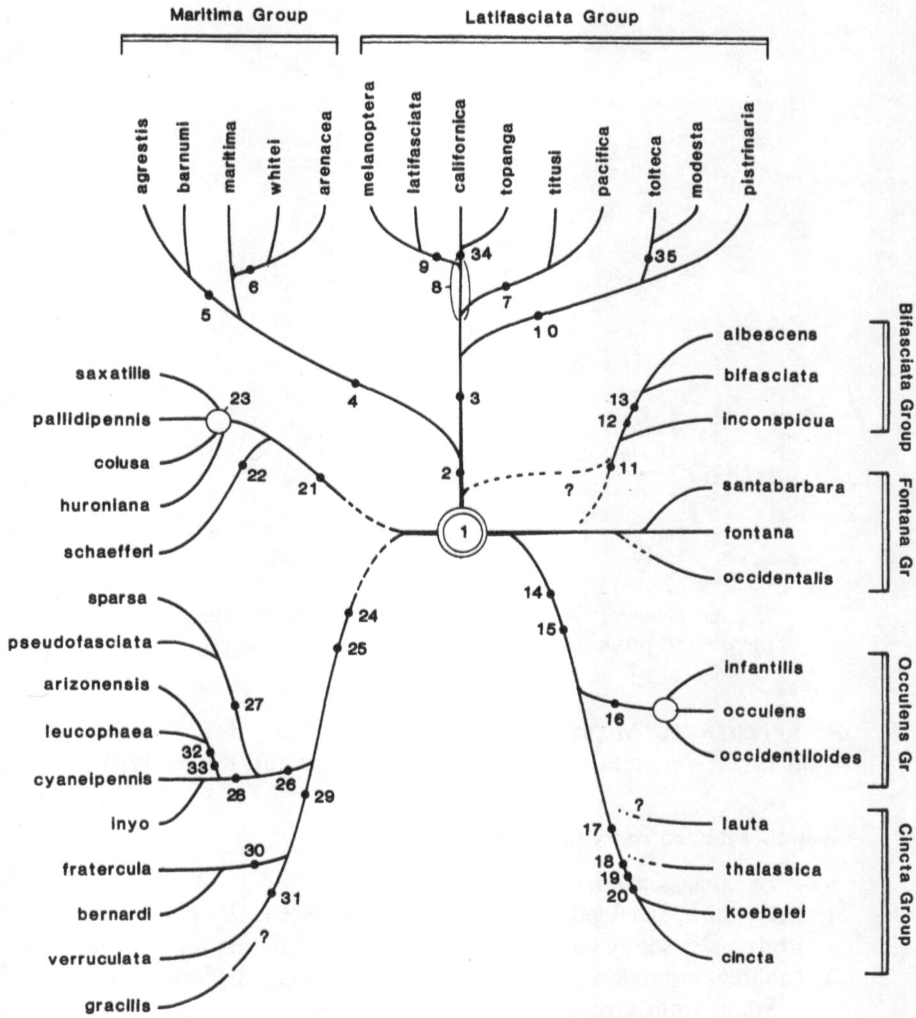

Fig. 48. Hypothetical phylogeny of *Trimerotropis* species (see Appendix 1 for explanation).

an elongate pronotum. Side of lateral lobe never has a continuous dark oblique band as in some Psinidia group species. No species is spotted as in Hippiscus group species. Hindwings usually with a clearly defined outer dark band. Basal area of hindwings yellow, yellow-green, or blue, never orange or red.

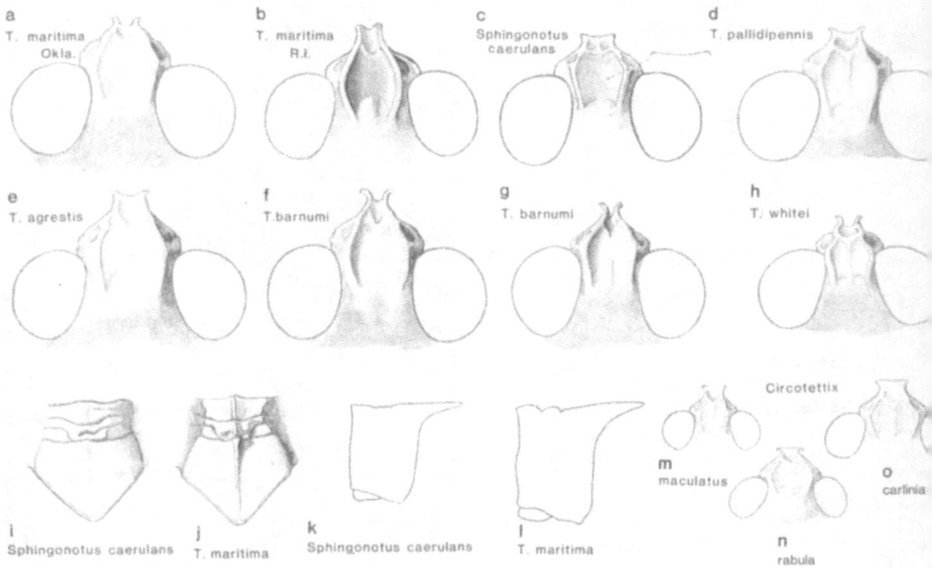

Fig. 49. *Trimerotropis, Sphingonotus,* and *Circotettix* species. Fastigium and pronotum.

REFERENCES. McNeill 1901, White 1949, 1951, Rentz and Weissman 1980, Weissman and Rentz 1980, Vickery and Kevan 1983.

▪ Key to Trimerotropis Species

EAST OF MISSISSIPPI RIVER

1. Body dark, speckled with pale spots. Inhabits rocky places: **2**
   Body pale, sandy colored or straw-yellow. Inhabits sandy areas: **3**
2. Inhabits mountain tops and rock outcroppings in southern United States from Georgia to Oklahoma (Pl. 18): **saxatilis**
   Inhabits lake shores and rocky areas across northern United States and Canada from Minnesota to Maine and from Alberta to New Brunswick and Nova Scotia (Pl. 18): **verruculata**
3. Inner face of hind femora with broad basal dark band. Crossbands of forewings faint but present in most specimens, visible on dorsal surface. Found only on dunes on northern shores of lakes Michigan, Huron, and Superior (Pl. 18): **huroniana**
   Inner face of hind femora with short basal dark band. Crossbands of forewings obsolete or fading out on dorsum, especially in spec-

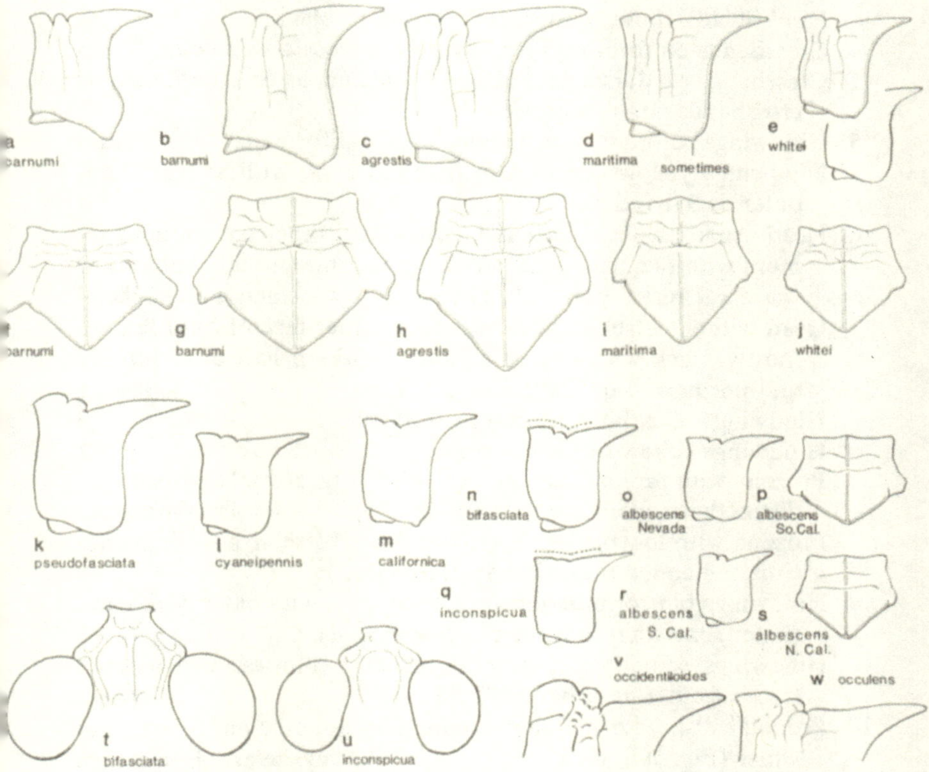

Fig. 50. *Trimerotropis* species. Fastigium and pronotum.

imens from lake and ocean beaches. Widespread along sandy shores; on Great Lakes, only on southern beaches (Pl. 15): **maritima**

WEST OF MISSISSIPPI RIVER

1. Lower posterior margin of lateral lobes without projection or tooth (Fig. 50*k, l, n–r*): 2
   Lower posterior margin of lateral lobes with projection or tooth (Fig. 50*a–d, m*): 21
2. Body dark gray or blackish. Forewings without crossbands. Anal veins thickened in males. Basal area of hindwings yellow, outer half smoky. Found in openings in mountain conifer forests (Pl. 18): **verruculata suffusa**
   Without above combination of characters: 3
3. Hind tibiae blue or blue-gray or slightly greenish: 4

Hind tibiae yellow, orange, red, brown, or black: **27**

4. Crossbands on forewings usually do not cross dorsal field: **5**
   Crossbands on dorsal field almost as distinct as on lateral field or crossbands largely obsolete: **7**
5. Hindwings colorless, transparent. Baja California (Pl. 18): **lauta**
   Hindwings yellow or greenish in basal area, with at least faint outer dark band: **6**
6. Head usually with black band across face. Inner face of hind femora with two pale bands. Widespread in mountain habitats of western United States (Pl. 17): **cincta** and **koebelei**
   Head without black band across face. Inner face of hind femora mostly black with one pale band. Southern half of California and northern Baja California (Pl. 17): **thalassica**
7. Hindwings blue, blue-green, or colorless: **8**
   Hindwings yellow or yellow-green: **12**
8. Prozona with pronounced median ridge (Fig. 50*k*). Hindwings indistinctly banded or unbanded (Pl. 18): **pseudofasciata**
   Prozona with low median ridge (see Fig. 50*l, m, n, o*). Hindwings usually banded (indistinct in *leucophaea*): **9**
9. Hindwings blue or blue-green in basal area, with outer dark band (latter somewhat indistinct in *leucophaea*): **10**
   Hindwings sometimes pale blue, usually colorless and without trace of outer dark band (Pl. 18): **sparsa**
10. Pronotal disk with definite shoulders at anterior end. Fastigium wider (Fig. 56*a;* Pl. 15): **cyaneipennis** and **inyo**
    Pronotal disk rounding onto lateral lobes at anterior end. Fastigium quite narrow (Fig. 56*b, c*): **11**
11. Outer dark band crosses hindwings. Forewings with distinct crossbands: **arizonensis**
    Outer dark band on hindwings distinct only on fan. Forewings without distinct crossbands (Pl. 15): **leucophaea**
12. Crossbands on forewings narrow, sometimes not reaching anterior margin. Small species from California coastal ranges: **13**
    Crossbands on forewings broader, reaching anterior margin, or crossbands obsolete. Size of specimens varies: **15**
13. San Mateo County: **infantilis**
    San Luis Obispo to Los Angeles: **14**
14. Los Angeles County (Pl. 17): **occidentiloides**
    San Luis Obispo and Santa Barbara counties (Pl. 17): **occulens**
15. Found only on Santa Barbara Island: **santabarbara**
    Found on mainland: **16**
16. Forewing crossbands broad and dark on lateral field, lighter on

dorsal field, sometimes almost disappearing at midline. Outer third of hindwings often smoky (Pl. 17): **fontana**

Forewing crossbands not much broader or darker on lateral field than on dorsal field, not disappearing at midline. Hindwings not smoky in distal third: **17**

17. Hindwings pale yellow-green (Pl. 17): **occidentalis**

Hindwings yellow: **18**

18. Body sand-colored. Inhabits sand dunes in Oregon, Nevada, and Idaho (Pl. 15): **arenacea**

Body with strong crossbands: **19**

19. Pronotal carina higher, shaped as in Fig. 50*n* (Pl. 17): **bifasciata**

Pronotal carina lower, as in Fig. 50*o, q, r:* **20**

20. Body white and black. Lateral lobes flared when viewed from above. Prozona often black or dark gray on dorsum. Lateral lobes with large black mark (Pl. 17): **albescens**

Body cream and gray, not white and black. Lateral lobes do not flare. Prozona not dark gray or black on dorsum. Lateral lobes without large black mark (Pl. 17): **inconspicua**

21. Lower posterior margin of pronotum as in Fig. 50*a, b, c,* not as in Fig. 50*d, e:* **22**

Lower posterior margin of pronotum as in Fig. 50*d, e,* not as in Fig. 50*a, b, c:* **23**

22. Pronotum as in Fig. 50*c, h.* Inner face of hind femora with black bands (Pl. 15): **agrestis**

Pronotum as in Fig. 50*a, b, f, g.* Inner face of hind femora without black bands (Pl. 15): **barnumi**

23. Hind tibiae blue or gray. Southern California and Baja California (Pl. 16): **topanga**

Hind tibiae white, yellow, orange, or red: **24**

24. Forewings without distinct darkened crossbands (Pl. 15): **maritima and whitei**

Forewings with distinct darkened crossbands: **25**

25. Hind tibiae orange or red. Widespread from New Mexico to southern California: **californica**

Hind tibiae white or yellow: **26**

26. Inner face of hind femora with weak or narrow dark bands (Pl. 16). Dark band on outer face of hind femora usually interrupted in medial area. Hind femora very narrow (Pl. 16). Central Valley of California and along rivers emerging from surrounding hills: **pacifica**

Inner face of hind femora with broad dark bands (Pl. 16). Dark band on outer face of hind femora continuous through medial area (Pl. 16). California coastal ranges between Monterey and Santa Barbara: **titusi**

27. Hind tibiae orange or red: **28**
Hind tibiae yellow, brown, or blackish: **31**

28. Dark band of hindwings relatively broad with very clearly defined margins (Pl. 16): **29**
Dark band of hindwings narrower, or with poorly defined margins (Pl. 16): **30**

29. Inner face of hind femora with single pale band. Dark band of hindwings usually very broad, not curved along posterior edge (Pl. 16). Dorsum of abdomen usually reddish: **melanoptera**
Inner face of hind femora with at least two pale bands. Dark band of hindwings variable in width; when very broad, it curves around posterior margin. Dorsum of abdomen rarely reddish (Pl. 16): **latifasciata**

30. Forewings with very sharply contrasting pale and dark bands. Narrow dark bands have clearly defined margins, pale bands have no small spots. Pronotal carina very low or obsolete (Pl. 16): **pistrinaria**
Forewings not contrastingly banded, pale band usually with darker spots. Pronotal carina distinct (Pl. 16): **tolteca** and **modesta**

31. Hindwings with very broad dark band (similar to some *melanoptera* and *latifasciata,* but tibiae not red or orange) Dunes on Gulf Coast from Tamaulipas, Mexico, to Louisiana (Pl. 18): **schaefferi**
Hindwings with narrower or indistinct dark band: **32**

32. Body pale to medium gray. Forewings without distinct darker crossbands, streaked. Dark band of hindwings varies from moderately distinct to obsolete. Inner face of hind femora largely black, with one pale band. Sagebrush areas in western United States (Pl. 15): **gracilis**
Body color variable; if pale or medium gray, then with darker crossbands. Dark band of hindwings distinct. Inner face of hind femora with two pale bands: **33**

33. Body generally with lighter and darker gray bands but never very dark. Inhabits dry grasslands, salt flats, and desert scrub: **35**
Body generally dark, sometimes nearly black, sometimes with strongly contrasting light and dark areas. Inhabits mountain forests: **34**

34. Inhabits forested slopes in San Bernardino Mountains (largely indistinguishable from *fratercula*): **bernardi**

Inhabits lightly forested slopes in front ranges of Rocky Mountains (Pl. 18): **fratercula**

35. Ratio of forewing length (measured from origin of R vein) to hind femur length 2.08–2.35 (Fig. 52, Pl. 18): **pallidipennis**

Ratio of forewing length to hind femur length 1.85–2.09: **36**

36. Head width (measured through eyes) 0.91–1.04 times posterior width of pronotal disk (Fig. 52): **salina**

Head width 0.82–0.92 times posterior width of pronotal disk (Fig. 52): **diversellus**

***Trimerotropis maritima* (Harris)** Pl. 15

TAXONOMY. In this work I treat *acta* Hebard and *citrina* Scudder as synonyms of *maritima.* On the basis of morphology, reproductive behavior, and hybridization experiments it is evident that a single geographically variable species is involved. Tibial color, width of the black hindwing band, degree of banding of the forewings, and color of the inner face of the hind femora were previously used to distinguish the three species. These characters do not vary concordantly, and intermediate conditions occur at some localities. In general, specimens from beaches are paler. Those along the northern Atlantic coast (New Jersey to Massachusetts) usually have yellow tibiae, those farther south have reddish tibiae. Carothers (unpublished data) successfully crossed specimens of *maritima* and *citrina,* getting fertile F1 and backcrosses (specimens in University of Michigan Museum of Zoology). Courtship behavior in Michigan, New York, South Carolina, and Oklahoma is virtually identical (Otte 1970). Blatchley (1920: 295) writes: "There is no doubt but that *acta* is only a southern form of *maritima,* replacing the latter along the South Atlantic coast. Davis has numerous specimens taken at Riverhead, N.J., and near Ft. Monroe, Va., that are intermediate in coloration between *maritima* and *acta.*"

DISTRIBUTION. Widespread in eastern, central, and southern United States, west to eastern Arizona, north to Wisconsin and Ontario; usually along sandy shores. Along the Atlantic coast it is the only *Trimerotropis* species. Along the Texas coast it lives with *T. schaefferi.*

RECOGNITION. Figs. 49*a, b, j, l,* 50*d, i.* In the east this is the only *Trimerotropis* species living in sandy areas. From Florida to New Jersey the hind tibiae are red or orange; from New Jersey to Massachusetts and along the Great Lakes they are yellowish; in most other regions they are orange to reddish. In the western portions of its range it differs from *pallidipennis* in lacking a strong crossband on dorsum of forewings and in having orange tibiae. Differs from *agrestis* in shape of lateral lobes. Along the Texas coast distinguished from *schaefferi* by having a narrower dark band on the hindwings.

HABITAT. Open, sandy shores of rivers, lakes, and oceans, and occasionally other exposed sandy areas. In Oklahoma Hebard (1938) notes: "This species was collected in greatest numbers in cotton fields but was also found in corn and along roadsides. The insect prefers bare or almost bare soil and so is rarely found in grassy pastures or in normal plains environment."

BEHAVIOR. Crepitation seems to be produced mainly by males approaching other individuals and by disturbed males. I have never seen isolated males crepitate. Courting males usually shake their femora, striking the ground at the same time, and then stridulate as they approach females. In male–male interactions males shake and tip their femora (Otte 1970).

LIFE CYCLE. Adults from midsummer to fall.

REFERENCES. Scudder 1876, Hebard 1915, Fox 1917, Blatchley 1920, Otte 1968, 1970.

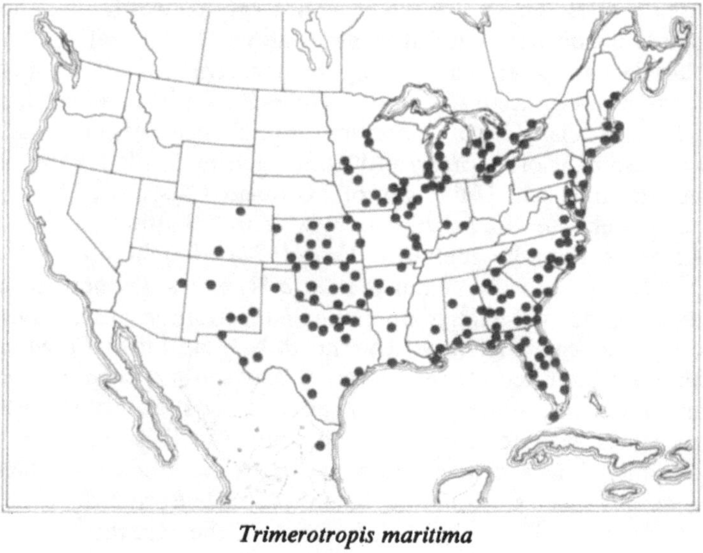

Trimerotropis maritima

***Trimerotropis agrestis* McNeill** Pl. 15

DISTRIBUTION. East of the Rocky Mountains; northeastern New Mexico and western Oklahoma north to Alberta, Saskatchewan, and

Manitoba. Specimens from west of the Rockies belong to *barnumi,* here considered to be a distinct species.

RECOGNITION. Figs. 49*e,* 50*c, h.* Lateral lobes strongly produced along lower posterior margin. Forewings without crossbands. Hindwings bright yellow in basal area and with strong outer dark band. Hind tibiae orange or red. Most similar to *barnumi,* but differs as follows: pronotum narrower; lateral lobes not so produced; inner face of hind femora with two black bands. Body length to end of forewings 24–36 mm in males, 31–43 mm in females.

HABITAT. Sandy tracts with scattered vegetation, dunes, and blowout areas. In the prairie provinces of Canada, "*T. agrestis* is collected only on bare yellow sand, and when disturbed will circle to rest on sand again" (Brooks, 1958).

LIFE CYCLE. Egg-overwintering. Adults from July to August. In Idaho the first nymphal instars appear in early June (Scoggan and Brusven 1972).

REFERENCES. Tinkham 1961, 1964, Scoggan and Brusven 1972.

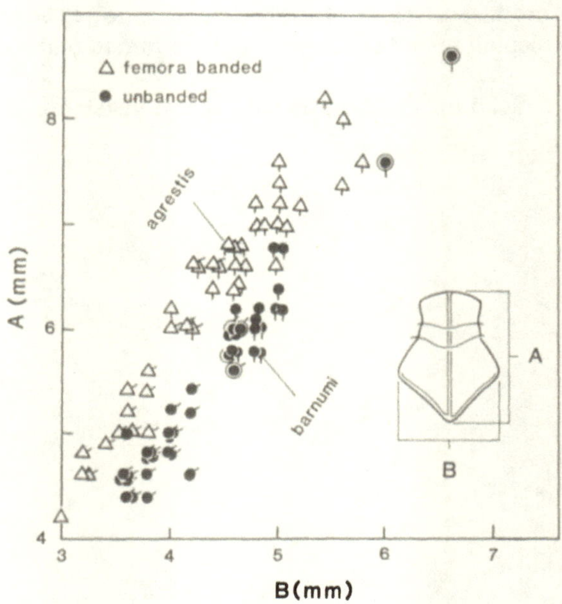

Fig. 51. *Trimerotropis.* Scatter diagram showing separation of *agrestis* and *barnumi.* Black rings with circles are specimens identified by Tinkham as *T. agrestis gracewileyae.*

***Trimerotropis barnumi* Tinkham** Pl. 15

TAXONOMY. This species was described as a subspecies of *agrestis,* but the two forms differ by a set of concordantly varying characters (see Fig. 51 and Recognition). *T. agrestis gracewileyae* is here treated as a synonym of *barnumi.* The type of this subspecies is somewhat different from *barnumi,* but the presence of intermediates between typical *barnumi* and *a. gracewileyae* reduces the usefulness of the subspecies category.

DISTRIBUTION. Central New Mexico to northern Nevada and southern Idaho.

RECOGNITION. Figs. 49*f,* 50*a, b, f, g.* Hindwings bright yellow in basal area and with a broad outer dark band. Forewings often with posteriorly converging pale stripes on dorsal surface (especially evident in Utah populations). Side of forewings speckled but without crossbands. Hind tibiae red or orange, inner face of hind femora orange without black bands. This species is most easily confused with *agrestis* but differs as follows: pronotum relatively shorter and wider; hind femora without two black bands on inner face; pronotal lateral lobe more produced; venter of thorax closer to lower edge of lateral lobes. Body length to end of forewings 31–35 mm in males, 35–44 mm in females.

HABITAT. Sand hills and dunes with sparse vegetation.

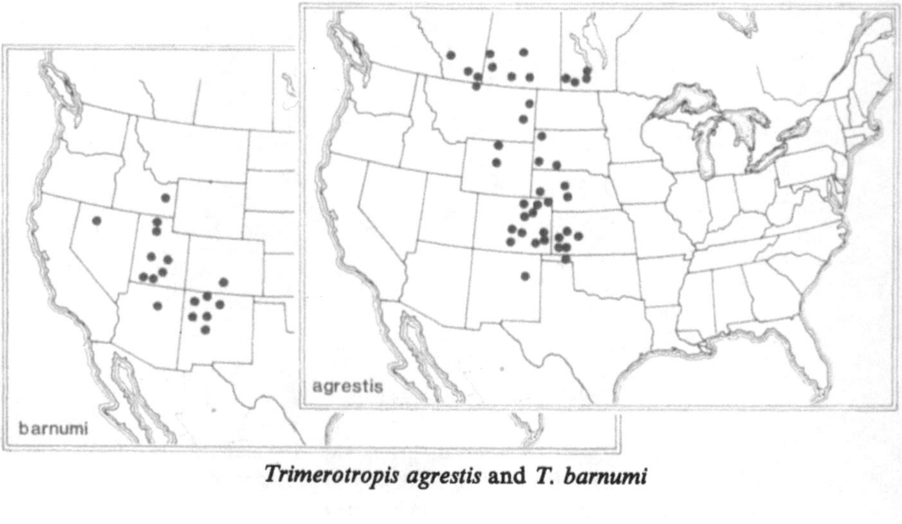

Trimerotropis agrestis and *T. barnumi*

BEHAVIOR. Isolated males and females both perform crepitation flights to attract the opposite sex (Otte 1970, Willey and Willey 1970). Males usually stridulate and occasionally tip their femora while approaching females. Femur-shaking with ground-striking and femur-tipping signals are performed in male–male interactions.

LIFE CYCLE. Adults from early July to September.

REFERENCES. Tinkham 1960, Willey and Willey 1970.

Trimerotropis arenacea Rehn — Pl. 15

DISTRIBUTION. Great Basin region of Idaho, Nevada, and Oregon. Perhaps *arenacea* and *whitei* belong to the same species.

RECOGNITION. A sand-colored species with poorly indicated crossbands on the forewings and with flaring lateral lobes. Most similar to *whitei,* but geographically separated from it. It may be confused with *sparsa* or *gracilis.* It differs from *sparsa* in hindwing banding and from *gracilis* in forewing banding and hind femur banding. Hind tibiae yellowish to gray or bluish. Body length to end of forewings 22–26 mm in males, 27–36 mm in females.

HABITAT. Loose, sandy soils, especially eroded areas including dunes, blowouts, or gullies (Scoggan and Brusven 1972).

LIFE CYCLE. Egg-overwintering. The first nymphs appear in late May and early June; adults from late June to September.

REFERENCES. Rehn 1910, Helfer 1963, Scoggan and Brusven 1972.

Trimerotropis whitei Rentz and Weissman — Pl. 15

DISTRIBUTION. Presently known only from southern Utah.

RECOGNITION. Figs. 49*h,* 50*e, j.* This species is most similar to *maritima* and *arenacea,* but it does not coexist with either species. Hind tibiae yellow (bluish in *arenacea,* red or orange in western *maritima*). It can be distinguished from the sympatric *barnumi* by its lateral lobes, which are not greatly produced. Basal area of hindwings bright yellow, outer dark band distinct. Crossbands on forewings indistinct and mottled. Body length to end of forewings 23–26 mm in males, 28–32 mm in females.

HABITAT. Pink sand hills and dunes with sparse vegetation.

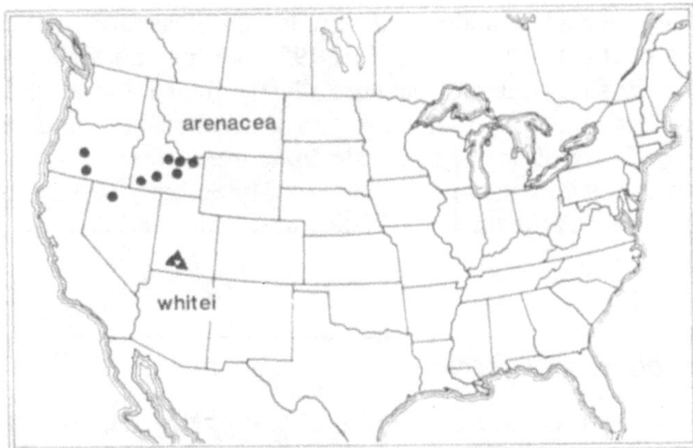

Trimerotropis arenacea and *T. whitei*

LIFE CYCLE. Adults have been collected between early August and early September.

REFERENCES. Rentz and Weissman 1984.

Trimerotropis latifasciata Pl. 16

TAXONOMY. This widely distributed species presents two major problems. First, geographic variation in size and width of hindwing band is considerable, raising the question whether more than one species is involved. The resemblance between *latifasciata* and *californica* is close, especially where the wing band of *latifasciata* approaches that of *californica* and where specimens of *californica* have a reduced tooth on the lateral lobes or lack this tooth altogether. I have little confidence that I can always distinguish the two species; therefore the distribution maps given for the two species may be inaccurate, especially in northern Chihuahua.

DISTRIBUTION. Western half of United States and southern Canada, northern Mexico.

RECOGNITION. Most easily confused with *californica, melanoptera, pistrinaria,* and *campestris.* It can be distinguished from *melanoptera* as follows: inner face of hind femora has at least two pale bands; dorsum of abdomen usually not orange or reddish in distal half; central dark band on hindwings usually not broad; when it is broad, as in some populations in Arizona, the black band curves around the posterior edge of the hindwing.

Specimens with a narrow dark band on the hindwings are likely to be confused with *californica*. The latter species has a small projection or tooth at the lower posterior margin of the lateral lobes. When this tooth is absent, I have identified specimens as *latifasciata*. This single difference would suffice if it were certain that all *californica* bear the tooth, but that is not certain, so we are left with the possibility that the two species are sometimes indistinguishable. Body length to end of forewings 30–41 mm in males, 35–49 mm in females.

HABITAT. Open, dry grasslands. In the prairies of western Canada it is found "on rather bare, sandy-alkaline flats where the chief conspicuous vegetation is grease wood (*Sarcobatus* sp) and sage brush" (Brooks 1958). Scoggan and Brusven (1972) report that in Idaho the species inhabits "disturbed areas in the shadscale or sage brush communities. At lower elevations [it is common] on fine loamy soil. At higher elevations preferred habitats are rocky, often with a wind-paved appearance. Vegetation cover varies from dense to sparse. Vegetation frequently includes *Salsola kali, Poa secunda, Chenopodium album, Agropyron spicatum* and *A. cristatum,* and several annual forbs." In western Colorado Hebard (1929) reports that the species is limited to alkali flats in the valleys at lower elevations. In northern Mexico I collected the species at the edges of salt pans.

LIFE CYCLE. Egg-overwintering. Adults in southern portion of the range from June to December. In Idaho the first instars appear from May to early June, adults from early July to August.

REFERENCES. Hebard 1929, Tinkham 1948, Brooks 1958, Scoggan and Brusven 1972.

Trimerotropis melanoptera McNeill Pl. 16

DISTRIBUTION. West Texas, New Mexico, Arizona, Colorado, and northern Mexico.

RECOGNITION. This species has a very broad black band on the hindwings, a reddish abdomen, and a single pale band on the inner face of the hind femora. Some Arizona populations of *latifasciata* also have a broad black band, but its shape is different. The two species also differ in the coloration of the hind femora (see *latifasciata*). Hind tibiae are reddish, and pale areas on inner face of hind femora are orange or reddish. Body length to end of forewings 32–40 mm in males, 37–50 mm in females.

HABITAT. Open grasslands and desert grasslands. Tinkham (1948) notes: "It is a very common species on the grasslands of the Davis

Mountain region [Texas], especially at about 6,000 feet elevation. In Arizona it prefers narrow gravelly ridges and the bare areas around the ant colonies of *Pogonomyrmex* for its special habitat." Hebard (1929) writes: "The distribution of *melanoptera* on the plains of eastern Colorado is discontinuous, as the insect is limited to the small bare alkaline areas there found." In northern Mexico I collected the species at the edges of salt pans.

LIFE CYCLE. Egg-overwintering. Adults from June to November.

REFERENCES. Ball et al. 1942, Tinkham 1948, Otte 1970.

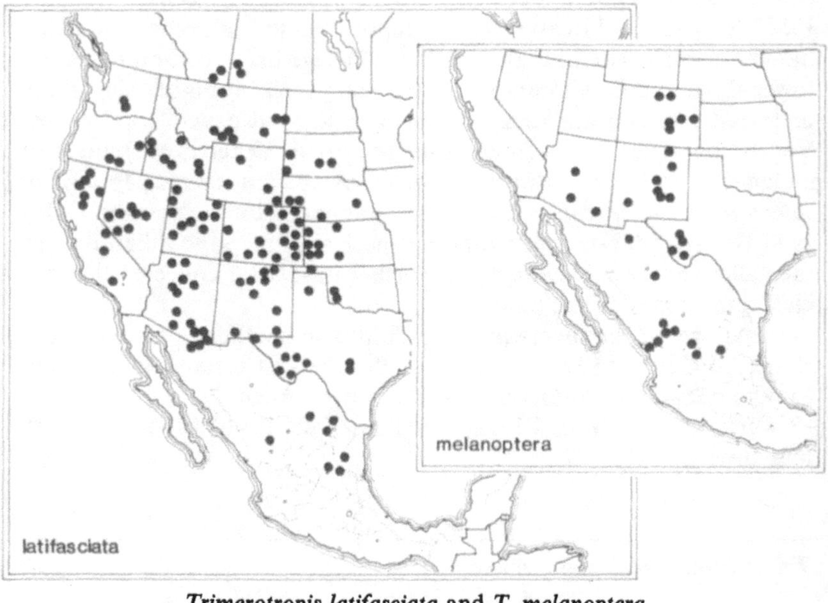

Trimerotropis latifasciata and *T. melanoptera*

Trimerotropis californica Bruner — Pl. 16

TAXONOMY. *Trimerotropis strenua* McNeill is here treated as a synonym of *californica.* Individuals collected near Benson and Gila Bend, Arizona, and in the San Gabriel Mountains and near Lancaster, California, have identical reproductive behavior. Specimens previously assigned to the two nominal species are morphologically indistinguishable. McNeill described *strenua* from a single male from Boise City, Idaho, and stated that the species was very closely related to *californica.* Rehn and Hebard (1905, 1910) thought that *strenua* was not distinct from *californica* on the basis of morphology. LaRivers (1948) states: "*T. strenua* is very close to *T. californica* and may eventually prove to be a variety of the latter."

DISTRIBUTION. Widespread in arid habitats over the western United States.

RECOGNITION. Fig. 50*m.* This species may be confused with *latifasciata,* especially those individuals with a narrower dark band on the hindwings. If all specimens of this species have the pronotal tooth, and the tooth is consistently absent in *latifasciata,* then there is no problem in separating them, but we suspect that *californica* occasionally lacks the tooth.

The color of this species usually resembles the prevailing ground color of its habitat. Lower posterior margin of the lateral lobes has a small tooth. Hind tibiae red or orange. Inner face of hind femora with a broad basal dark band and two narrower distal ones. *T. topanga* is similar but has blue hind tibiae. *T. pacifica* is similar but often has yellow tibiae (reddish or orange in some individuals), and the inner and outer dark bands on the hind femora are different. *T. titusi* is also similar but has yellow hind tibiae, and the dark crossbands on the forewings are narrower and more conspicuous. Body length to end of forewings 25–35 mm in males, 30–43 mm in females.

HABITAT. Open stony ground and dry river washes. In Arizona this species is fairly common from the Lower Sonoran Zone into the Transition Zone on rocky soil with scattered clumps of grass (Ball et al. 1942). LaRivers (1948) writes: "In Paradise Valley [Nevada] the species was commonly found feeding and resting in the numerous clumps of sand grass, *Oryzopsis hymenoides.* In areas where dried, almost blood-red leaves of an unidentified plant lay scattered about after the death of the plant, a nearly blood-red color phase of *T. strenua* predominated, achieving considerable protection from the color resemblance. This weed did not occur at Sand Spring, and no red individuals were seen there." Scoggan and Brusven (1972) report the species only from the shadscale community in Idaho. "Typical areas inhabited . . . are mixed shrub areas with perennial bunchgrass and numerous forbs. Shrubs include *Atriplex confertifolia, Artemisia spinescens, Erayia spinosa. Oryzopsis hymenoides* is the predominant grass but *Bromus tectorum* may be present."

BEHAVIOR. Crepitation by solitary males apparently serves to attract females. All crepitation consists of a single sequence of snapping sounds. Femur-shaking with ground-striking is the only courtship signal produced. No stridulation has been observed. In male–male encounters Arizona males only tipped their femora. Those from California tipped and shook femora.

LIFE CYCLE. Egg-overwintering. Adults in California from June to October. Adults in Arizona and New Mexico from June to September. In Idaho adults appear in mid-July.

REFERENCES. McNeill 1901, Rehn and Hebard 1905, 1910, Ball et al. 1942, LaRivers 1948, Otte 1970, Scoggan and Brusven 1972.

***Trimerotropis topanga* Rentz and Weissman** Pl. 16

DISTRIBUTION. Southern California and Baja California.

RECOGNITION. This is the only species that has all three of the following characters: a tooth at the lower posterior edge of the lateral lobes, blue tibiae, and yellow wings. It is superficially similar to *pallidipennis* but may be more closely related to *californica, pacifica,* and *titusi.* Body length to end of forewings 23–31 mm in males, 30–38 mm in females.

HABITAT. Open stony ground with bare patches and scattered clumps of grass and bushes.

LIFE CYCLE. Adults from June to September.

REFERENCES. Rentz and Weissman 1981.

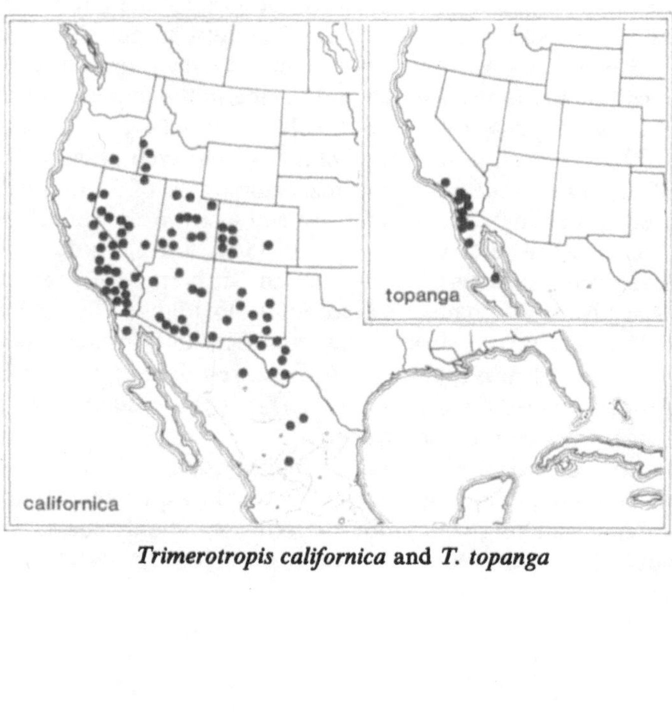

Trimerotropis californica and *T. topanga*

Trimerotropis pacifica Bruner — Pl. 16

TAXONOMY. I am unable to properly distinguish Scudder's *Trimerotropis munda* from *californica* and am therefore treating his species as a junior synonym. Strohecker, Middlekauff, and Rentz (1968) state: "The dark gray ground color of this form seems to be the only feature separating it from *pacifica.*"

DISTRIBUTION. Central Valley of California and surrounding ranges.

RECOGNITION. Like *californica,* this species has a small tooth on the bottom of the lateral lobe. Most specimens are gray to straw-yellow and have very distinct dark crossbands on the forewings. But some specimens are almsot uniformly straw-yellow and show only traces of dark crossbands. The pale individuals also have very indistinct dark bands on the hindwings. This species is most easily confused with *californica* and *titusi.* It differs from *californica* in having whitish or yellowish hind tibiae and more slender hind femora. It differs from *titusi* in the wider forewing crossbands, in the shape of the hind femora, and in body size. The dark crossbands on the forewings are often interrupted along the margins. Body length to end of forewings 26–30 mm in males, 33–39 mm in females.

HABITAT. Dry grasslands at lower elevations and dry washes.

LIFE CYCLE. Adults from July to October.

REFERENCES. Strohecker, Middlekauff, and Rentz 1968.

Trimerotropis titusi Caudell — Pl. 16

DISTRIBUTION. Coastal ranges of California between Monterey and Santa Barbara.

RECOGNITION. Very pale with dark narrow crossbands on the forewings; some specimens from Pinnacles National Monument lack crossbands. Most similar to *pacifica* and *californica.* It differs from *pacifica* in the degree of banding on the hind femora, in body size, and in the width of the forewing crossbands. It differs from *californica* in having yellow tibiae, a less distinct dark band on the hindwings, narrower forewing crossbands, and a lower prozontal ridge. Dark band on hindwings quite variable, distinct in some specimens, obsolete in others. Body length to end of forewings 21–27 mm in males, 27–33 mm in females.

HABITAT. Dry washes with sparse vegetation.

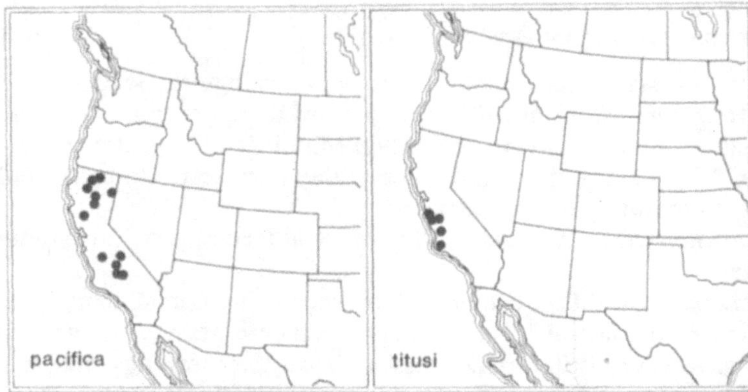

Trimerotropis pacifica and *T. titusi*

LIFE CYCLE. Adults from June to September.

REFERENCES. Caudell 1906, Strohecker, Middlekauff, and Rentz 1968.

Trimerotropis pistrinaria Pl. 16

TAXONOMY. This species displays geographic variation in the color of the hind femora and in forewing banding, which requires further study. Perhaps several species will be shown to exist. David Ferguson (personal communication) notes a strong resemblance and perhaps an intergradation between *pistrinaria* and *modesta* in New Mexico. In their most common configuration the two are morphologically distinct, and they occupy different habitats. But their relationship merits further study.

DISTRIBUTION. Great Plains from Alberta and Saskatchewan south to Arizona and central Mexico.

RECOGNITION. Pale and dark crossbands on forewings very conspicuous and with clearly defined edges. Pronotal carina very low or obsolete. Hind tibiae reddish or orange. Inner face of hind femora variable. Most likely to be confused with *latifasciata,* but differs from that species as follows: forewings not speckled anterior to first dark band; second dark band rather solid and lines up nicely with dark band on hind femora; dark band of hindwings does not extend into proximal half of wing in the second lobe. Likely to be confused with *Hadrotettix,* but without oblique dark band on hind femora.

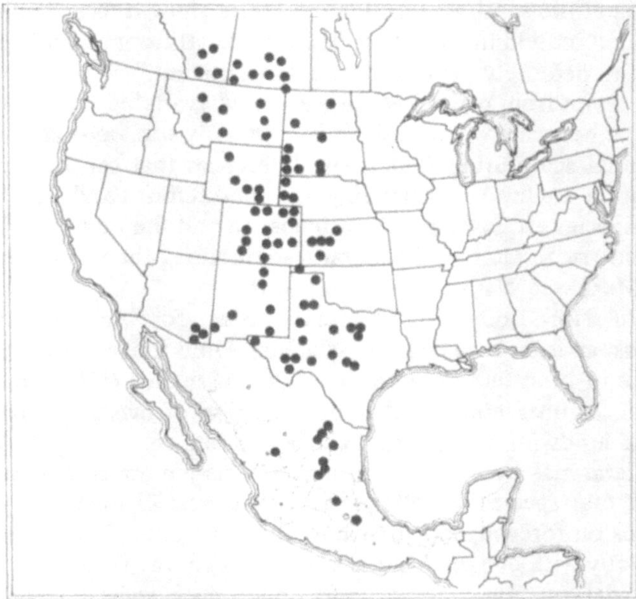

Trimerotropis pistrinaria

HABITAT. Open, eroded, stony habitats. Helfer (1963) states: "Occurs on dry gravelly plains and hillsides and on eroded barren lands." In the upper Rio Grande Valley Hebard (1935) found it at 7,-000 feet in "bare pebbly areas among junipers on upper slopes and ridges." Brooks (1958) reports that in the Canadian prairies "*T. pistrinaria* inhabits gravelly situations and hillsides. It is a moderately strong and erratic circle-back flier; the crackling noise is moderately loud but of short duration. The species is a mixed feeder, preferring forbs such as *Astragalus* spp."

LIFE CYCLE. Adults from June to October.

REFERENCES. Hebard 1935c, Ball et al. 1942, Tinkham 1948, Brooks 1958, Helfer 1963.

***Trimerotropis modesta* Bruner** Pl. 16

TAXONOMY. Hebard (1932) synonymized this species under *tolteca*. He wrote (p. 260): "Comparison with *tolteca* shows it to be so closely related that we believe that nothing more than racial recognition is warranted. This is emphasized by the fact that large series from the mountains of the southwestern United States show it to be subject to very decided individual variation. More material from central Mexico is needed to definitely prove this racial assignment."

Good series from central Mexico are now available and show discontinuities between *tolteca* and *modesta* in body size, body coloration, and pronotal sculpturing. The main differences that can be used for identification are hindwing banding and hind femur banding (Pl. 16). The two species are sympatric in Michoacán and the Federal District.

DISTRIBUTION. Utah and Colorado south along the western mountains of Mexico to Nayarit and Michoacán.

RECOGNITION. Body medium to dark brown, forewings with two or three crossbands (*tolteca* lacks crossbands). Hind tibiae orange to red. Inner face of hind femora with two pale crossbands (*tolteca* has one pale band, as does *pistrinaria* from the region of overlap). Proximal margin of hindwing bands are distinct (gradual in *tolteca*). In Colorado *modesta* may be confused with *Spharagemon campestris,* but differs from that species as follows: prozonal crest distinctively lower; crossbands on forewings have distinct margins; outer face of hind femora usually with one distinct dark band; if two, then the distal one is distinctly darker; inner face of hind tibiae black at base (reddish in *campestris*). Differs from *latifasciata* and *melanoptera* in hindwing banding and from *californica* in that it lacks a spur on lateral lobes. Body length to end of forewings 23–32 mm in males, 29–37 mm in females.

HABITAT. In Arizona Ball et al. (1942) report: "One of the most common grasshoppers of the grasslands of the Transition Zone extending into the Lower Sonoran Zone. Northern and eastern Arizona to Oak Creek Canyon and the Baboquivari mountains." In New Mexico Hebard (1935c) notes: "Found only on volcanic sand and gravel under western yellow pines at 8,000 feet on a mesa two miles north of Frijoles Canyon, Sandoval County. It was the most abundant grasshopper there present, occurring in fairly frequent small colonies. Individuals crackle when alighting, the males doing this more strongly than the females."

LIFE CYCLE. Egg-overwintering. Adults from July to September.

REFERENCES. Ball et al. 1942, Hebard 1932, 1935c.

Trimerotropis tolteca Saussure — Pl. 16

TAXONOMY. *T. tolteca* is here treated as distinct from *modesta* (see *modesta* Taxonomy).

DISTRIBUTION. Highlands of central Mexico.

RECOGNITION. Body dark brown, sometimes almost black. Forewings without crossbands. Proximal margins of hindwing band usually gradual. Inner face of hind femora largely black, with one pale crossband. Hind tibiae deep orange to bright red. Disk of pronotum roughly sculptured. Body length to end of forewings 18–25 mm in males, 20–30 mm in females.

HABITAT. Not reported in literature; probably eroded hillsides and openings in pine forests.

LIFE CYCLE. Adults July through September.

REFERENCES. Hebard 1932.

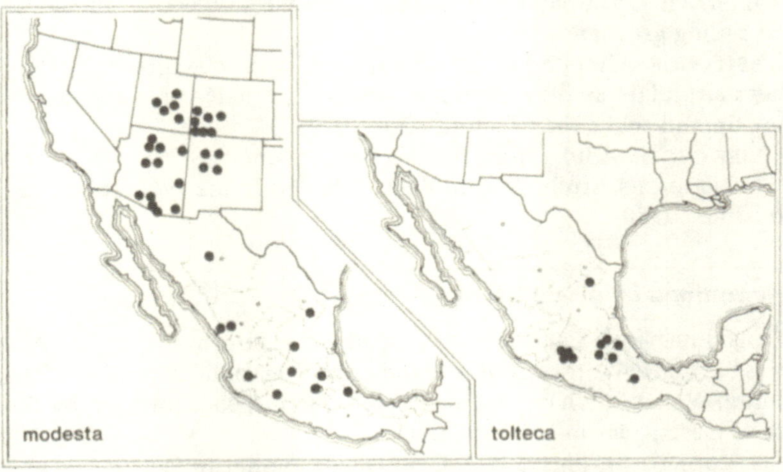

Trimerotropis tolteca and *T. modesta*

Trimerotropis albescens McNeill Pl. 17

DISTRIBUTION. Eastern and southern California, western Nevada, and Baja California.

RECOGNITION. Fig. 50*o, p, r, s.* A small, contrastingly marked grasshopper. Background color varies from white to pale reddish brown or blue-gray, depending on the color of the ground on which it lives. Hind tibiae bluish to gray. In southern California this species may be confused with *bifasciata,* especially with contrastingly marked specimens of the latter, but in *bifasciata* the lateral lobes never flare out, males are usually more than 25 mm long (to ends of forewings), and females are usually more than 29 mm. Occasional specimens of *fontana* are conspicuously marked and may be confused with this species. *T. inconspicua* is also very similar, but in this species the area preceding the first crossband of the forewings is usually darker than the area behind the first two crossbands. East of the Sierras and in Nevada the connection between pronotal disk and the lateral lobes on the prozona is more rounded. The pronotal lateral lobes flare out most noticeably in Southern California (Los Angeles and San Diego Counties) and Baja California, and the median pronotal carina is higher and more crestlike.

Body length to end of forewings 19–28 mm in males, 25–30 mm in females.

HABITAT. Open patches of gravelly ground on wooded hill slopes and rolling grasslands.

BEHAVIOR. Crepitation by solitary males is common. Courting males stridulate as they approach females; in male–male encounters they tip and shake their femora (Otte 1970).

LIFE CYCLE. Adults from June to September.

REFERENCES. Strohecker, Middlekauff, and Rentz 1968, Rentz and Weissman 1980.

Trimerotropis bifasciata **Bruner** Pl. 17

DISTRIBUTION. Central Valley; southern California from Los Angeles to Hemet. At present it is found only around the periphery of the Central Valley, perhaps having been extinguished elsewhere by the use of insecticides in irrigated areas.

RECOGNITION. Figs. 50*n, t.* Confusingly similar to *inconspicua* and sometimes to *albescens.* I have used the following characters to distinguish it from *inconspicua:* frontal ridge when viewed from above almost always slightly concave (straight or convex in *inconspicua*); fastigium usually with a rather definite low median carina (absent or quite indistinct in *inconspicua*); fan of hindwings with a greenish or bluish tinge; center of hindwings usually within yellow area (outside the yellow area in *inconspicua*).

Sometimes very similar to *albescens.* I misidentified specimens of *bifasciata* collected in San Diego County as *albescens* in my 1970 work on communication. The resemblance to *albescens* is especially close because of the strongly contrasting colors and the nearly identical dark bands on the lateral lobes. It differs from *albescens* in having a lower pronotal carina and in lacking flaring lateral lobes. Body length to end of forewings 25–29 mm in males, 30–36 mm in females.

HABITAT. Bare eroded soil of the Central Valley. I have collected the species along a gravelly man-made lake shore in Colusa County. The Los Angeles County specimens were collected on "grass prairie land" by Poling in 1907. Near Warner Springs, San Diego County, I collected the species on thinly grassed, gravelly, rolling prairie.

BEHAVIOR. In San Diego County males of this species produced short bursts of crepitation with a wing-stroke rate of about 44 per second and about five bursts of crepitation per second. Courting males performed only O-stridulation. One femur is used in each stridulation series, consisting of two to three leg strokes.

LIFE CYCLE. Adults from late June to mid-September.

REFERENCES. Strohecker, Middlekauff, and Rentz 1968, Rentz and Weissman 1980.

Trimerotropis inconspicua **Bruner** Pl. 17

DISTRIBUTION. All western states, but not yet known from Montana and Washington.

RECOGNITION. Figs. 50*q*, *u*. Basal area of hindwings yellow; dark band distinct to indistinct. Hind tibiae often bluish or gray, sometimes greenish, rarely yellowish. Disk of pronotum rounds imperceptibly onto lateral lobes. Dark crossbands of forewings well defined throughout, and area between bands one and two is noticeably pale.

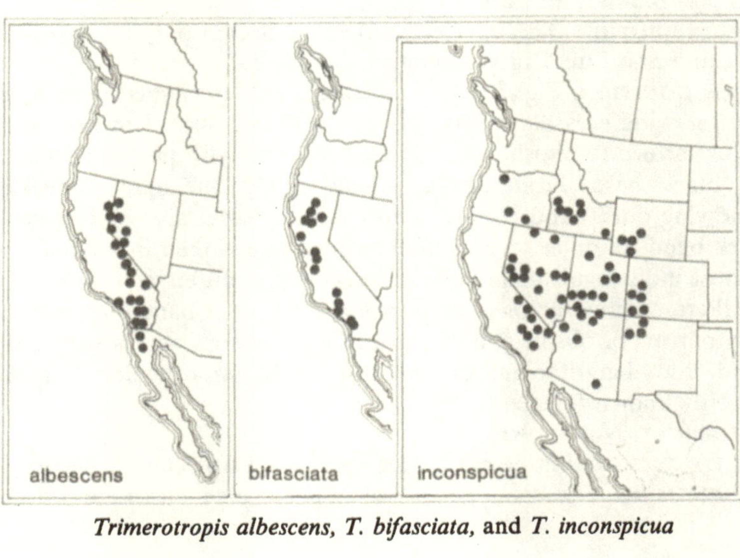

Trimerotropis albescens, T. bifasciata, and *T. inconspicua*

This species may be confused with *bifasciata,* but their ranges are largely nonoverlapping. To distinguish them see Recognition under *bifasciata.* May also be confused with *albescens,* which can usually be recognized by its more contrasting coloration and less distinct hindwing band. Body length to end of forewings 22–27 mm in males, 28–33 mm in females.

HABITAT. Open ground on hillsides or flats, often on stony ground. I collected the species in the western foothills of the White Mountains, Inyo County, California, where it lived on a stony hillside with scattered piñon pine and little other vegetation.

BEHAVIOR. I described the reproductive behavior of this species under the name *T. koebelei* Bruner (Otte 1970). I heard males crepitate only during escape flights. Courtship behavior included femur-tipping and ordinary stridulation signals. Both tipping and shaking signals were performed in male–male encounters.

LIFE CYCLE. Adults from July to September.

REFERENCES. Strohecker, Middlekauff, and Rentz 1968, Otte 1970 (under *T. koebelei*).

Trimerotropis occulens Otte, n. sp. Pl. 17

TAXONOMY. This species appears to be most closely related to *T. occidentiloides* from the Los Angeles area and *infantilis* from the San Francisco area. Because the three species have allopatric distributions, it is possible that they are not specifically distinct.

DISTRIBUTION. This species has been collected only in Santa Barbara and San Luis Obispo counties, California.

RECOGNITION. Fig. 50*w.* A very small, pale gray species with narrow forewing crossbands interrupted on the dorsum. Hind tibiae are yellow-brown to bluish, black at the base, and with pale band next to the black base. Pronotum and head finely but lightly speckled. Hindwing dark band is narrow to obsolete; basal area yellow-green. Dark band on outer face of hind femora interrupted in medial area. Almost indistinguishable from *occidentiloides;* differs as follows: pronotal crest distinctly lower on prozona; outer dark band on hindwings very narrow or obsolete; outer face of hind femora with a single dark band. Body length to end of forewings 19–22 mm in males, 27 mm in the only known female.

HABITAT. Not known.

LIFE CYCLE. All nine adult specimens were collected in August.

SPECIMENS. The holotype and seven paratypic males were collected at Lompoc, Santa Barbara County, California, on August 6, 1938, by L. W. Hepner. A paratypic female was collected at Paso Robles, San Luis Obispo County, California, on August 21, 1909, at 750–1,100 feet. All specimens are in the Philadelphia Academy of Natural Sciences collection.

Trimerotropis infantilis Rentz and Weissman

DISTRIBUTION. Mountains inland from Santa Cruz, California.

RECOGNITION. Similar to *occulens* (Pl. 17) in coloration and body size. Basal area of hindwings pale yellow and with a rather faint band, remainder of wing transparent. Body pale gray, dark crossbands on forewings. Hind tibiae blue-gray.

HABITAT. At the type locality, *infantilis* was found "on a sparsely covered sandy substrate at the base of pines above a rock quarry" (Rentz and Weissman 1984).

LIFE CYCLE. All known adults were collected between August 13 and September 28. Approximately 25 percent of the population was in the fourth or fifth instar on July 16, 1977 (Rentz and Weissman 1984).

REFERENCES. Rentz and Weissman 1984.

Trimerotropis occidentiloides Rentz and Weissman — Pl. 17

DISTRIBUTION. Known only from Los Angeles County, California.

RECOGNITION. Fig. 50*v*. Basal area of hindwings greenish yellow, dark band distinct but with faded margins. Forewings with narrow crossbands on lateral field; bands are indistinct or obsolete on dorsal field. Hind tibiae blue. This species may be confused with *topanga* and *thalassica* from southern California. Unlike *topanga* it lacks a small tooth on the lower margin of the lateral lobes. Unlike *thalassica* it has

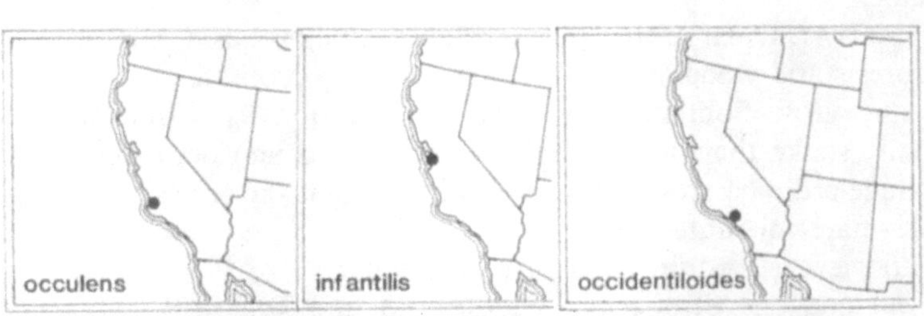

Trimerotropis occulens, T. infantilis, and *T. occidentiloides*

two pale bands on inner face of hind femora. It differs from its presumed close relative, *occulens,* as follows: pronotal crest relatively higher; posterior margins of pronotal disk form a right angle in males. Body length 22–24 mm in males and about 32 mm in females.

HABITAT. The only specimens in collections were collected along a roadside in southern California.

LIFE CYCLE. Adults from June 24 to August 19.

REFERENCES. Rentz and Weissman 1981.

***Trimerotropis fontana* Thomas** Pl. 17

DISTRIBUTION. Widespread over the western United States from Rocky Mountains west to the Pacific and north to British Columbia.

RECOGNITION. This is a highly variable species. Its color varies from dark brown or dark gray to light brown or gray. Some individuals are nearly black, others are almost as contrastingly marked as *albescens.* The dark crossbands on the forewings are usually quite characteristic, being darkest along the lateral surface and becoming fainter on the dorsum. Basal area of hindwings varies from nearly transparent to greenish yellow. Central dark band usually lacks strongly defined margins. Distal areas of hindwings vary from transparent to smoky.

May be confused with *koebelei* and *cincta,* but differs as follows: transition from the darker lateral fields of the forewings to the dorsal field not sudden as in *koebelei, cincta,* and *thalassica. T. santabarbara* is similar but inhabits only Santa Barbara Island, southern California. Also confusingly similar to *occidentalis,* but in *fontana* there is, over most of the range, no greenish or blue-green cast to the hindwings. The crossbands on the forewings of *occidentalis* usually cross the dorsal field.

Body length to end of forewings 22–29 mm in males, 30–38 mm in females.

HABITAT. Open sunny spots on slopes, along roadsides, in openings in forests and woodlands, and other dry, hilly habitats.

BEHAVIOR. Solitary males crepitate occasionally. Courting males usually shake their femora against the ground and occasionally also produce ordinary stridulation. Femur-tipping is the only signal seen in male–male encounters.

LIFE CYCLE. Adults from July to October.

REFERENCES. Strohecker, Middlekauff, and Rentz 1968, Otte 1970, Rentz and Weissman 1980.

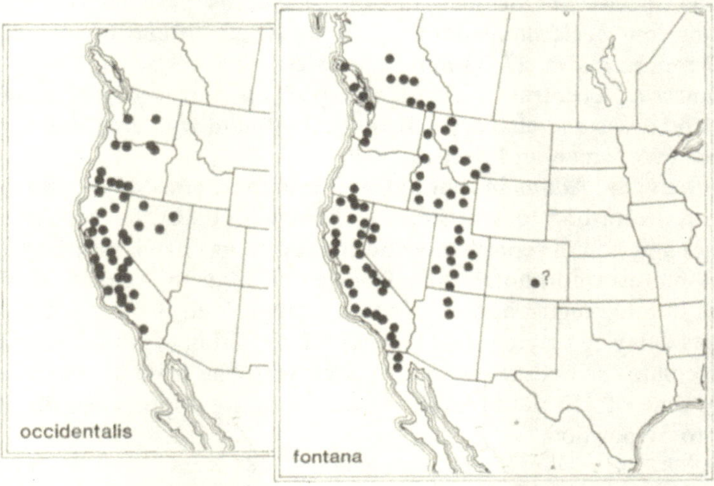

Trimerotropis fontana and *T. occidentalis*

Trimerotropis santabarbara Rentz and Weissman

DISTRIBUTION. Santa Barbara Island, Los Angeles County, California.

RECOGNITION. Very similar to—almost indistinguishable from—*fontana.* Hindwings greenish yellow in basal area with a very indistinct outer dark band. Forewing crossbands indistinct to obsolete on dorsal field; distinct on lateral fields. Hind tibiae banded—with a pale proximal band, then a dark gray band, then a blue or gray band, and at distal end a dark gray band. Body length to end of forewings about 22 mm in males, 30–33 mm in females.

HABITAT. Lives on the ground among chaparral vegetation.

LIFE CYCLE. All adults collected in May.

REFERENCES. Rentz and Weissman 1981.

Trimerotropis occidentalis (Bruner) Pl. 17

DISTRIBUTION. California, northern Nevada, Oregon, and Washington.

RECOGNITION. Very difficult to separate from *fontana.* Usually the bluish or blue-green color of the hindwings in *occidentalis* can be used to separate them. Forewing crossbands usually are fairly distinct on the dorsal field and more discontinuous than in *fontana.* Behaviorally the two species are distinct: *fontana* shakes its hind femora while courting, but *occidentalis* does not. Body length to end of forewings 20–27 mm in males, 27–33 mm in females.

HABITAT. In central California the species is restricted to serpentine grassland areas and chaparral. It is usually found with *Dissosteira pictipennis* (Weissman and French 1980).

LIFE CYCLE. Adults in central California from mid-May to late November. According to Weissman and French (1980) this species undergoes a period of reproductive dormancy: "Eggs hatch in spring and adults mature morphologically from early June to August. Mating begins in late August and oviposition starts in mid-September. Some females delay egg laying for 14 weeks after molting to adults. Average survivorship for both sexes was 70 days, with one male having a residency time of 153 days. Adult female mortality approached 40–50% prior to oviposition."

REFERENCES. Weissman and Rentz 1978, Rentz and Weissman 1980, Weissman and French 1980.

Trimerotropis lauta Scudder — Pl. 18

DISTRIBUTION. Presently known only from nine specimens, all from Guadalupe Island, Baja California.

RECOGNITION. Hindwings transparent but sometimes with a faint bluish cast. Forewings indistinctly banded on the lateral field, sometimes only speckled. Pronotum with lateral carinae distinct on the prozona. Top front of head rather angulate. Face has dark band running between lower front margins of the eyes; occasionally black beneath the lateral foveolae. Lateral pronotal lobes occasionally with a dorsal dark band, a continuation of the postocular dark band. Inner face of hind femora with one and a half or two pale bands. Body length to end of forewings 20–26 mm in males, 28–32 mm in females.

HABITAT. Not known. Guadalupe island is covered with desert scrub vegetation.

LIFE CYCLE. All nine adult specimens in the Philadelphia Adademy of Natural Sciences collection were taken in June and July.

REFERENCES. Scudder 1876c, Rentz and Weissman 1980.

Trimerotropis cincta (Thomas) — Pl. 17

TAXONOMY. *T. cincta* and *T. koebelei* (Bruner) are almost indistinguishable. They are entirely allopatric; *koebelei* is restricted to California and Oregon, while *cincta* is distributed east of these states.

DISTRIBUTION. Arizona and New Mexico north to Idaho, Montana and British Columbia.

RECOGNITION. This species almost always has a black band across the face between the lower inner margins of the eyes and beneath the antennal sockets. The area directly above the antennal sockets and beneath the lateral foveolae is often black as well. It is mainly this feature that separates it from *fontana.* As in *Conozoa* species, the dark forewing crossbands almost never cross the dorsal field. Dorsal field often distinctly lighter than lateral field, but in rare specimens traces of crossbands are visible. Unlike *Conozoa* species, in *cincta* the tibiae are blue or gray, never red, and seldom yellow. Basal area of hindwings yellow, pale yellow, or greenish yellow.

Body length to end of forewings 21–28 mm in males, 28–36 mm in females.

HABITAT. Commonly found on bare patches of ground in grasslands at moderate elevations (2,000–9,000 feet), often in open pine woodlands. I collected this species in open pine woods on Rocky Mountains slopes and foothills. In Colorado Alexander and Hilliard (1969) report: "We have found it primarily in close association with pines and common only in small clearings. Its resident range in our area is from 6,000 to 10,000 feet in altitude."

BEHAVIOR. Crepitation by solitary males consists of one to several bursts of irregular length. Silent femur-shaking and stridulation signals are produced by courting males. The shaking signal is more common than stridulation. In male–male interactions femur-tipping is common and femur-shaking rare (Otte 1970).

LIFE CYCLE. Adults from late June to September. In Colorado at 7,200 feet, hatching begins by June 1 or earlier; at 8,500 feet by the second week in June. Adults appear the third week of July and the first week of August. Adults are most abundant through August and the first half of September.

REFERENCES. Hebard 1929, Hebard 1935c, Strohecker, Middlekauff, and Rentz 1968, Alexander and Hilliard 1969, Otte 1970.

Trimerotropis koebeli (Bruner) Pl. 17

DISTRIBUTION. California to British Columbia.

RECOGNITION. Most similar to *cincta;* some individuals easily confused with *fontana;* also similar to *thalassica.* Basal area of hindwings pale yellow (in *thalassica,* bluish or greenish blue). Dark bands on forewings extend up to but do not cross dorsal surface; transition from darker lateral field to lighter dorsal field is sudden (usually gradual in *fontana*). Outer third of hindwings beyond dark band often suffused with smoky brown (as in *fontana* and *cincta* but not as in *thalassica*). Inner face of hind femora with at least two pale bands (one in *thalassica*). Head often with a black band crossing the face below the level of the antennae (also present in most *cincta* and some *fontana,* always lacking in *thalassica*).

HABITAT. Grassy clearings on mountain slopes in the Sierra Nevada and Cascade ranges. Seems to prefer mid- to high-altitude conifer woodland.

LIFE CYCLE. Adults from midsummer to early fall.

REFERENCES. McNeill 1901, Strohecker, Middlekauff, and Rentz 1968.

Trimerotropis thalassica **Bruner** Pl. 17

DISTRIBUTION. Southern two-thirds of California, mainly in hills and mountain foothills.

RECOGNITION. Body color pale gray to leaf brown, occasionally darker gray or brown. Crossbands on forewings sometimes indistinct, present only on the lateral field; dorsal field usually slightly lighter. Posterior margin of pronotal disk often pale and with black dots. Basal area of hindwings usually pale blue or blue-green, occasionally very pale green to transparent; central dark band usually indistinct, sometimes almost invisible; when it is dark, the margins are indefinite. Inner face of hind femora usually with a single pale band, rarely with

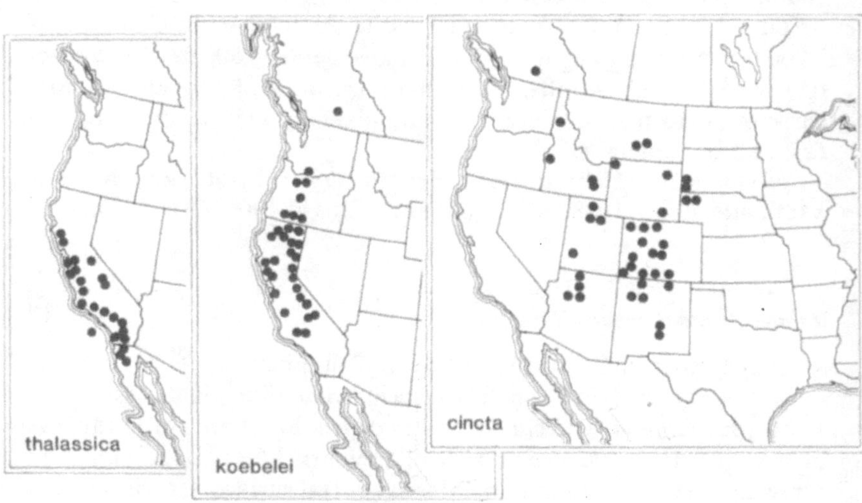

Trimerotropis cincta, T. koebelei, and *T. thalassica*

two. Hind tibiae blue. Body length to end of forewings 21–32 mm in males, 26–36 mm in females.

HABITAT. Coastal ranges in chaparral zone, often among manzanita bushes and in live oak woodlands, often found in the leaf litter of live oaks. When disturbed, they often alight on bushes.

BEHAVIOR. No crepitation has been reported for this species. Males stridulate during courtship and tip and shake their femora in male–male interactions.

LIFE CYCLE. Adults from June to September.

REFERENCES. Rehn and Hebard 1910, Strohecker, Middlekauff, and Rentz 1968, Otte 1970, Rentz and Weissman 1980, 1981.

***Trimerotropis pallidipennis* (Burmeister)** Pl. 18

TAXONOMY. A study of thousands of specimens identified as *T. pallidipennis* from all of North America in the Academy of Natural

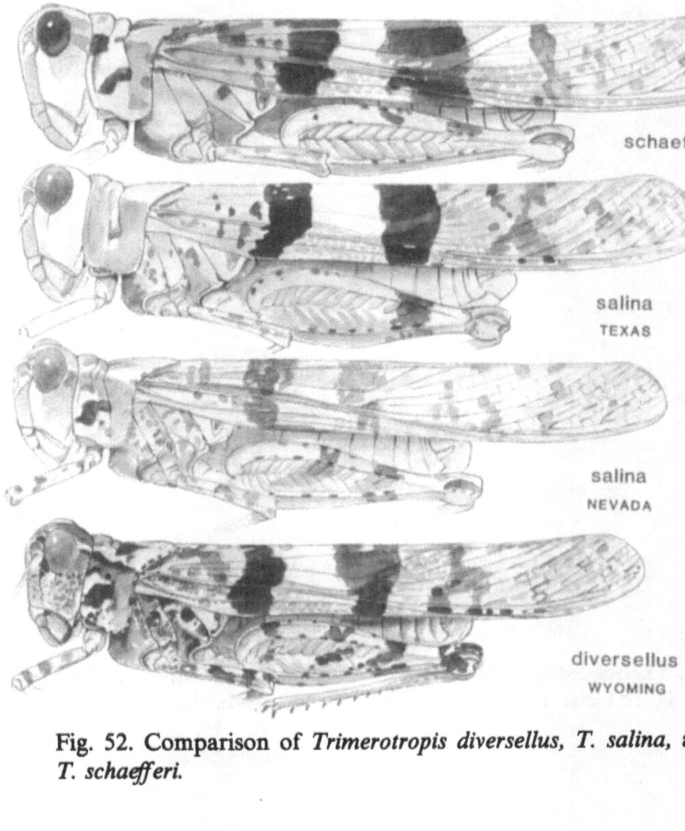

Fig. 52. Comparison of *Trimerotropis diversellus, T. salina,* and *T. schaefferi.*

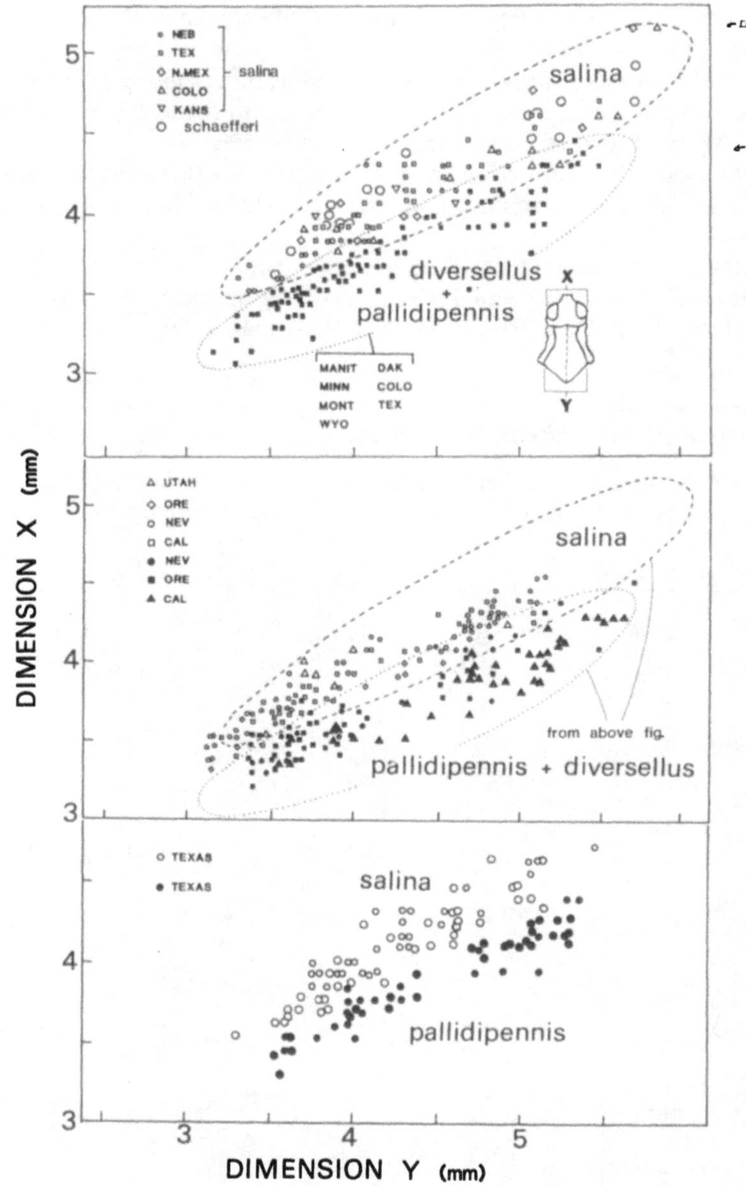

Fig. 53. Scatter diagrams showing separation of *salina, pallidipennis,* and *diversellus,* using head and pronotal proportions.

Sciences collection has shown that at least three distinct species are represented. We can summarize the status of this complex as follows: (1) McNeill's *salina* is a valid species and not merely a subspecies of *pallidipennis;* it is widespread over the western states from Oregon southeastward to Texas and then northward to Colorado and Nebraska. Both *salina* and *pallidipennis* are occasionally collected at the same localities. They differ principally in the relative proportions of the head and pronotum and the lengths of the forewings and hind femora (see Recognition and Figs. 52–54). (2) Specimens previously thought by Hebard and others to be *T. pallidipennis salina* from the Dakotas, Minnesota, Montana, and western Canada are members of Hebard's *T. diversellus.* This species, previously known only from Wyoming and Montana, is widespread from Manitoba and Minnesota west to British Columbia, Washington, Oregon, and California. *T. diversellus* can be separated from *pallidipennis* by its relatively shorter forewings and its smaller size. And it can be separated from *salina* by its head and pronotal proportions (Fig. 53). (3) Specimens from the central valley of California previously thought to belong to *pallidipennis* are tentatively included under *diversellus.* The courtship patterns of males from Colusa and Kern counties, California, first suggested that they belonged to an unrecognized species, very similar to *pallidipennis.* In my 1970 paper on grasshopper communication I referred to it as *T.* (near *pallidipennis*). Additional behavioral observations on this species were made in Colusa County in 1982. I was initially inclined to treat this species as a new species, until I found that I was unable to separate it from *diversellus* (see *diversellus* Recognition and Behavior).

DISTRIBUTION. This is the most widely distributed oedipodine in the New World. It ranges from southwestern Canada at least to central Argentina. Within this range, however, there are large areas through Central America where it does not occur. It inhabits for the most part dry grasslands, such as the western prairies, desert grasslands, and desert scrub. In North America it is generally found at lower elevations, being replaced at higher elevations by other species of *Trimerotropis.* In Argentina it inhabits a wide altitudinal range, from low salt flats to mountain grasslands at 10,000 feet or more. Its wide distribution is no doubt a result of its considerable flight capabilities and its habit of dispersing at night. In the southwest species is commonly found near electric lights in places it does not normally inhabit, such as city parks and streets.

RECOGNITION. Figs. 49*d,* 52, 53, 54, 55*a–f, m, n.* This species may be confused with many other *Trimerotropis* species. It possesses the

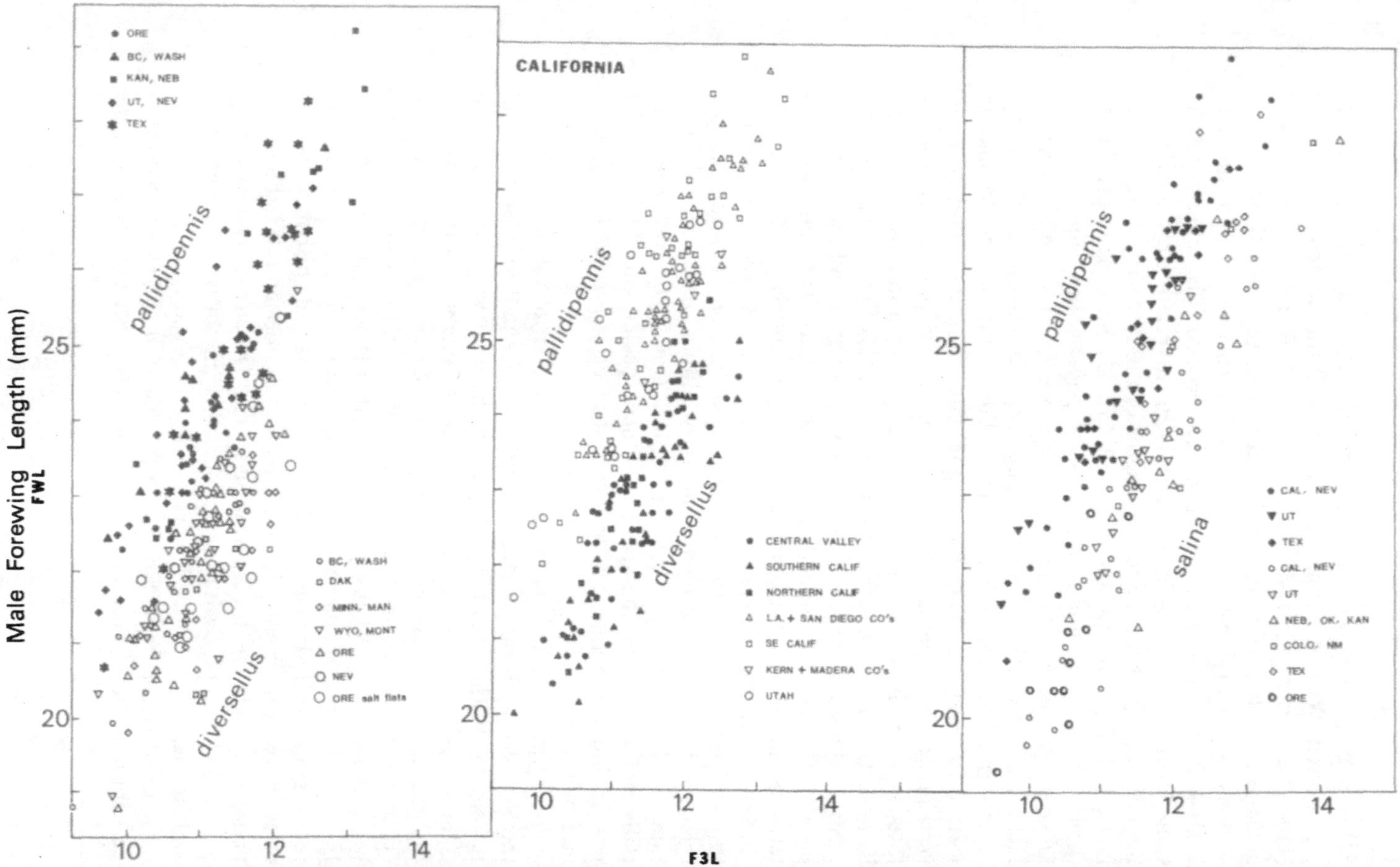

Fig. 54. Scatter diagrams showing separation of *pallidipennis, salina,* and *diversellus* using forewing length (FWL measured from the beginning of the R vein) and male hind femur length (mm) (F3L).

following combination of characters: dark crossbands of forewings distinct on dorsum; basal area of hindwings pale whitish yellow to yellow, always with a distinct black band; hind tibiae yellowish to pale brown; and posterior margin of lateral lobes never with a small tooth. Through parts of Oregon, Nevada, Utah, New Mexico, and the Great Plains it may be confused with *salina,* from which it differs as follows: head is narrower (X/Y = 0.82–0.92; in *salina,* 0.91–1.04); hindwings always with a distinct dark band (band often faint or obsolete in *salina*); forewings relatively longer (FWL/F3L = 2.08–2.35; in *salina* 1.86–2.09). It differs from *diversellus* as follows: forewing crossbands always distinct and dark (often pale or broken up in *diversellus*); forewings shorter in *diversellus* (FWL/F3L = 1.85–2.09); dark band on hindwings highly variable in *diversellus*—from very broad and dark to faint or obsolete. *T. inconspicua* is also very similar to *pallidipennis,* but this species is smaller and its pronotal disk usually rounds off onto the lateral lobes more gradually.

Body length to end of forewings 25–35 mm in males, 30–45 mm in females.

HABITAT. This species inhabits the western prairies, desert grasslands, and deserts. It breeds in areas of thin soil and sparse vegetation. Tinkham (1948) noted that it was the most ubiquitous of desert grasshoppers and was abundant in every desert locality he visited in the Southwest. David Ferguson (in litt.) writes that *T. pallidipennis* "is common in about any bare dry spot in the Lower and Upper Sonoran

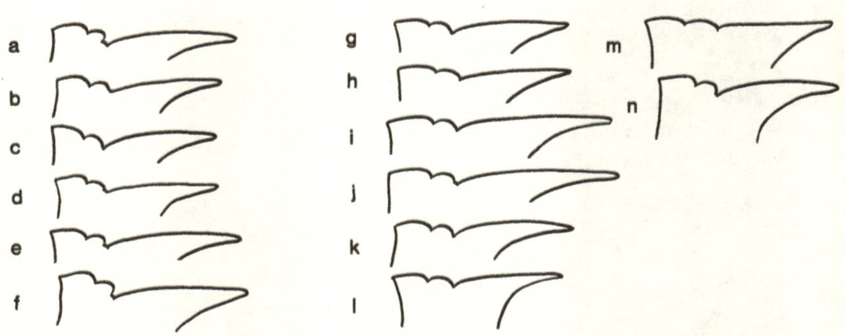

Fig. 55. *Trimerotropis.* Profile of pronotal crest in *pallidipennis* (*a-f, m, n*) and *diversellus* (*g-l*) from various localities. *a,* Blyth, Calif.; *b,* Portland, Ore.; *c,* Harney Co., Ore.; *d,* Prescott, Ariz.; *e,* Baja California Norte; *f,* Baker, Calif.; *g, j, k, l,* Colusa Co., Calif.; *h, i,* Kern Co., Calif.; *m,* Benedict Island; and *n,* Clarion Island, both southwest of the tip of Baja California.

zones and through the Transition Zone. It may be found in alkali-encrusted habitats, but usually not as the favored habitat. Everywhere I have found *salina* I have also found *pallidipennis*".

BEHAVIOR. *T. pallidipennis* feeds on a wide variety of plant species but often concentrates on grasses. (Mulkern et al. 1969, Otte and Joern 1977). Otte (1970) describes the reproductive behavior. Crepitation consists of snapping sounds in groups of three to six snaps. Usually a number of such groups are produced during a single flight. Crepitation flights in Argentina are similar to those in North America.

LIFE CYCLE. This species produces at least two generations a year in the southwestern United States. Rentz and Weissman (1980) state that three generations a year may occur in southern California, with first-generation specimens smaller on the average than second- or third-generation ones. In Idaho the life cycle is complex. "In many areas it overwinters in the egg state, with a single generation per year. In some

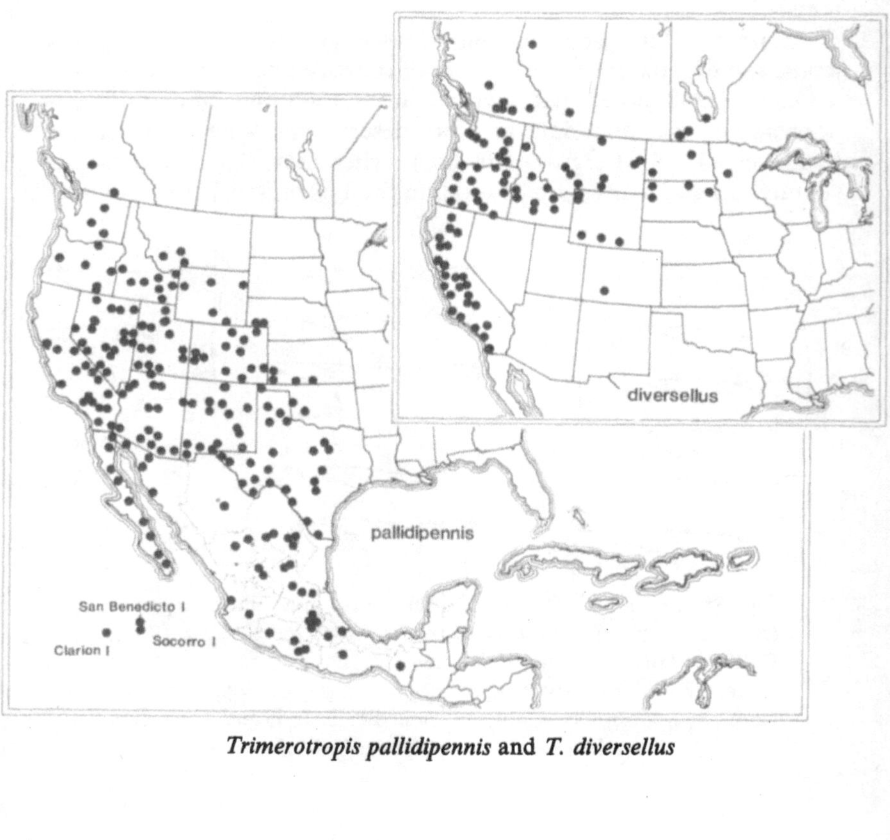

Trimerotropis pallidipennis and *T. diversellus*

areas there is either a complete or partial second generation, or perhaps continuous generations. As a result, early instars may be collected from early June to late August, and adults from early June to late October" (Scoggan and Brusven 1972). Note: These latter statements may also refer to *diversellus.*

REFERENCES. Hebard 1928, 1931a, Tinkham 1948, Otte 1970, 1976, Scoggan and Brusven 1972, Ball et al. 1972, Otte and Joern 1977.

Trimerotropis diversellus Hebard

TAXONOMY. *T. diversellus* was described from northwestern Wyoming and southern Montana. But the range of this species was never thought to be as extensive as shown here. All specimens from the Dakotas, Minnesota, and Manitoba, identified by Hebard as *T. pallidipennis salina,* belong to this species. Specimens from Montana believed by Hebard to be intermediate between *salina* and *pallidipennis* all belong to *diversellus; T. salina* is presently not known from north of Nebraska. *T. diversellus* can be distinguished from *salina* by head width (in relation to pronotum width) and from *pallidipennis* by wing length, body size, forewing banding, behavior, and usually by habitat (see Recognition). These features are quite consistent within the three species. Specimens from central and southern California are tentatively assigned to this species; what is certain is that they are distinct from *pallidipennis.*

DISTRIBUTION. Minnesota and Manitoba, westward to British Columbia and Washington, and along the west coast southward to California.

RECOGNITION. Figs. 52, 53, 54, 55*g–i,* 57. *T. diversellus* has shorter forewings (FWL/F3L = 1.85–2.09) than *pallidipennis* (2.08–2.35). In this character it is similar to *salina* (1.86–2.09). It has a measurably narrower head (X/Y = 0.82–0.92) than *salina* (0.91–1.04).

VARIATION. Specimens from mountain habitats are usually darker than those from valleys. Specimens from salt flats and lake margins in Oregon resemble *T. salina* rather closely, being pale in color and having a faint hindwing band or no bands at all. Lightfoot (personal communication) has found that along alkaline lake margins all degrees of variation, from completely unbanded, sand-colored insects to wide-banded, gray insects, exist within cruising range of one another.

In Wyoming and Montana the dark hindwing band is also quite variable in width, sometimes extending toward the base beyond the midpoint of the wings.

In California's Central Valley and other valleys specimens are pale gray with indistinct crossbands. The separation between *pallidipennis* and *diversellus* in California is shown in Fig. 54.

BEHAVIOR (based only on California specimens). As males approach females, they produce widely and irregularly spaced pulses of stridulation, using single strokes of the hind femora. These are interspersed with bursts of hind femur-shaking, one leg at a time. At East Park Reservoir in Colusa County, I could hear no sound during shaking, but in Kern County the femora struck the forewings during this movement (see Fig. 10*b, c* in Otte 1970). Another minor difference was noted: in Colusa County the male did not produce a burst of O-stridulation just before mounting the female, but such variation is common among species that shake their femora during courtship.

HABITAT. From open pine woodland up to 7,500 feet and down to grassy plains and alkaline depressions. In California it ranges from low-flying flatlands of the Central Valley to rolling foothills. In southern California *pallidipennis* seems to live at higher altitudes than *diversellus.*

LIFE CYCLE. Adults from July to October.

REFERENCES. Hebard 1928, Otte 1970.

Trimerotropis saxatilis McNeill — Pl. 18

DISTRIBUTION. Mountain tops from eastern Oklahoma and Kansas to Illinois and southern Georgia. In the western portion of the range the presence of specimens intermediate between *saxatilis* and *pallidipennis* suggests that gene flow between the two forms persists.

RECOGNITION. Most similar to *pallidipennis.* Specimens from Kansas are intermediate between typical *saxatilis* and *pallidipennis.* Over most of its range this species is highly speckled, often closely resembling lichen-covered rocks. The darker pigments are often black, the lighter ones pale green. Black areas sometimes much larger than pale green areas. Hind tibiae yellowish to dark brown, usually with a pale band near base. Body length to end of forewings 23–30 mm in males, 28–38 mm in females.

HABITAT. Exposed, lichen-covered rocks at the summits of mountains. Hart (1906) notes: "Restricted to the larger sandstone ledges of the Ozark hills in southern Illinois, the insects closely resemble in color the lichen-covered rock surfaces on which they rest." Morse (1907) writes: "*T. saxatilis* has been taken only on rock surfaces, sometimes lichen-encrusted, sometimes bare and sun-scorched, either wholly unprotected from sun and storm or amid a considerable growth of forest which has sprung up since the station was colonized by the locust. In color and markings it so closely resembles the background of its habitat as to be indistinguishable there when at rest. It is a sluggish, unsuspicious species, flying but a short distance when disturbed, often with a distinct but relatively faint crepitation, and is easily captured."

LIFE CYCLE. Adults from July to October.

REFERENCES. McNeill 1901, Morse 1907, Blatchley 1920.

Trimerotropis huroniana **Walker** Pl. 18

DISTRIBUTION. Northern dunes of Lake Michigan, Lake Huron, and Lake Superior. The only other *Trimerotropis* in this region is *maritima.* Although the two species have identical habitats, they are never found at the same place. They displace one another in Michigan along the shores of Lakes Michigan and Huron, with *huroniana* occupying the northern part and *maritima* the south (Otte 1968). The boundary between them is likely the result of some incompatibility, since there are no evident natural boundaries to their northward or southward spread. The boundary has shifted in the last few decades. In 1941, according to the University of Michigan collection, the boundary on the eastern shore of Lake Michigan was between Pierport, Manistee County, and Platte River Point, Benzie County. My collecting in 1965 and 1966 indicates a boundary fifty miles farther south, between Pentwater and Muskegon. On the western shore of Huron *maritima* has apparently pushed north into areas formerly occupied by *huroniana.* Hubbell (1929) recorded *huroniana* from two beaches in Huron County and *maritima* from one beach in St. Clair County. In 1966 I collected *maritima* in St. Clair, Huron, and Bay counties and *huroniana* only as far south as Iosco County. These two species may well be competitive on four different lake fronts, on the north and south shores of Lake Michigan and Lake Huron.

RECOGNITION. Fig. 52. Separated from *maritima* as follows: inner face of hind femora with a broad basal dark band (*maritima* has a short band). Crossbands of forewings faint but present in most specimens and visible on the dorsal surface (in *maritima* from the Great Lakes and bands are either obsolete or fade out on the dorsum). Body length to end of forewings 24–30 mm in males, 29–40 mm in females.

HABITAT. Sand dunes along lakes and in back of lakes where vegetation is sparse.

BEHAVIOR. Flight crepitation is very similar to that of *T. pallidipennis,* as is courtship behavior (Otte 1970).

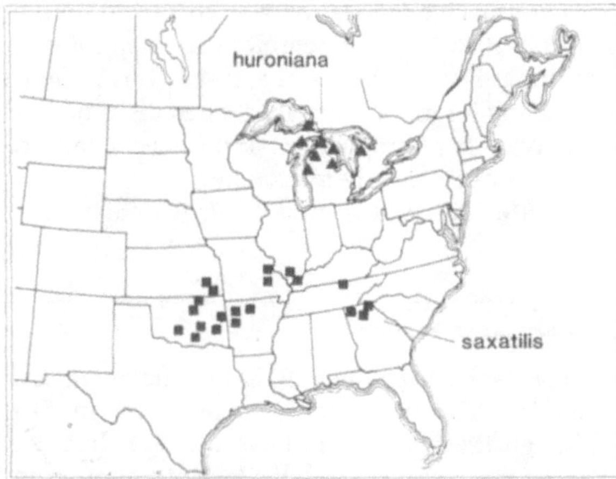

Trimerotropis saxatilis and *T. huroniana*

LIFE CYCLE. Adults from July to September.

REFERENCES. Walker 1902, Blatchley 1920, Hubbell 1929, Otte 1968, 1970.

Trimerotropis salina Thomas

TAXONOMY. *T. salina* was described from Lincoln, Nebraska. Hebard (1928) first treated it as a synonym of *T. pallidipennis* mainly on the basis of specimens from the Bozeman area of Montana (but where *salina* does not occur). The species was known from alkaline depressions in Kansas, Nebraska, and Colorado and was thought until now to extend north to Manitoba and Saskatchewan. Hebard's (1928) Montana specimens of "salina" graded imperceptibly into *T. pallidipennis,* which caused him to treat *salina* as a subspecies. An examination of Hebard's specimens shows that he was confusing specimens of *T. diversellus* with *salina.* In Hebard's collection (Philadelphia Academy of Natural Sciences collection) *salina* is not known from north of Nebraska. The species is easily confused with *pallidipennis* and *diversellus* but is readily distinguished from each by measuring head and pronotal dimensions.

T. salina appears to be most closely related to *schaefferi,* which is known only from beaches on the Gulf of Mexico in Texas. Indeed, when the hindwings are closed, specimens of *schaefferi* are virtually indistinguishable from *salina* specimens from Nevada, Utah, and New Mexico. Hebard (1928) also states that *salina* is not known from west of the Rocky Mountains, but we now know that it ranges through New Mexico west to California and Oregon. Hebard examined specimens from Oklahoma and Texas and labeled them "Trimerotropis, species near *pallidipennis.*" These specimens have darker bands on the forewing than those from Nebraska and were therefore not recognized by Hebard as *salina.*

DISTRIBUTION. Nebraska southward to Texas, then westward and northwestward to California and Oregon.

RECOGNITION. Figs. 52–54, 57. Body coloration in *salina* varies from pale sand-colored with indistinct forewing crossbands in the salt flats of California, Nevada, Oregon, and Utah to darker, grayer specimens with blackish crossbands in central Texas, Oklahoma, and Kansas. *T. salina* has a wider head (X/Y = 0.91–1.04) than *pallidipennis* or *diversellus* (0.82–0.92), and shorter forewings (FWL/F3L = 1.85–2.09) than *pallidipennis* (2.08–2.35) (see Fig. 53). The width and darkness of the hindwing band is highly variable, although on the average it is wider than *pallidipennis* in the Great Plains and Texas. In the prairies and open woodlands of central Texas the hindwing band is very dark and varies in width. In the salt flats of Utah, Nevada, California, and Oregon the band is occasionally narrow and distinct but is usually faded or entirely absent. In Nevada, California, and Oregon *salina* may be confused with *sparsa,* but they differ in the ratio of length to width of pronotal disk (Fig. 57). Basal area of hindwings always yellow. Body length to end of forewings 23–34 mm in males, 32–45 mm in females.

HABITAT. Over most of its range this species inhabits sparsely vegetated margins of salt flats and dry river beds. David Ferguson (in litt.) sent me the following notes on behavior and habitats in Colorado and Wyoming: "*T. salina* is strictly limited to white alkali-encrusted areas which are often wet and very muddy and also often subject to flooding. It seems associated with salt grass (*Distichlis stricta*) and alkali sacaton (*Sporobolus airoides*), which seem the most likely food plants. In some areas greasewood (*Sarcobatus vermiculatus*) is common, and the hoppers hide under them on hot afternoons. *T. salina* seems to be quicker and more erratic in flight [than *pallidipennis*]. When disturbed they circle back to a nearby place (probably an adaptation to usually small habitat patches)." The species also occurs in the rolling hills of central Texas. What its habitat is like in this region is not known.

LIFE CYCLE. Adults from July to September.

REFERENCES. Hebard 1928, 1931.

Trimerotropis schaefferi Caudell Pl. 18

DISTRIBUTION. This species is known only from the coastal dunes of Texas from Brownsville to Galveston.

RECOGNITION. Fig. 52. Likely to be confused only with *pallidipennis.* Recognized by its very broad dark hindwing band and its generally higher pronotal crest. Hind tibiae yellowish, sometimes dirty yellow to brownish. Body length to end of forewings 30–37 mm in males; 39–52 mm in females.

HABITAT. Coastal dunes and areas behind the dunes.

LIFE CYCLE. Adults in the Philadelphia Academy of Natural Sciences collection were taken from May to August, mainly in July and August.

REFERENCES. Caudell 1904.

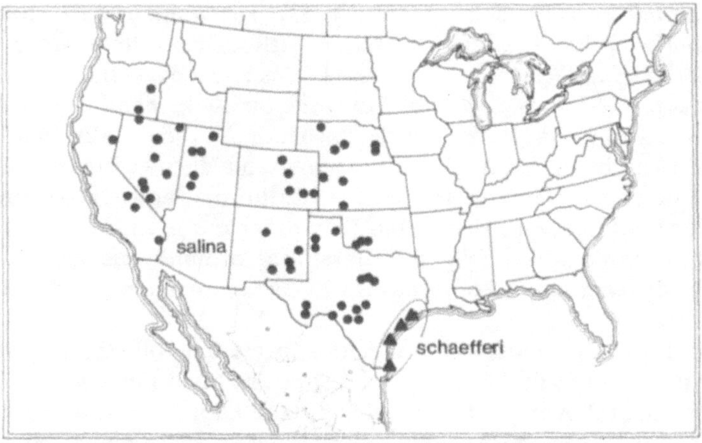

Trimerotropis salina and *T. schaefferi*

Trimerotropis cyaneipennis Bruner

TAXONOMY. This species belongs to a group that includes *inyo, leucophaea, fratercula,* and *bernardi.* The entire complex requires further study, because even the status of well-established species such as *cyaneipennis* and *fratercula* have become questionable with the discovery of what appears to be a morphological cline between them in New Mexico.

David Ferguson (in litt.) believes that *fratercula* and *cyaneipennis* may belong to the same species. This belief, or that they hybridize with one another, is based on information on behavior and morphology.

BEHAVIOR. The crepitation patterns of New Mexico *cyaneipennis* (Sandia Mountains—Ferguson recording) and of *fratercula* (Otte 1970) are largely indistinguishable. Only *T. verruculata* has a similar pattern.

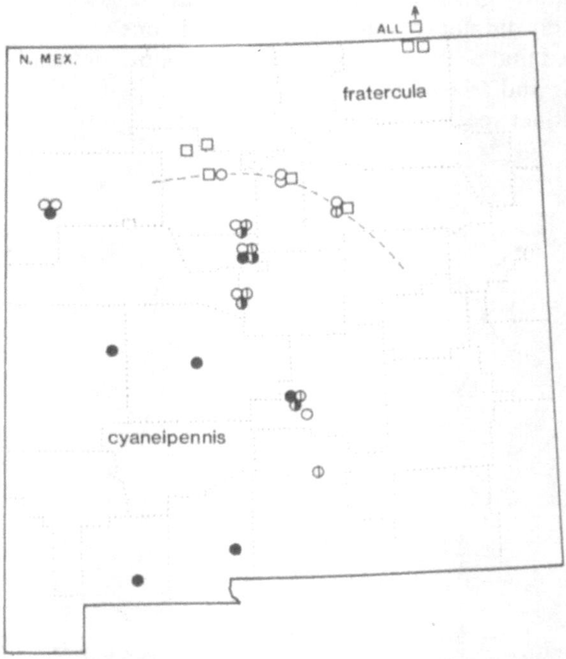

Transition zone between *T. cyaneipennis* and *T. fratercula* in New Mexico. Square: brownish, yellowish, or pale greenish blue tibiae (*fratercula*). Circle: blue or blue-green tibiae (*cyaneipennis*). Open symbol, yellow wings; circle with line, yellow-green wings; half darkened circle, blue-green wings; black circle, blue wings.

MORPHOLOGY. The hindwings of *fratercula* are always yellow in the basal area; in *cyaneipennis* they are usually blue. In Arizona, New Mexico, and west Texas *cyaneipennis* with blue-green wings can be found, and in New Mexico yellow-winged specimens are found increasingly toward the central and northeastern portions of the range, but all have a slightly greenish cast near the base (see map based on

Philadelphia specimens and Ferguson notes). The hind tibiae of *cyaneipennis* are blue, but the degree of blueness diminishes as the hindwings become yellower. Most specimens in the Philadelphia collection from New Mexico previously identified as *fratercula* have yellowish-brown tibiae; occasional specimens have faintly greenish yellow or pale, very slightly bluish tibiae. The zone of transition from pure yellow to slightly greenish wings coincides with the transition of yellow-brown or faintly greenish tibiae to blue tibiae. If the change of these two characters did not coincide geographically one could be more easily convinced that a single species is involved. Until detailed studies on *cyaneipennis* and *fratercula* are conducted it is perhaps best to leave them as distinct species.

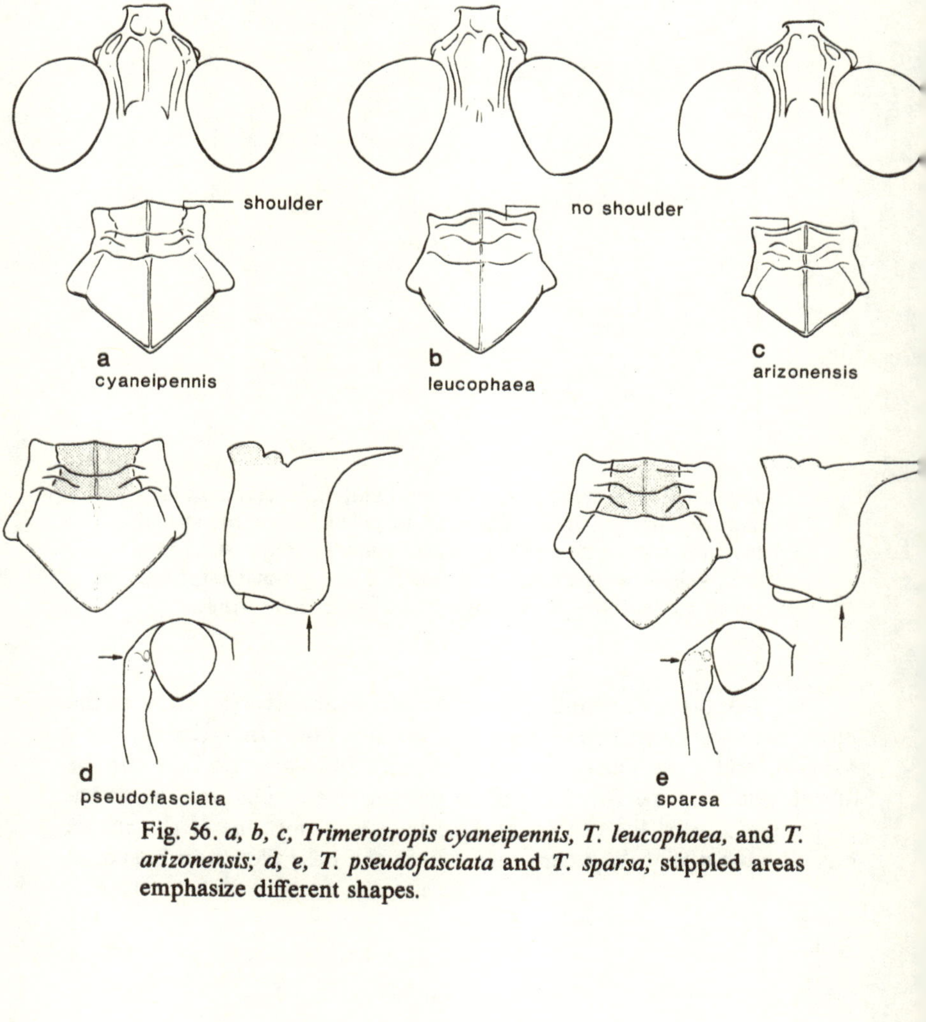

Fig. 56. *a, b, c, Trimerotropis cyaneipennis, T. leucophaea,* and *T. arizonensis; d, e, T. pseudofasciata* and *T. sparsa;* stippled areas emphasize different shapes.

DISTRIBUTION. Dry mountain habitats from Texas west to California and north to Idaho.

RECOGNITION. Figs. 50*l*, 56*a*. Hindwings blue to greenish blue in basal area with a dark outer band. Very similar to *leucophaea*, but differs as follows: pronotal disk with definite shoulders at anterior end (rounded in *leucophaea*); hindwings with a distinct to indistinct dark band; when indistinct it is present on the anterior fold (absent on anterior fold of *leucophaea*). Also similar to *inyo*, but differs in wing color (see *inyo*).

In New Mexico and Arizona the hindwings are often greenish blue and sometimes even greenish yellow.

Body color varies considerably in *cyaneipennis*. Individuals collected on park rocks and lava flows are dark gray, those collected on pale rocks are pale gray. Dark individuals have darker and broader dark bands on the forewings and hindwings. Hind tibiae are almost always blue or blue-gray; in rare specimens they are yellowish brown. This species differs from the similar *pseudofasciata* and *sparsa* as follows: prononal carina lower; prozona relatively longer; hindwings with a more distinct dark band (but rarely *pseudofasciata* has a distinct dark band); hindwings bluer; hind tibiae blue, never yellow, rarely brown; frontal ridge of head less concave; fastigium without a very prominent median carinula (as in *pseudofasciata*). Lower margin of pronotum without a blunt tooth (tooth is present in *pseudofasciata*). Body length to end of forewings 25–35 mm in males, 28–40 mm in females.

HABITAT. Dry rocky slopes in wooded areas, often on slopes among junipers and piñon pine. Usually found above 3,000 feet up to 8,000 feet. Common along trails in the Grand Canyon. "It was not plentiful along the canyon edge, but lower on the Bright Angel Trail it was found almost everywhere, most plentiful, however, about bare places near precipices" (Rehn and Hebard 1908). In Idaho Scoggan and Brusven (1972) write:

> *T. cyaneipennis* is a rock-loving species. It can be found on lava flows, talus, and basaltic cliffs in either the shadscale or sage brush communities. Vegetation often consists of *Bromus tectorum, Oryzopsis hymenoides, Artemisia tridentata, Juniperus* spp., or *Atriplex confertifolia,* along with scattered annual forbs. It may be found in association with *Trimerotropis fontana* and *Cordillacris occipitalis*. Nymphs are found in open areas among rocks or pumice, and respond lethargically when disturbed. *T. cyaneipennis* is not an economic pest in Idaho. It inhabits rough, generally ungrazable terrain and seldom reaches numbers greater than one per square yard as nymphs.

LIFE CYCLE. Adults from July to October in Arizona. In Idaho it overwinters in the egg stage. First nymphal instars appear in early June, and adults appear in mid-July.

REFERENCES. Rehn and Hebard 1908, 1909; Hewitt and Barr 1967; Scoggan and Brusven 1972.

Trimerotropis inyo Rentz and Weissman Pl. 15

DISTRIBUTION. Known only from the eastern slopes of the Sierra Nevada Mountains above the Owens Valley, Inyo County, California. It has been collected at the following places: 12.8 and 17.6 km west of Lone Pine on the Whitney Portal Road, and 9.6 km west of Big Pine.

RECOGNITION. Most similar to *cyaneipennis* in appearance, but somewhat paler than typical *cyaneipennis.* Hindwings yellow-green in basal area; greenish wings occur in *cyaneipennis* populations in New Mexico and Arizona, but in all populations near the distribution of *inyo,* hindwings are blue. Band variable: sometimes almost as broad as the forewings and uninterrupted; sometimes less than half as wide and interrupted. Forewings have at least three blotchy crossbands. Shape of fastigium and pronotum as in *cyaneipennis* (Fig. 53*a*); prozona with definite shoulders (unlike *leucophaea* or *arizonensis*). Hind tibiae blue. Body length to end of forewings 26–29 mm in males, 33–36 mm in females.

HABITAT. Bare gravelly ground in the piñon-sage community of the eastern slopes of the Sierra Nevada Mountains. It was found along with *T. albescens.*

LIFE CYCLE. Adults were collected between July 19 and September 1.

REFERENCES. Rentz and Weissman 1984.

Trimerotropis leucophaea Rentz and Weissman Pl. 15

TAXONOMY. This species differs from *cyaneipennis* in chromosome number and slightly in morphology. Whether it is distinct from *arizonensis* is questionable; the karyotype of the latter is not known, but morphologically they are very similar. The two species have not been collected at the same locality, therefore populations from intervening areas should be studied to understand the relationships between them.

DISTRIBUTION. Specimens have been collected at the following places in California: Silver Canyon in White Mountains, Wild Rose

Canyon in Panamint Range, and Boundary Canyon in Amargosa Range, all in Inyo County; Trona, San Bernardino County; and Sherwin Hill, Mono County.

RECOGNITION. Fig. 56*b*. Differs from *cyaneipennis* in that the outer dark band on the hindwings is more or less obsolete; when present, the band is most conspicuous along the posterior margin. Pronotal disk rounds gradually onto lateral lobes at anterior end (*cyaneipennis* has definite shoulders). Pronotal disk narrows steadily from rear to front (in *cyaneipennis* it straightens out or widens anteriorly). See *arizonensis* for differences with that species.

HABITAT. Gravelly mountain slopes.

LIFE CYCLE. Adults in late summer and fall.

REFERENCES. Rentz and Weissman, 1984.

Trimerotropis arizonensis **Tinkham**

TAXONOMY. This species is almost indistinguishable from *cyaneipennis* and *leucophaea*. They are here treated as three distinct species, but I believe further work on these species is needed.

DISTRIBUTION. Collected at Dome, Yuma County, Arizona; Volcanic, Lincoln County, and Logandale, Clark County, Nevada; Goffs and Providence Mountains, San Bernardino County, California.

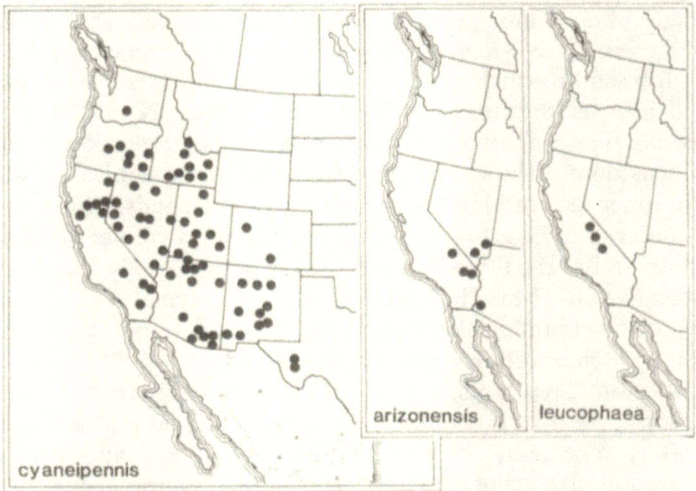

Trimerotropis cyaneipennis, T. arizonensis, and *T. leucophaea.*

RECOGNITION. Fig. 56*c*. Differs from *cyaneipennis* as follows: fastigium narrower; lateral foveolae sometimes not concave; disk of pronotum rounds gradually onto lateral lobes so prozona has no shoulders; pronotum relatively short. Differs from *leucophaea* as follows: forewings with distinct crossband; outer dark band of forewings distinct (largely obsolete in *leucophaea*); pronotum somewhat shorter. Body length to end of forewings 22.5–27.5 mm in males, 31–33 mm in females.

HABITAT. Dry, rocky hillsides.

LIFE CYCLE. Adults have been collected from early August to mid-September.

REFERENCES. Tinkham 1947.

Trimerotropis fratercula McNeill Pl. 18

TAXONOMY. An analysis of the contact zone between *fratercula* and *cyaneipennis* in New Mexico suggests that these two nominal species may belong to the same species (see discussion under *cyaneipennis* Taxonomy). *T. bernardi* may also be conspecific.

DISTRIBUTION. Mountains of New Mexico, Colorado, and Wyoming.

RECOGNITION. Most similar to *cyaneipennis* in north central New Mexico where yellow-winged *cyaneipennis* are common. At present the two species can be separated by the color of the hind tibiae, which are distinctly blue or blue-green in *cyaneipennis* and yellowish brown or faintly greenish in *fratercula.* Basal area of hindwings yellow; dark band distinct, brownish; outer third of hindwings largely transparent. Crossbands on forewings usually present, especially on lateral field; sometimes they are formed only by a clustering of dark spots, but some specimens do not show this. Hind tibiae yellow, greenish, or slightly bluish. Inner face of hind femora with two pale bands. This species is also similar to *suffusa,* but in *suffusa* the forewings are not banded and the outer half of the forewings is brown or smoky. Body length to end of forewings 28–33 mm in males, 32–38 mm in females.

HABITAT. Mountain slopes and eroded banks. In the Rocky Mountains it is often found in open pine woodlands.

BEHAVIOR. Crepitation consists of pairs of closely spaced snapping sounds, *t-k, t-k, t-k.* In Colorado I heard individual males flying and crepitating repeatedly; between flights they walked about over the rocky terrain producing courtship signals. Males approaching females produced only ordinary stridulation. Femur-tipping and shaking signals were performed in male–male encounters.

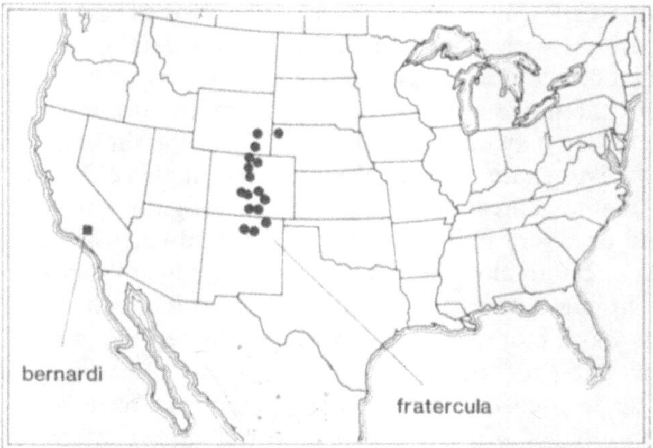

Trimerotropis fratercula and *T. bernardi*

LIFE CYCLE. Adults from July to September.

REFERENCES. McNeill 1901, Hebard 1929, Otote 1970.

Trimerotropis bernardi Rentz and Weissman

DISTRIBUTION. Known only from between 7,200 and 9,000 feet at Vivian Creek and High Creek in the San Bernardino Mountains of southern California.

RECOGNITION. Largely indistinguishable from *fratercula* of New Mexico and Colorado, but separated by a gap of 600 miles. In southern California it is most likely to be confused with *Circotettix shastanus,* but in *bernardi* the hindwings have a well-defined outer dark band, and veins 2A and the anterior branch of $2A_a$ are not fused. Hind tibiae blue. Forewings with barely indicated crossbands formed by aggregations of smaller dark spots. Body length to end of forewings 30–32 mm in males, 36–39 mm in females.

HABITAT. The labels on the insects indicate heavy forest.

LIFE CYCLE. All Philadelphia Academy of Natural Sciences specimens were collected in late August.

REFERENCES. Rentz and Weissman 1984.

Trimerotropis pseudofasciata Scudder Pl. 18

DISTRIBUTION. Washington south to Baja California, east to Idaho and Utah.

RECOGNITION. Fig. 50*k*. Basal area of hindwings pale blue to colorless, sometimes with a very indistinct central dark band, sometimes represented only by darkened veins. In specimens with slightly bluish hindwings, the tibiae are blue or gray; in specimens with transparent hindwings, they may be yellow. It differs from the similar *cyaneipennis* mainly in having a rather high pronotal crest; it also differs in having a more pronounced median carinula on the fastigium and in having no dark band or a very indistinct one on the hindwings and in having a low tooth or bump along the lower posterior edge of the lateral lobes. It differs from *sparsa* as follows: prozona is much higher and shorter; forewings have more distinct crossbands (especially where the two species overlap geographically); hind tibiae are often blue (yellowish, never blue in *sparsa*); lateral lobes sometimes have a very small projection along the lower posterior corner. Body length to end of forewings 24–35 mm in males, 32–42 mm in females.

HABITAT. Low-elevation deserts and semideserts. Rehn and Hebard (1909) report the species from beach vegetation among sand dunes.

LIFE CYCLE. Adults from July to September.

REFERENCES. Rehn and Hebard 1909, Strohecker, Middlekauff, and Rentz 1968.

***Trimerotropis sparsa* (Thomas)** Pl. 18

TAXONOMY. According to White (1951), *sparsa* is a member of section B of the genus. "It is a cytologically polymorphic species in which several of the autosomes (up to seven in one population) may be represented by acrocentric or metacentric chromosomes. [It] appears to be a complex species in which much local cytological differentiation has taken place."

DISTRIBUTION. Northern New Mexico and Arizona north to Alberta and from western Nebraska and the Dakotas west to eastern California.

RECOGNITION. Fig. 57. Hindwings variable: sometimes completely clear, sometimes pale blue, sometimes yellowish; dark band usually barely indicated or absent, but occasionally quite smoky; sometimes only the veins are darkened. Hind tibiae yellowish. Veins 3A, 4A, and 5A usually dark and somewhat thickened. Darkened crossbands on forewings absent or indistinct. Body generally grayish or pale brown and finely speckled in appearance. This species differs from the similar *pseudofasciata* as follows: prozona relatively lower and longer; forewings with less distinct crossbands (the difference is especially pro-

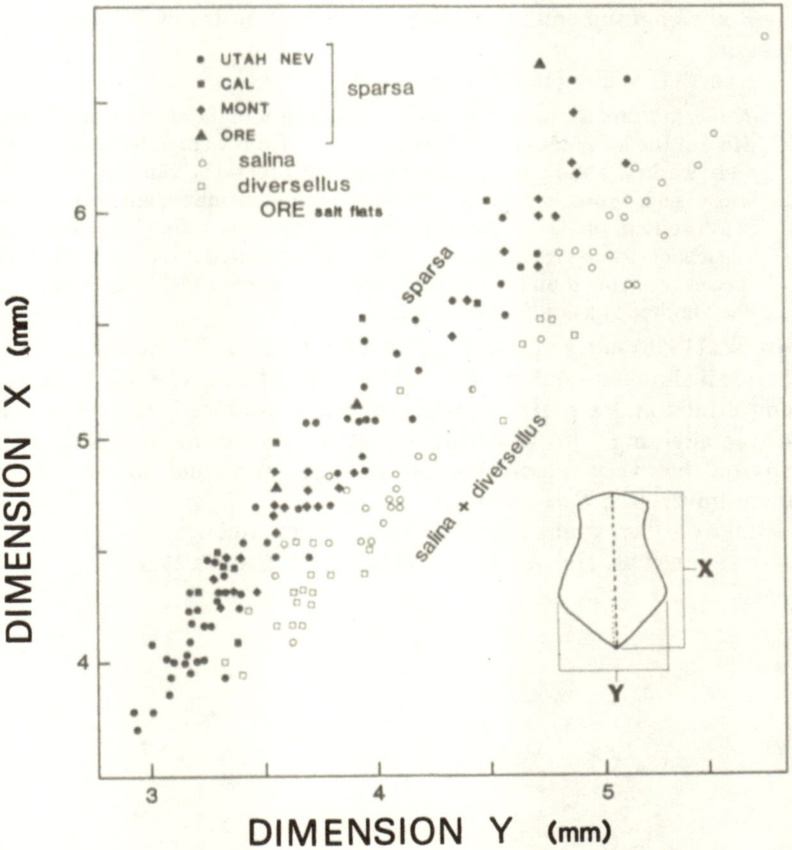

Fig. 57. Scatter diagram showing separation of *sparsa* from *salina* and *diversellus*.

nounced where the two species overlap); hind tibiae yellowish; lower posterior side of lateral lobe never with a small projection. This species is difficult to distinguish from *gracilis* in Nevada and Utah. It differs as follows: Hind femora with two pale bands (*gracilis* almost always has one pale band); basal area of hindwings transparent to bluish (transparent to greenish or yellowish in *gracilis*); pronotal crest lower on the metazona; forewings not as streaky, crossbands sometimes barely visible. In some *cyaneipennis* the dark band on the forewings is rather indistinct, but unlike *sparsa*, the hind tibiae are always blue. May also be confused with *salina* and some *diversellus* (see Fig. 57 for separation of these species).

Body length to end of forewings 24–32 mm in males, 30–42 mm in females.

HABITAT. White (1951) reports:

> *T. sparsa* is not a true desert species, although found only in arid regions. Most of the localities at which we have collected it were on alluvial soil in wide washes; although we have on several occasions found it in places where sage brush is present, it is not a common inhabitant of the sage brush zone in the Great Basin, in the way that *T. gracilis, T. fontana,* and *T. pseudofasciata* are. In spite of its wide distribution, it is a relatively rare species . . . being found in small colonies which are probably isolated from one another to a considerable extent.

Brooks (1958) notes that in the Canadian prairies: "*T. sparsa* inhabits bare, alkaline flats and hillsides in the eroded areas. The flight is short and erratic and a sharp crackling sound is produced, especially just before alighting." In New Mexico Hebard (1935c) found a large colony on the "very smooth bare adobe floor of a small arroyo which there traversed a *Sarcobatus* flat."

LIFE CYCLE. Adults from late July to September.

REFERENCES. Hebard 1935c, White 1951, Brooks 1958.

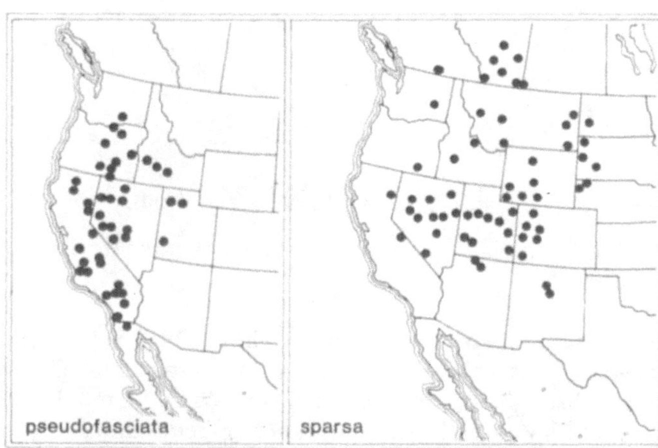

Trimerotropis sparsa and *T. pseudofasciata*

***Trimerotropis gracilis* McNeill** Pl. 15

TAXONOMY. E. M. Walker (1902) described *T. sordida* from Assiniboia, Saskatchewan, a species with a distinct dark band on the hindwings. Subsequently the species was considered to be merely a subspecies of *gracilis* (Hebard, 1928, 1931, and others). Vickery and Kevan (1983) give the impression that the two can be separated on the basis of wing banding, and they consider the intermediates found by Hebard (1928) to be "atypical" *gracilis.* I have analyzed the wing banding across the western United States (see map) and find a trend from pale wings to banded wings as one goes from west to east. I conclude that two species cannot be recognized on the basis of wing color. Vickery's (1979) cytological studies suggest that there are two species, but Rentz and Weissman (1980) and Weissman and Rentz (1980) treat them as subspecies, even though they have different chromosome numbers. Additional work could shed light on the problem.

DISTRIBUTION. All states west of Kansas and Nebraska; southwestern Canada.

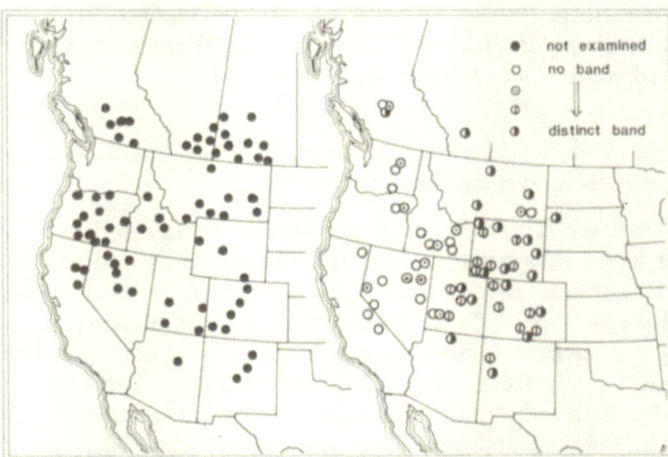

Trimerotropis gracilis. Symbols indicate condition of the dark band on the hindwings.

RECOGNITION. A gray or pale brown species without distinct crossbands on the forewings. Body streaked in appearance. Hindwings yellowish in basal area and with or without a central dark band. In the far west the band is indistinct or absent; it is most strongly developed in the eastern portion of the range. The map shows the location of unbanded wings, banded wings, and intermediate forms. Hind tibiae yellowish or brownish. *T. gracilis* is most easily confused with *sparsa* (see *sparsa* for differences).

HABITAT. This species is most commonly found on bare ground in sagebrush areas. Scoggan and Brusven (1972) report it only from sagebrush areas in Idaho: "Favored habitats have a dense shrub cover of *Artemisia tridentata,* and abundant perennial grasses, e.g., *Agropyron* sp. and *Poa secunda.* Associated grasshopper species include *Ageneotettix deorum, Trachyrhachys kiowa, Aulocara elliotti,* and on more rocky terrain, *Trimerotropis bilobata.* Nymphs are active and usually seek vegetation, particularly sagebrush, when disturbed. Their coloration closely matches that of sagebrush bark."

BEHAVIOR. Creptitation flights by solitary males are frequent, and males run about on the bare ground between crepitations. During crepitation a snapping sound is produced on every third or fourth beat of the wings; consequently, the crepitation rate is comparatively slow. Males stridulate and silently tip their femora while approaching females. In male–male interactions they tip and shake their femora.

LIFE CYCLE. Adults from late June through September.

REFERENCES. Hebard 1925, 1928, 1929, 1935c, Otte 1970, Scoggan and Brusven 1972, Vickery 1979, Rentz and Weissman 1980, Weissman and Rentz 1980, Vickery and Kevan 1983.

***Trimerotropis verruculata verruculata* (Kirby)** Pl. 18

TAXONOMY. This subspecies is morphologically and behaviorally similar to the subspecies *T. v. suffusa.* The courtship and pair-forming behaviors of these two forms are nearly identical. Hebard (1929) notes that the two "species" are confusingly similar; Vickery and Nagy (1973) and Vickery, Johnstone, and Kevan (1974) considered the two forms to be subspecies. Helwig (1955) was able to hybridize the two forms but believed that they belonged to different species because they have different chromosome numbers. Rentz and Weissman (1981) have treated the two forms as separate species. In this work I am following the scheme of Vickery, Johnstone, and Kevan (1974).

DISTRIBUTION. The distribution closely follows the southern edge of the boreal forest region from western Canada to Nova Scotia. It meets that of *T. v. suffusa* in Alberta and British Columbia.

RECOGNITION. This is the only dark-colored *Trimerotropis* species ranging across Canada and into the Great Lakes region and the northeastern United States; it cannot be confused with any other. Its preference for rocky shores and outcroppings separates it from the only other two northern or northeastern *Trimerotropis* species, *huroniana* and *maritima,* which live on sand dunes. The body is often strongly speckled with black and small pale markings, so it is well camouflaged on lichen-encrusted and other darkly speckled rocks. The hindwings are similar to those of *v. suffusa* but less opaque in the distal third. Forewings often have dark markings grouped into indefinite crossbands. Hind tibiae are brownish to blackish, usually with a pale band near the base. Body length to end of forewings 25–35 mm in males, 28–40 mm in females.

HABITAT. Exposed gravel, rock, and clay banks along river courses, lake shores, and roadsides. Sometimes found in forest clearings and on exposed rocky hillsides. Morse (1897: 113) writes: "Its favorite haunts [in New England] are bare ledges on elevated land and low mountains." In Minnesota, according to Somes (1914): "It is primarily an insect of dry or even dusty places and is somewhat saxicolous, loving rocky exposures. On the granite hills about Lake Vermillion it occurs in countless numbers and on a bright day the [crepitation] of the males in flight fills the air . . . In Tower and some of the other northern towns of Minnesota, this usually shy insect takes the place of *Dissosteira* in the streets and appears as thoroughly at home there as the Carolina Locust in the southern cities."

BEHAVIOR. Flight displays by the males are spectacular. They hover over one general area and dance one to fifteen feet above the ground while producing pairs of loud snaps. The interval between the two snaps is constant but the interval between pairs is more variable. As males rise in their flight the intervals between pairs become shorter. Flights last a minute or more. Males stridulate while approaching females and shake and tip their femora when interacting with males.

LIFE CYCLE. Adults from July to September.

REFERENCES. Morse 1897, Somes 1914, Blatchley 1920, Otte 1970.

Trimerotropis verruculata suffusa Scudder — Pl. 18

DISTRIBUTION. Mountains from British Columbia and Alberta south to Colorado, Utah, and Arizona.

RECOGNITION. Body dark brown or dark gray to black. Forewings nearly always without darker crossveins. Hindwings have enlarged posterior lobes; basal area is yellow, and in males the veins are enlarged and yellow; outer half of hindwings dark brown to smoky brown, sometimes with a lighter, more translucent area in the outer third. Hind tibiae variable: brownish, yellow-brown, or bluish. This species differs from the related *fratercula* mainly in being much darker and in lacking forewing crossbands. Inner face of hind femora with one or two complete pale bands. Body length to end of forewings 28–35 mm in males, 32–38 mm in females.

HABITAT. Pine forest openings in mountains, often on ridges, eroded slopes, and rocky courses of mountain streams.

BEHAVIOR. Solitary males often fly up, and during an up-and-down zigzag flight produce pairs of loud, closely spaced snapping sounds. Flights last from a few seconds to a minute or more. The signal is very similar to that of *fratercula* and almost identical to that of *T. v. verruculata.* Courting males produce bursts of ordinary stridulation with one leg at a time as they approach females. In male–male interactions they shake and tip their femora. There is some variation in courtship signaling over the range of the species (Otte 1970).

LIFE CYCLE. Adults from June to September.

REFERENCES. Ball et al. 1942, Helwig 1955, Otte 1970, Vickery and Nagy 1973, Vickery, Johnstone, and Kevan 1974, Rentz and Weissman 1980.

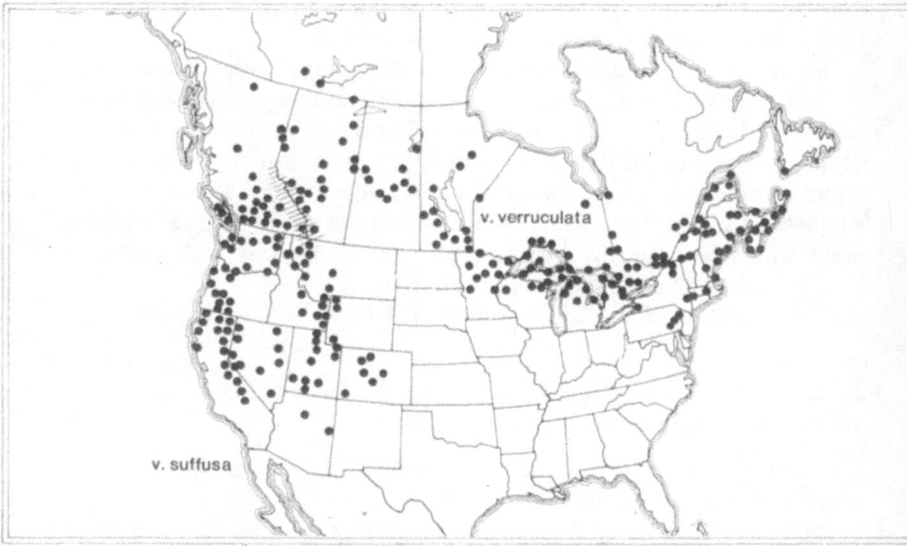

Trimerotropis verruculata. Crosshatching indicates possible contact zone between subspecies.

Bryodema Genus Group

One North American and five Eurasian genera belong to this group (Group Bryodemae of Bei-Bienko 1930). These are *Circotettix* (includes *Aerochoreutes*) from North America, and *Bryodema, Angaracris, Helioscirtus, Uvaroviola,* and *Cosmorrhipis* from Eurasia. This grouping of Bei-Bienko and Rehn (1921) is, I believe, essentially correct. White (1951) has challenged this view, believing that *Circottetix* is not closely related to *Bryodema* but shares an exclusive common ancestry with section B of *Trimerotropis* (see Fig. 54). He argues as follows: the genus *Trimerotropis* contains two groups of species, those with only acrocentric chromosomes (section A) and those with one or more metacentric chromosomes (section B, White 1949, 1951; Weissman and Rentz 1980). *Circotettix* species have metacentric chromosomes, and *Bryodema tuberculatum bavaricum* has acrocentric chromosomes. Could the *Bryodema* acrocentric condition be derived from the *Circotettix* metacentric condition? White maintains that such a change is unlikely (see his 1954 paper for reasons). But could the *Circotettix* metacentric condition be derived from the *Bryodema* acrocentric condition? In general, such a change is feasible, but *B. tuberculatum bavaricum* shows an extreme type of chiasma localization that makes the change from acrocentric to metacentric difficult, if not impossible, in that lineage.

The difficulties with White's scheme are these: (1) If *Trimerotropis* section B (metacentric) is derived from section A (acrocentric), then why could not *Circotettix* be derived similarly from *Bryodema?* Not all *Bryodema* species have been studied; other species may not have the extreme form of chiasma localization seen in *B. t. bavaricum,* in which case the metacentric condition could be derived from the acrocentric condition. (2) White says it is unlikely, *but not impossible,* to derive the acrocentric *Bryodema* condition from the metacentric *Circotettix* condition. (3) If *Trimerotropis* sections A and B are more closely related to *Bryodema* and *Circotettix,* respectively, than to one another, the genus is polyphyletic. (4) Chromosome differences, like other morphological differences, must be correlated with other characters in determining phylogenetic relationships. Further studies of *Bryodema* and its relatives, including ones that do not show extreme forms of chiasma localization, are required to show the relationships among *Bryodema, Trimerotropis,* and *Circotettix.*

RECOGNITION. Fig. 49 *m, n, o.* Hindwings with a greatly enlarged posterior lobe and anal veins usually thickened, especially in males. Forewings relatively broad. Hind femora relatively small.

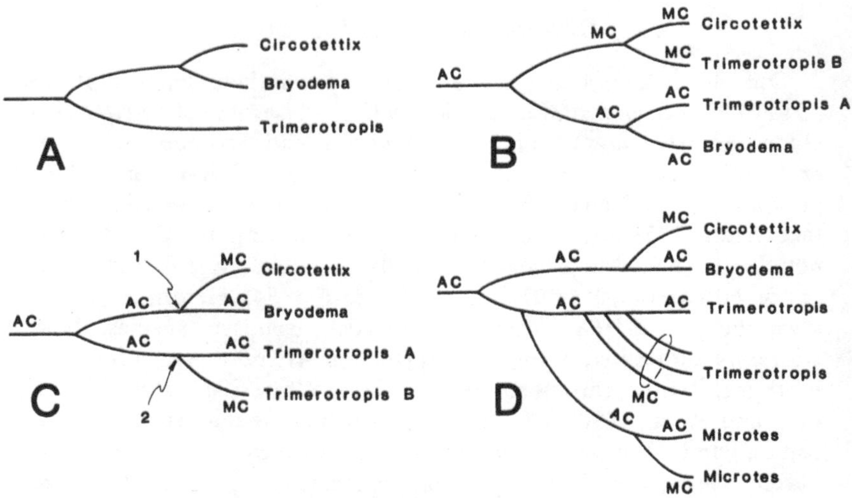

Fig. 58. *A,* phylogeny implied by Rehn and Bei-Bienko envisages a common stem for *Bryodema* and *Circotettix. B,* White's implied phylogeny envisages an acrocentric (AC) common ancestor for *Trimerotropis, Circotettix,* and *Bryodema.* His argument implies a biphiletic origin of *Trimerotropis,* with section B sharing a common stem with *Circotettix,* and section A sharing a common stem with *Bryodema. C,* combination of Rehn–Bei-Bienko, and White schemes can be envisaged in which the *Bryodema* lineage gave rise to metacentric (MC) species *Circotettix*), and the *Trimerotropis* lineage also gave rise to a metacentric species. This scheme satisfies the implied requirement that acrocentrics can give rise to metacentrics but not the other way around. White rejects the AC to MC change in the *Bryodema* lineage, saying this could not happen given the peculiar type of chiasma localization in *B. tuberculatus bavaricum,* the only member of the Eurasian Bryodema group that was studied. Obviously AC can give rise to MC as his scheme shows. To argue that *Circotettix* and *Bryodema* are unrelated on the basis of chiasma localization in one population of *Bryodema* and to ignore substantial morphological similarities between *Circotettix* and *Bryodema* seems less than parsimonious. *D,* a scheme based on morphological, behavioral, and cytological data envisages the acrocentric condition giving rise to the metacentric condition more than once—at least once in the Bryodema group, at least three times in *Trimerotropis,* and once in *Microtes.* Change in chromosome number from 23 to 21 (2n in m) has also occurred independently in at least three different lineages within *Trimerotropis,* in *cyaneipennis, gracilis, verruculata, sparsa,* and *occidentalis.*

Genus CIRCOTETTIX Scudder

DISTRIBUTION. Rocky Mountains and Great Plains in United States and Canada to West Coast.

TAXONOMY. The genus *Aerochoreutes* is here treated as a synonym of *Circotettix.* Previously the distinctness of these two genera was based on the nature of the anal veins, most of which were said to be swollen in *Aerochoreutes* and only a few of which were swollen in *Circotettix.* But in *A. maculatus* there is no pronounced swelling of the anal veins. Most *Circotettix,* including *carlinianus,* have twenty-one chromosomes (2 N♂); only *maculatus* and *stenometopus* have twenty-three.

RECOGNITION. Figs. 49*m, n, o,* 60*a–d.* Forewings with strongly developed posterior lobe, lateral margin concave. Some or most anal veins of hindwings are thickened in males (except in *maculatus*).

BIOLOGY. Males of all species perform noisy flight displays, evidently to attract females. In some species (*undulatus, rabula*) males make wide circling flights; in others (*carlinianus, maculatus*) they hover over one site. Most species live on mountain slopes, road or railroad cuts, or sloping eroded ground, but *carlinianus* often inhabits dry rolling prairies and wheat fields.

REFERENCES. Rehn 1921, White 1951, Rentz and Weissman 1980, Weissman and Rentz 1980.

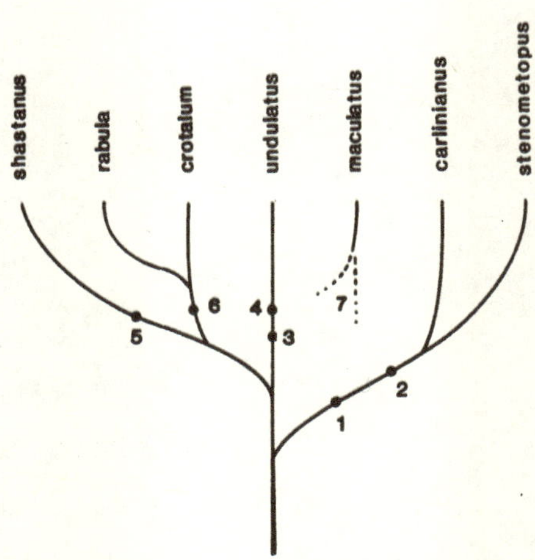

Fig. 59. Hypothetical phylogeny of *Circotettix* species (see Appendix 4 for explanation).

• Identification of Circotettix Species

maculatus (higher elevations of Sierra Nevada and San Bernardino mountains, California) Fig. 60*f*

1. Hindwings mostly blackish, sometimes clear in distal quarter in females
2. Vein $2A_a$ of hindwing: anterior branch present
3. Male hindwings without thickened anal veins
4. Hind tibiae yellowish in distal half, two dark bands in proximal half, pale ring near base

carlinianus (Colorado to British Columbia) Fig. 60*b*

1. Hindwings transparent or with black basal area

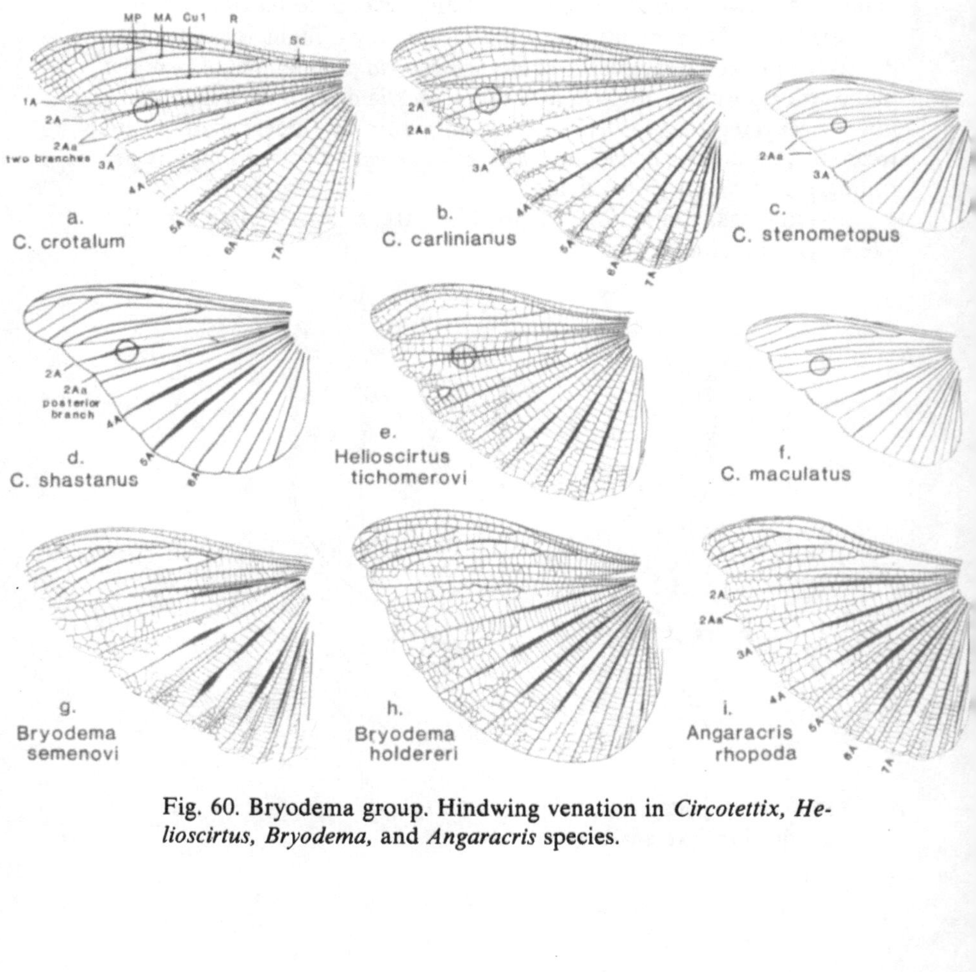

Fig. 60. Bryodema group. Hindwing venation in *Circotettix, Helioscirtus, Bryodema,* and *Angaracris* species.

2. Vein $2A_a$ of hindwing: anterior branch present
3. Male hindwings with veins 3A through 8A black and thickened
4. Hind tibiae yellowish

stenometopus (California: mountains west of northern Central Valley) Fig. 60*c*

1. Hindwings yellowish in basal area, dark band visible only in anterior lobe
2. Vein $2A_a$ of hindwing: anterior branch present
3. Male hindwings with veins 6A, 7A, and 8A thicker than 3A, 4A, and 5A
4. Hind tibiae yellow, black on inner face at base

rabula (Rocky Mountains from New Mexico to British Columbia)

1. Hindwings yellow in basal area with distinct to indistinct central dark band. (Vein Cu_1 closer to MP than to 1A, compared to *undulatus*)
2. Vein $2A_a$ of hindwing: anterior branch present
3. Male hindwings with veins 4A and 5A thickened
4. Hind tibiae yellowish or pale brown over most of range, bluish in northern Arizona

crotalum (Nevada) Fig. 60*a*

1. Hindwings yellow in basal area and with distinct central dark band. Cu_1 vein closer to MP vein than to 1A
2. Vein $2A_a$ of hindwing: anterior branch present
3. Male hindwings with veins 4A and 5A somewhat thickened
4. Hind tibiae bluish

shastanus (Sierra Nevada range from California to Oregon) Fig. 60*d*

1. Hindwings yellowish in basal area, dark central band somewhat indistinct, especially near middle. Cu_1 vein closer to MP than to1A
2. Vein $2A_a$ of hindwing: anterior branch absent (or very short)
3. Male hindwings with veins 4A and 5A thickened, 5A thicker than 4A
4. Hind tibiae blue

undulatus (Utah and Nevada north to Washington, Idaho, and Montana)

1. hindwings greenish yellow or faintly greenish blue, central dark band very indistinct or obsolete. Vein Cu_1 about equidistant between MP and 1A
2. Vein $2A_a$ of hindwing: anterior branch well developed, often as thick as 2A
3. Male hindwings with veins 4A and 5A thickened
4. Hind tibiae blue or gray

Circotettix rabula Rehn and Hebard — Pl. 19

TAXONOMY. Rehn's *Circotettix coconino* is here treated as a synonym of *rabula.* The differences between these nominal species are very slight, and I am not able to separate them on morphological bases. Specimens assigned to *coconino* are all from northern Arizona, and the majority of them have blue tibiae (*rabula* usually has yellowish tibiae). But the presence of a few northern Arizona specimens with yellowish tibiae and the lack of other strong distinguishing features leads me to the present synonymy. Rehn (1921) recognized three subspecies of *rabula:* the nominal subspecies, *rabula nigrafasciata* Beamer, and *rabula altior* Rehn. Interested readers should consult Rehn (1921) for his discussion of differences among the three and of geographic variation within each. I have examined all of the specimens Rehn studied, as well as others collected since. Some geographic variation can be seen, but many specimens cannot be assigned to one of the three names with certainty. Consequently I have not used subspecies names.

DISTRIBUTION. Widespread over the Rocky Mountain states from New Mexico and Arizona north to Alberta and Saskatchewan, extending eastward into the short-grass prairies.

RECOGNITION. Body light gray. Forewing crossbands usually faintly indicated, sometimes obsolete. Hindwings yellow in basal area and with a moderately distinct central dark band. Hind tibiae usually yellowish, but bluish in northern Arizona. Male hindwings with veins 4A and 5A enlarged. Differs from *crotalum* in the proportions of the prozona and metazona and in the relative length of the forewings; *crotalum* always has blue tibiae. Differs from *undulatus* in hindwing coloration. Differs from *shastanus* in wing venation. Body length to end of forewings 30–40 mm in males, 34–49 mm in females.

HABITAT. Sparsely vegetated rocky hillsides, stream and river banks, road cuts, and eroded hills. Ranges from low-lying plains into pine and spruce zone.

BEHAVIOR. *C. rabula* produces one of the loudest flight crepitations known in grasshoppers. Males generally fly loudly back and forth, circling over stretches of eroded or rocky hillsides. Crepitation is broken up into bursts lasting up to several seconds. The males fly from a few feet up to thirty feet above the ground. Although generally hovering over one area, they occasionally cover as much as 150 to 300 feet during one flight. Males seem to stimulate one another to fly. Long silent periods alternate with periods when several males fly simultaneously (Otte 1970).

LIFE CYCLE. Egg-overwintering. Adults from late June to September.

REFERENCES. Rehn 1921, Brooks 1958, Otte 1970, Scoggan and Brusven 1972.

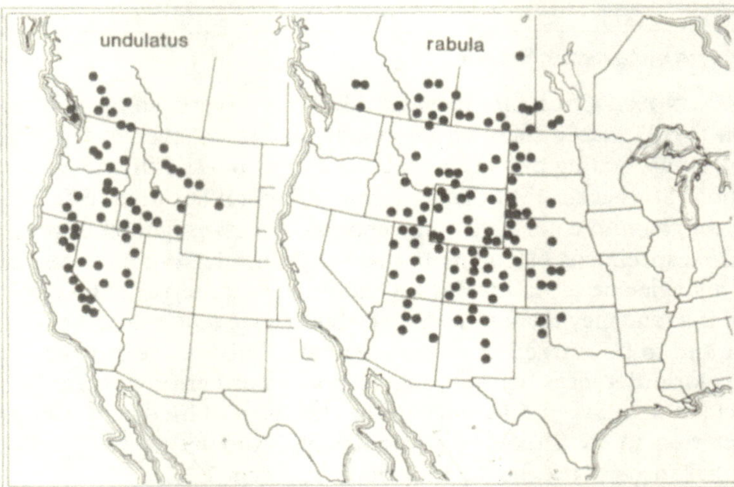

Circotettix rabula and *C. undulatus*

***Circotettix crotalum* Rehn** Pl. 19

DISTRIBUTION. Presently known only from extreme southern Nevada (from Lee Canyon, Spring Mountains, north of Las Vegas, and from Charleston Peak, west of Las Vegas).

RECOGNITION. Fig. 60*a*. Body gray to dark gray. Forewings without darker crossbands. Hindwings yellow in basal area and with a distinct dark central band. Hind tibiae blue. Differs from *rabula* in the proportions of the prozonal and metazonal lengths and wing length. Differs from *shastanus* by having a branched $2A_a$ vein on the hindwing. Differs from *undulatus* in having yellow hindwings. Body length to end of forewings 34–39 mm in males, 37–40 mm in females.

HABITAT. Mountain slopes with pine, 8,000–11,000 feet. Rehn describes several habitats as follows: "at 7,200 feet . . . a park-like region of juniper and pinyon with sage brush, and open areas showing some bare gray limestone; at 8,000 to 8,500 feet a region of yellow pine forest with small glades and some cut-over areas, but as a whole one of shadows and reduced light, with limestone of the mountains largely mantled with needles and dry soil."

LIFE CYCLE. All adults were collected between June 22 and August 19.

REFERENCES. Rehn 1921.

***Circotettix shastanus* Bruner** Pl. 19

TAXONOMY. *Circotettix splendidus* Rehn is a synonym of *shastanus.* Rehn (1921) placed these two nominal species together in the Shastanus group, stating that their exact relationship was still under investigation. Strohecker, Middlekauff, and Rentz (1968) note differences between them in color and wing venation. An investigation of several hundred specimens belonging to the two groups leads me to conclude that a single species is involved. Coloration is quite variable, though there is a tendency for southern specimens (*splendidus*) to be lighter in color and to be marked more contrastingly. This can be attributed to soil color differences between southern and northern mountain habitats. I see no marked difference in the thickness of hindwing vein 2A, as referred to by Strohecker, Middlekauff, and Rentz (1968). Both nominal species lack the anterior branch of vein $2A_a$.

DISTRIBUTION. San Bernardino and Sierra Nevada ranges in California, north to Oregon.

RECOGNITION. Fig. 60*d.* This is the only *Circotettix* species that lacks the anterior branch of vein $2A_a$ on the hindwing. Body pale gray

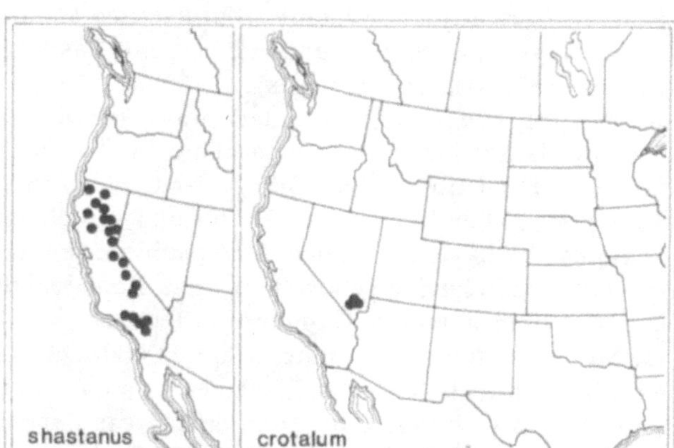

Circotettix crotalum and *C. shastanus*

and spotted (southern region) to darker gray (northern region). Forewings without noticeable dark crossbands. Basal area of hindwings yellow, dark band somewhat indistinct, especially in central region. Male hindwings with veins 4A and 5A thickened. Hind tibiae blue. Body length to end of forewings 31–37 mm in males, 32–40 mm in females.

HABITAT. Sunny clearings with exposed soil in mountain woodland from 4,000 to 10,000 feet.

LIFE CYCLE. Egg-overwintering. Adults from June to September, most abundant in July and August.

REFERENCES. Rehn and Hebard 1909, Rehn 1921, Strohecker, Middlekauff, and Rentz 68, Otte 1970.

Circotettix undulatus **(Thomas)** Pl. 19

TAXONOMY. Strohecker, Middlekauff, and Rentz (1968) placed *thalassinus* as a subspecies of *undulatus,* stating that *undulatus* presents a cline from Montana through southern Idaho, northern California, and Nevada to Inyo County, California. Here *thalassinus* is treated as a synonym.

DISTRIBUTION. Western California and Nevada north to British Columbia, east to Montana and Wyoming.

RECOGNITION. Body pale gray to dark gray-brown, often strongly speckled. Forewings without darker crossbands. Hindwings pale greenish yellow through most of the northwestern states, faintly greenish blue in parts of Idaho, Oregon, northern California, and Nevada, sky-blue in parts of eastern California and western Nevada. Dark central band of hindwings barely indicated or obsolete. Vein $2A_a$ of hindwings has well-developed anterior branch, almost as thick as 2A and often joining 2A distally. Hind tibiae blue, gray, or greenish. Usually bluest in blue-winged individuals. Body length to end of forewings 35–40 mm in males, 35–44 mm in females.

HABITAT. Stony and gravelly hillsides, often among juniper, piñon pine and other conifers. In Idaho Scoggan and Brusven (1972) note: "*C. undulatus* is a cliff-loving species. It thrives on extremely rocky areas or steep talus slopes. It can be found in nearly all major vegetational zones of the state. Preferred habitats are scantily vegetated, usually with *Festuca idahoensis, Carex* sp. and numerous forbs including *Phlox* sp., *Penstemon* sp., and *Lupinus* sp. Few grasshopper species occur in these areas."

BEHAVIOR. Unlike other species of *Circotettix, undulatus* males perform a two-part crepitation flight consisting of a series of widely spaced snapping sounds (estimated at three to five per second) followed by a rapid buzz (estimated at fifteen to twenty-five per second). During a single flight a male may alternate between two crepitations several times as he flies and circles about an area. Flights may last from a few seconds to half a minute or more and cover hundreds of square meters.

REFERENCES. Strohecker, Middlekauff, and Rentz 1968, Scoggan and Brusven 1972.

***Circotettix maculatus* (Scudder)** Pl. 19

DISTRIBUTION. Higher elevations of the Sierra Nevada and San Bernardino ranges of California.

RECOGNITION. Fig. 60*f*. This species differs from other *Circotettix* in having largely black hindwings; they are entirely blackish in males, transparent in the outer third in females. Unlike all other *Circotettix* species, in *maculatus* the hindwing anal veins are not thickened. Body generally dark, strongly speckled with light and dark gray; spots on forewings often aggregated to form indistinct crossbands. Hind tibiae banded; beginning basally, the bands are black, then pale, then blackish in basal half; mostly yellowish in distal half. Surface of body covered with fine hairs. Body length to end of forewings 26–29 mm in males, 30–34 mm in females.

HABITAT. Strohecker, Middlekauff, and Rentz (1968) report: "This interesting black-winged grasshopper is commonly heard in the higher

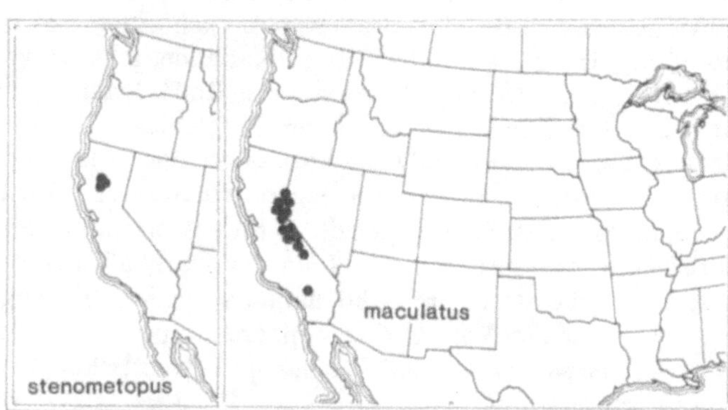

Circotettix maculatus and *C. stenometopus*

elevations of Sierra Nevada and the San Bernardino Mountains. It frequents granitic slopes and its loud rattling-type stridulation is startling at first to the unsuspecting hiker since the grasshopper's sound is not greatly unlike that of a rattlesnake. Males can be seen hovering for short periods in stridulation above resting females."

LIFE CYCLE. Adults from early July into September.

REFERENCES. Rehn 1921, Strohecker, Middlekauff, and Rentz 1968.

***Circotettix carlinianus* (Thomas)** Pl. 19

DISTRIBUTION. Colorado north to Manitoba, Saskatchewan, and Alberta and west to California, Oregon, and Washington.

TAXONOMY. Subspecies names are not useful. Two subspecies of *carlinianus* were recognized by Rehn (1921) and subsequent authors. An eastern form with shorter wings, *carlinianus carlinianus,* occurs in the higher Great Plains region and extends into the Rocky Mountains in Montana and western Colorado. A western form, *carlinianus strepitus,* was described from the Great Basin and Green River regions. Rehn describes specimens intermediate between these two forms from parts of Washington, Idaho, Montana, British Columbia, and Salt Lake Valley, Utah. In Canada Brooks (1958) notes: "In southern Saskatchewan and Alberta the two forms occur together, the minor form on bare, black-mud hillsides by preference, and the major form on flat, sandy-alkaline outwashes where sage brush and greasewood are abundant. Intermediates between the two are very common, and usually more numerous than the parent types in areas where the two occur." Obviously, the species does not divide into two distinct, geographically separated, units.

RECOGNITION. Fig. 60*b*. Hindwings largely transparent or with a black basal area. Male hindwings with veins 3A through 8A black and thickened. Crossbands of forewings faint or obsolete. Hind tibiae yellowish. Body length to end of forewings 29–40 mm in males, 33–46 mm in females.

HABITAT. Open dry prairies, desert grasslands, and eroded areas. Often found along roads and in wheat stubble fields. In Nevada, Rehn (1921) reports the species from "bare ground between scattered sage and other bushes on flats and talus slopes of valley rim," "on alkali encrusted adobe flat with scattered *Sarcobatus* bushes," and "on bare soil with small rock fragments on low rounded sage brush clothed hills"; in western Colorado from "bare ground with scattered chenopodaceous bushes and very sparse grass." In Idaho the species "prefers rocky soils in arid to semiarid habitats. It occurs commonly in the sage brush and shadscale communities, particularly in the foothills bordering the Snake River plains where it is most commonly associated with alluvial plains and moraines. Preferred habitats are sparsely vegetated and contain either *Atriplex confertifolia* or *Artemisia tridentata* as the major shrub. Principal herbaceous species in these areas are *Poa secunda, Oryzopsis hymenoides, Agropyron* sp. and numerous Chenopodiaceae" (Scoggan and Brusven 1972).

BEHAVIOR. Males have a dramatic flight display in which they hover over one spot for up to a minute. The display consists of long bursts of crepitation (0.7–1.3 seconds) interspersed with one to three short bursts (about 0.07 seconds). During long bursts the males ascend, and during silent intervals they descend, thus producing a continuous up-and-down dancing movement (Otte 1970). Occasionally one hears massive choruses of performing males. In August 1973 a stubble wheat field on the Colorado plains was silent for nearly an hour after I arrived. Then the whole field became active with hundreds of males performing their hovering displays. When I attempted to approach a male, he would drop to the ground.

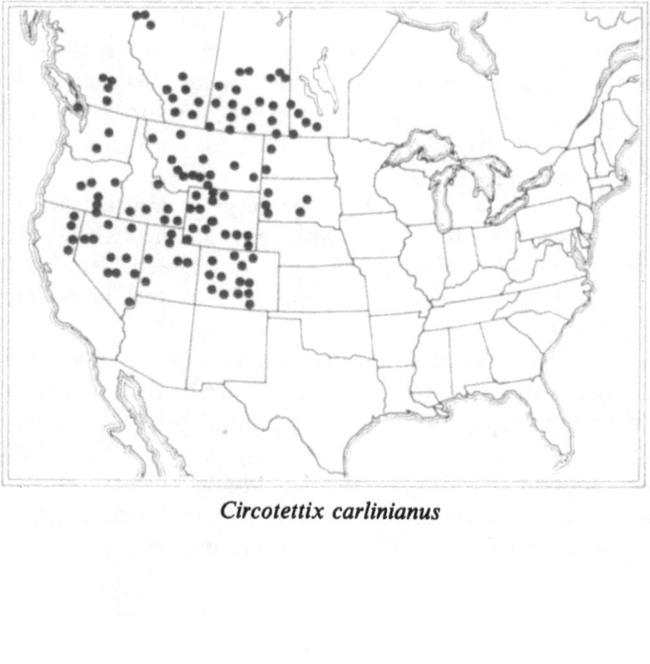

Circotettix carlinianus

LIFE CYCLE. Egg-overwintering. Adults from June to September. In Idaho first nymphal instars appear from late May to mid-June, and adults from late July to mid-August.

REFERENCES. Rehn 1921, Brooks 1958, Otte 1970, Scoggan and Brusven 1972.

***Circotettix stenometopus* (Strohecker and Buxton)** Pl. 19

DISTRIBUTION. Glen, Mendocino, and Trinity counties in the coastal ranges of northern California.

RECOGNITION. Fig. 60*c*. Hindwings pale yellowish in basal area; anal veins yellowish basally and becoming brown distally. Male hindwings with veins 6A, 7A, and 8A thicker than 3A, 4A, and 5A. Hind tibiae yellow but black at base on inner face. Body length to end of forewings about 29 mm in males.

HABITAT. Strohecker, Middlekauff, and Rentz (1968) report this species from mountain meadows. They also note: "*stenometopus* possesses a buzzing type of stridulation not greatly unlike that of *C. maculatus*. It flies rather close to the ground, seldom exceeding ten to twenty feet."

LIFE CYCLE. Adults in August and September.

REFERENCES. Strohecker and Buxton 1963, Strohecker, Middlekauff, and Rentz 1968.

Anconia Genus Group

Hebard placed this genus in the Group Anconiae with *Xeracris, Coniana,* and *Spaniacris* (unpublished synonymic checklist). Subsequently *Xeracris* and *Coniana* were placed in the Gomphocerinae (Jago 1971, Otte 1981), and *Spaniacris* was placed in the subfamily Romaleinae (Rehn and Grant 1961).

The position of the genus *Anconia* remains uncertain. By the male genitalia it is perhaps closest to *Trimerotropis* and *Sphingonotus,* but its peculiar long pronotum and posteriorly rounded pronotal disk are distinctive. With its smallish head and obsolete median pronotal carina, it is more like *Hadrotettix* than *Trimerotropis.* Here I follow Hebard and leave it in a separate group, but by itself.

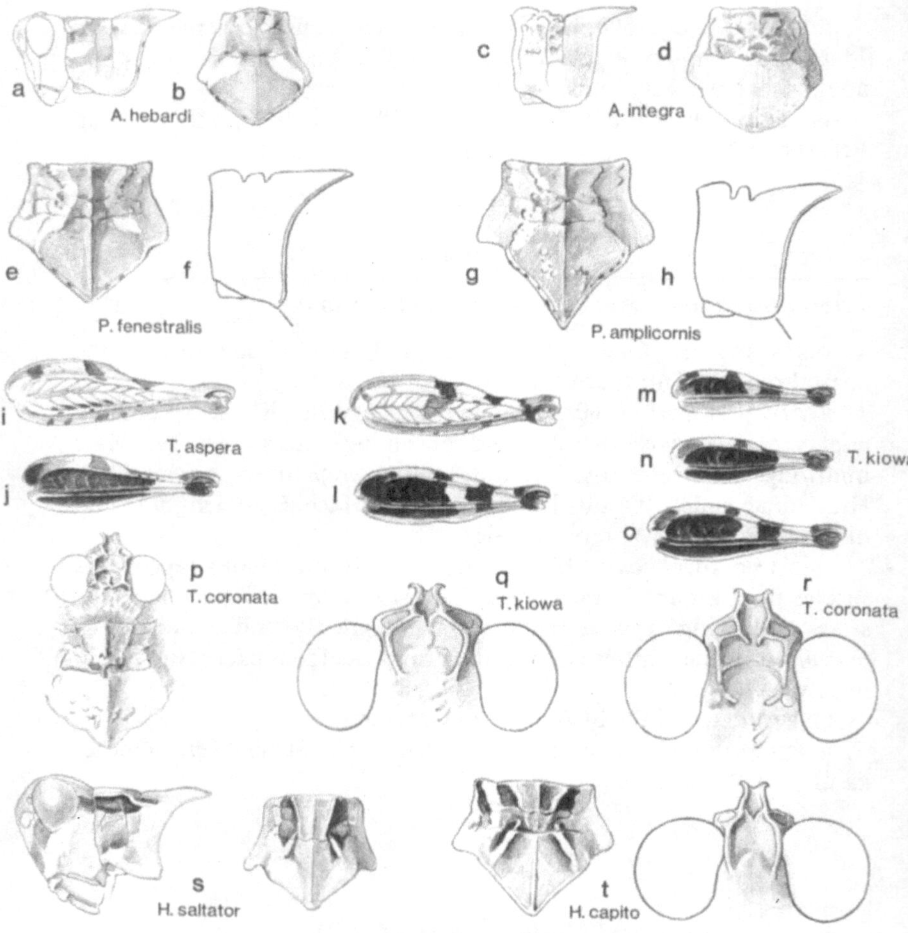

Fig. 61. ***Anconia, Psinidia, Trachyrhachys,*** and ***Hippopedon.*** Head, hind femur, and pronotum.

Genus ANCONIA Scudder

DISTRIBUTION. Deserts of southwestern United States and northwestern Mexico.

RECOGNITION. Fig. 61*a–d.* Pronotum elongated. Pronotal disk rounded posteriorly. Hindwings clear or bluish and without a dark band. Forewings with or without faint crossbands.

REFERENCES. Ball et al. 1942, Tinkham 1948.

• Identification of Anconia Species

integra (Sonoran Mojave deserts) Fig. 61*c, d*

1. Hindwings transparent, sometimes very faintly blue
2. Pronotal disk smooth, not lumpy
3. Inner face of hind femora without strong dark bands

hebardi (Chihuahuan desert) Fig. 61*a, b*

1. Hindwings sky-blue
2. Pronotal disk lumpy
3. Inner face of hind femora with strong dark bands

Anconia integra Scudder Pl. 22

TAXONOMY. This species was described and named three times. I have examined the types of *grisea* Bruner and *caeruleipennis* Bruner and all specimens of the latter determined by Hebard; these names are here considered synonyms because of difficulties in separating specimens from *integra.* The features Bruner uses to distinguish the species are based on a male of *grisea* and a female of *caeruleipennis;* one of the characters used was body size! "Possibly *A. grisea* is the male of [*caeruleipennis*], but if so the discrepancy in size between the sexes is much greater than it is in *A. integra.*"

DISTRIBUTION. Nevada and southeastern California to western Arizona and tip of southern California.

RECOGNITION. Fig. 61*c, d.* Hindwings transparent and without a dark band; basal area sometimes slightly bluish. The following color morphs occur: entirely pale green (uncommon); gray or pale gray and ivory with darker gray spots (common); entirely pale brown or straw-yellow (uncommon). Forewings without darker crossbands. Inner face of hind femora usually without dark band, sometimes lightly banded. Dorsum of pronotum smooth. Hind femora distinctly narrower than in *hebardi.* Body length to end of forewings 25–33 mm in males, 36–45 mm in females.

HABITAT. Ball et al. (1942) called this species the alkali grasshopper because it inhabits shrubs such as *Suaeda, Atriplex canescens, A. polycarpa,* and *Sarcobatus vermiculatus* on Arizona alkali flats. It is one of the few oedipodine species that does not live on the ground.

LIFE CYCLE. In Arizona nymphs from October to June and adults from March to July. In southern California adults from March to October.

REFERENCES. Ball et al. 1942, Strohecker, Middlekauff, and Rentz 1968.

***Anconia hebardi* Rehn** Pl. 22

DISTRIBUTION. Northern Mexico, western Texas, and southern New Mexico.

RECOGNITION. Fig. 61*a, b.* Hindwings sky-blue and becoming transparent distally. Body very pale brown to pale gray and mottled with darker markings. Pronotum rougher or more bumpy than in *integra.* Inner face of hind femora with two dark bands (excluding knees). Hind tibiae banded with gray and ivory, especially on inner face. Body length to end of forewings 27–30 mm in males, 39–42 mm in females.

HABITAT. Unlike *integra,* this species lives on the desert floor and on adobe flats. Tinkham (1948) describes the habitat in the Rio Grande Valley as follows: "The habitat was restricted to a dry arroyo bed and the adjacent gravelly areas on the adobe flat at the base of the mesa. The few plants in the immediate vicinity were mainly dense clumps of low mesquite and small isolated bushes of *Atriplex,* probably *canescens.*"

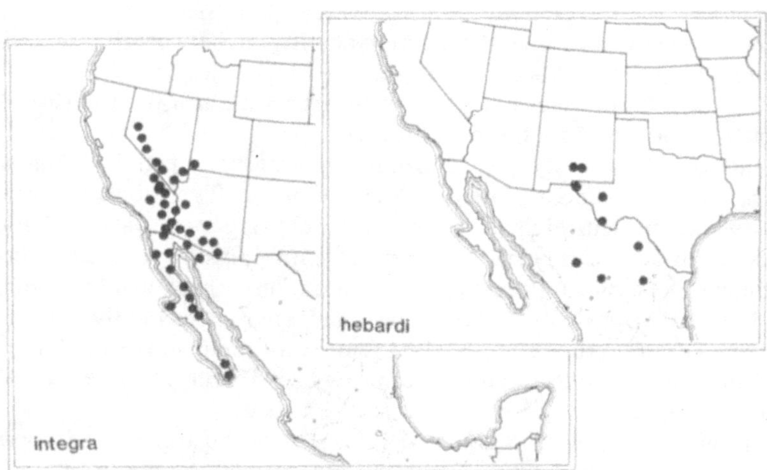

Anconia integra and *A. hebardi*

BEHAVIOR. Tinkham (1948) describes escape and reproductive behavior: "Many males were taken in the arroyo bed where their bluish gray bodies blended perfectly with the bluish gray stones of the arroyo and only by disturbing them and approaching the exact spot where they had alighted were they captured. The males were not in the habit of flying far when disturbed, usually flying ten or fifteen feet then alighting. This species when alighting makes a faint crackling noise. Any males within a radius of eight or ten feet of an alighted female usually approached her by a series of several short hopping flights and then at a distance of one or two feet more warily approached her."

LIFE CYCLE. Overwinters in the late nymphal stages. Adults from late April to July.

REFERENCES. Rehn 1919, Tinkham 1948.

Psinidia Genus Group

Eight North American genera are placed in this group: *Psinidia, Trachyrhachys, Hippopedon, Metator, Mestobregma, Trepidulus,* and *Derotmema.* From the Old World at least *Acrotylus* also belongs to the group. The position of *Derotmema* remains somewhat in doubt because its species do not possess trilobed posterior epiphallic lophi, as all the other genera do (Fig. 67). Pronotal and wing characters, however, suggest a close relationship to this group.

Within the group several apparently monophyletic subgroupings can be discerned (Fig. 62). *Trachyrhachys, Hippopedon,* and *Metator* form one subgroup; *Mestobregma, Trepidulus,* and perhaps *Derotmema* form a second subgroup; and *Psinidia* forms a third group.

RECOGNITION. Figs. 61, 71. Male epiphallus with trilobed lophi (except *Derotmema*). Pronotal crest cut by two sulci. Head held distinctly higher than pronotum. Frontal ridge of head usually deeply cut. Eyes large and globular. Top of head often with a black streak between the eyes. Middle lobe of pronotum often with three protuberances near median line. Forewings very narrow in some species. Hindwings narrow, with or without dark bands.

▪ Key to Psinidia Group Genera

1. Antennae ensiform. Forewings without crossbands. Hindwings with broad dark bands (Pl. 20): **Psinidia**
   Antennae filiform. Forewings with dark bands on lateral face. Hindwings with narrower dark bands: **2**
2. Cerci spoon-shaped (Fig. 64*e, f*): **Metator**
   Cerci conical: **3**

3. Color pattern on lateral lobe as in Fig. 8*h, i, j:* **4**
   Color pattern on lateral lobe as in Fig. 8*k, l:* **6**
4. Posterior margin of lateral lobe rounded (Fig. 8*j*): **Derotmema**
   Posterior margin of lateral lobe angulate (Fig. 8*h, i*): **5**
5. Pronotal color pattern as in Fig. 8*h*. Hind femora narrow (Pl. 20): **Hippopedon**
   Pronotal color pattern as in Fig. 8*i*. Hind femora wider (Pl. 20): **Trachyrhachys**
6. Hindwings without dark bands (Fig. 8*n*): **Trepidulus**
7. Hindwings with dark bands (Fig. 8*m*): **Mestobregma**

Genus PSINIDIA Stål

DISTRIBUTION. Atlantic and Gulf coasts, along some interior rivers and southern shores of the Great Lakes.

RECOGNITION. Figs. 8*f*, 61*e–h*. Antennae ensiform. Forewings without crossbands. Hindwings orange to rosy red in basal area and with a broad outer dark band.

REFERENCES. See under each species.

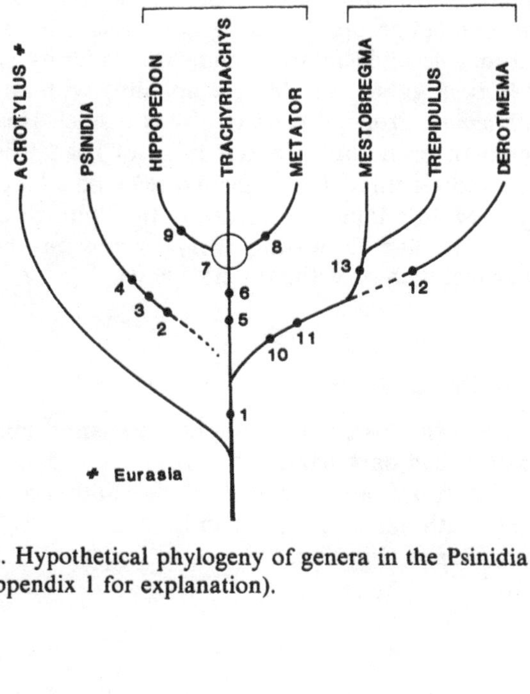

Fig. 62. Hypothetical phylogeny of genera in the Psinidia group (see Appendix 1 for explanation).

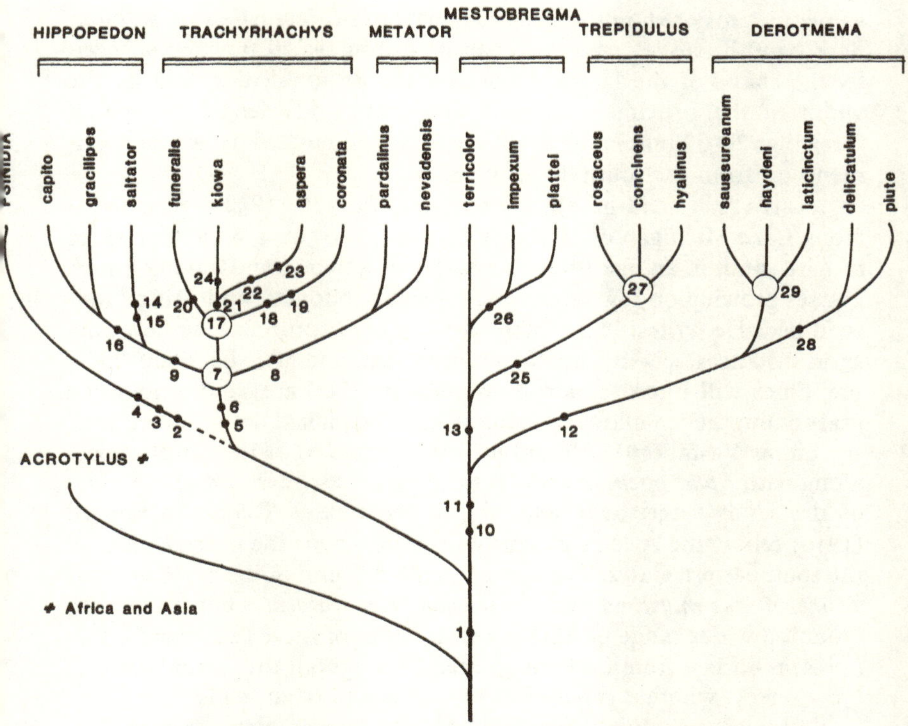

Fig. 63. Hypothetical phylogeny of species in the Psinidia group.

• Identification of Psinidia Species

fenestralis (Texas to New England) Fig. 61*e, f*

1. Posterior margins of pronotal disk straight
2. Lower posterior margin of lateral lobe comes to a point

amplicornis (eastern half of Texas) Fig. 61*g, h*

1. Posterior margin of pronotal disk concave, posterior angle of disk acute-angulate
2. Lower posterior margin of lateral lobe rounded

Psinidia fenestralis (Serville) Pl. 20

DISTRIBUTION. Eastern third of United States, along the shores of lakes and rivers.

RECOGNITION. Fig. 61*e, f.* Basal area of hindwings usually orange, sometimes rosy red; in the northeastern states rarely pale yellow; outer dark band broad. *P. fenestralis* can be distinguished from *amplicornis* by the shape of the lateral lobes and the pronotal disk, and by the width of the proximal antennal segments. Body length to end of forewings 20–32 mm in males, 26–36 mm in females. The largest specimens are from the south Texas coast.

HABITAT. A sand-inhabiting species. Hubbell (1922b) reported it along Lake Michigan on the edges of sandy blowouts, on a steep slope of bare sand, in an oak forest bordering a blowout, and among bunch grasses growing on low dunes and on an abandoned sand-drifted railroad bed. He writes: "Specimens were easily captured; they flew only short distances, usually less than twenty feet, and after being flushed a few times will often remain motionless on the bare sand among the grass clumps, even allowing themselves to be picked up in the fingers." In Illinois Vestal (1913) found it in bunchgrasses of the sand prairies along with *Spharagemon collare.* In Iowa Froeschner (1954) reports it on dry sandy ridges near lakes and larger streams. Rehn and Hebard (1916) report the species in sandy situations over the coastal plain of the southeastern states. The species is often found in the same areas as *Trimerotropis maritima* and *Spharagemon marmorata,* but it seems to inhabit a wider range of habitats than either of these two species. Fox (1914) reports it from bare sandy areas throughout the coastal plain of New Jersey, where it inhabits both woods and open fields.

LIFE CYCLE. Adults from early July to middle November in Alabama, from June to October in Georgia, from March to October in Florida (but I suspect adults occur at least sporadically throughout the year), from May to December along the Texas coast, and from July to September in the northern and northeastern states.

REFERENCES. Vesta 1913, Fox 1914, Hubbell 1922b, Froeschner 1954, Otte 1970.

***Psinidia amplicornis* Caudell** Pl. 20

DISTRIBUTION. Presently known only from Texas.

RECOGNITION. Fig. 61 *g, h*.. Basal area of hindwings orange with a broad outer dark band, nearly identical to *fenestralis.* This species differs from *fenestralis* in the shape of the lateral lobes and the pronotal disk and in having wider antennae. Body length to end of forewings 22–27 mm in males, 23–34 mm in females.

HABITAT. Open woodland with sandy soil. This species lives on exposed sand among scattered vegetation.

LIFE CYCLE. Adults from April to November.
REFERENCES. Caudell 1903.

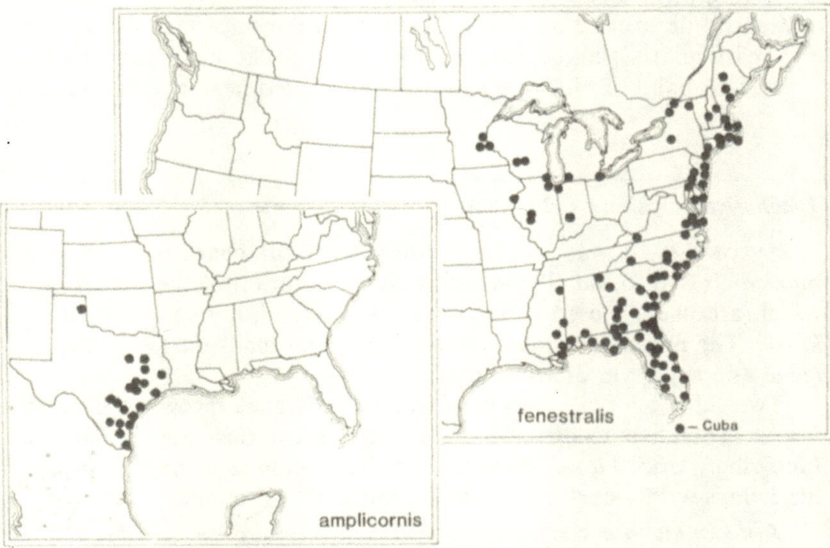

Psinidia fenestralis and *P. amplicornis*

Genus TRACHYRHACHYS Scudder

DISTRIBUTION. Central Mexico to southern Canada and most of the United States.

RECOGNITION. Figs. 8*i*, 61*i–r*. Most likely to be confused with *Metator* and *Hippopedon*. It lacks the spoon-shaped cerci of *Metator* and differs from *Hippopedon* as follows: FDE and FDI not always distinct; lateral carinae on metazona do not stand out as sharply defined white lines; fastigium distinctly wider; hind femora wider and without an oblique dark band on the medial area; middle lobe of pronotal crest distinctly crestlike.

▪ Key to Trachyrhachys Species

1. Hindwings almost entirely black (Michoacán, Mexico): **funeralis**
   Hindwings variable, never mostly black: **2**
2. Forewings with at least six rounded dark spots: **coronata**
   Forewings with at most three dark markings; these not round: **3**

3. Central dark band on upper face of hind femora distinctly slanting (Fig. 61*i*). Hind tibiae black at distal end. Dark band on hindwings, if present, located at periphery: **aspera**
   Central dark band on upper face of hind femora triangular or not particularly slanted (Fig. 61*j, k*). Hind tibiae not black at end. Dark band of hindwings, if present, located near center: **kiowa**

***Trachyrhachys kiowa* (Thomas)** Pl. 20

TAXONOMY. I was unable to discover what characters separate *mexicanus* and *kiowa.* The banding and coloration of the hindwings, which are usually used to separate them, are highly variable within *kiowa.* The presence of intermediate conditions leads me to treat *mexicana* as a synonym of *kiowa* (see Variation).

Two subspecies of *kiowa* have traditionally been recognized: *kiowa kiowa* (Thomas), recognized mainly by its colorless and unbanded hindwings, and *kiowa fuscifrons* (Stål), recognized by its yellow hindwings with a dark outer band. Smith (1981) reports:

> *T. kiowa kiowa* is common in the western two-thirds of the state . . . *T. kiowa fuscifrons* occurs in the eastern one-third of Kansas and has been recorded in Butler County. Hebard (1931) also reported intermediate forms from Butler County and several western counties. Froeschner (1954) designated such intermediates as the subspecies *kiowa thomasi* Caudell. Intermediate forms were collected in Washington, Riley, Chase, and Butler counties. All three forms were found together at two sites and two forms were found together at three sites, suggesting that the Flint Hills are a zone of transition and intergradation for the two subspecies.

I have not applied the subspecific names here because in numerous cases it is impossible to assign a specimen to one name or another. Courtship behavior is different in Michigan and in Kansas, suggesting that more than one species is indeed involved. The matter requires further study.

DISTRIBUTION. Southwestern Canada south to central Mexico and, in the United States, from California east to Virginia. This is one of the most widely distributed species of the subfamily in North America.

RECOGNITION. Fig. 61*m–o, q.* Hindwings variable in color and shading. The following conditions can be recognized, but all intermediate conditions exist: (1) entirely colorless and transparent; (2) transparent but yellowish near base; (3) colorless in basal area and with an outer dark band that varies in intensity; (4) yellowish in basal area and with an outer dark band that varies in intensity. Dorsal field of forewings sometimes with posteriorly converging pale streaks, sometimes entirely pale, and sometimes no lighter than lateral field. Inner face of hind femora with either one black band proximal to the narrowest part (southeastern states and Mexico) or two black bands (most of the United States); both conditions exist in some populations. Hind tibiae are pale blue or blue-gray.

T. kiowa differs from *aspera* as follows: head relatively large; hind femora narrower; lower margin of lateral lobe more strongly slanted; median pronotal carina of pronotum distinctly lower. Body length to end of forewings, 20–25 mm in males, 23–30 mm in females.

VARIATION. Variation in hindwing color and pattern is striking. In general this variation is clinal from north and west to south and east. From Nebraska northward the hindwings are colorless. Westward, in Utah, Idaho, and California, traces of yellow color appear in the basal area, and remnants of an outer dark band can be discerned. The transition to yellow wings with an outer dark band occurs in Kansas, Oklahoma, and northern Texas; here the full range of variation in color and pattern can be seen. East of Nebraska and Kansas the hindwing is almost always yellow with a dark outer band, although transitional forms exist in Iowa as well. From central Texas south to Mexico the hindwings are usually yellow with an outer dark band, but in numerous specimens the basal area is more or less colorless and the outer dark band is retained. In northern Arizona the outer dark band is indistinct or obsolete, but in southern Arizona it is distinct.

BEHAVIOR. Courtship in yellow-winged specimens from the eastern states differs from that seen in Kansas specimens, leading to the notion that courtship also varies clinally or that two species are indeed involved. Males observed in an old stone quarry in southeastern Michigan (Otte 1970) made no conspicuous flight displays. Instead they ran about on the gravelly ground and at times stood motionless for many minutes. Occasionally males made short, low flights, seemingly spontaneously. As males walked about they raised (tipped) their hind femora. Males approached other individuals by walking and hopping; they were never seen to approach by flying. Courtship in Michigan and Maryland specimens was similar; during approaches, males produced single strokes of stridulation using both femora. Michigan males approached other males while producing a fast version of leg tipping (duration 0.2 seconds); the males being approached, meanwhile, produced a much slower tipping movement (duration more than 1.0 second); these were then followed by equally slow tipping by approaching males and separation of the males. Evidently stridulation and rapid tipping are courtship signals, while slow tipping is a male recognition signal.

Courtship in males from Sherman County, Kansas, was quite different. Four males shook their femora as they approached other individuals. Twice these males mounted. Each male approached and mounted a female more than five times without producing any acoustical signals. Although males did not stridulate during approaches, one of them produced three bursts of vibratory stridulation as he walked about the observation chamber. As males walk about they also make rapid tipping movements like those of the Michigan specimens, but I observed no slow tipping movements.

HABITAT. Bare gravelly ground. Blatchley (1920: 291) reports: "[In Indiana] it frequents old abandoned fields, roadsides and bare rocky slopes . . . it was found to be quite common on a slope leading up to a hemlock covered bluff in Owen County. This slope was nearly bare and thickly covered with small weathered fragments of grayish-brown sandstone. Among these the locust was resting, and visible only when it moved." In Michigan this species was found only on gravelly roadsides, gravel pits, and stone quarrys. In eastern Kansas, according to Smith (1981): "Most of the sites at which *T. kiowa* was collected were

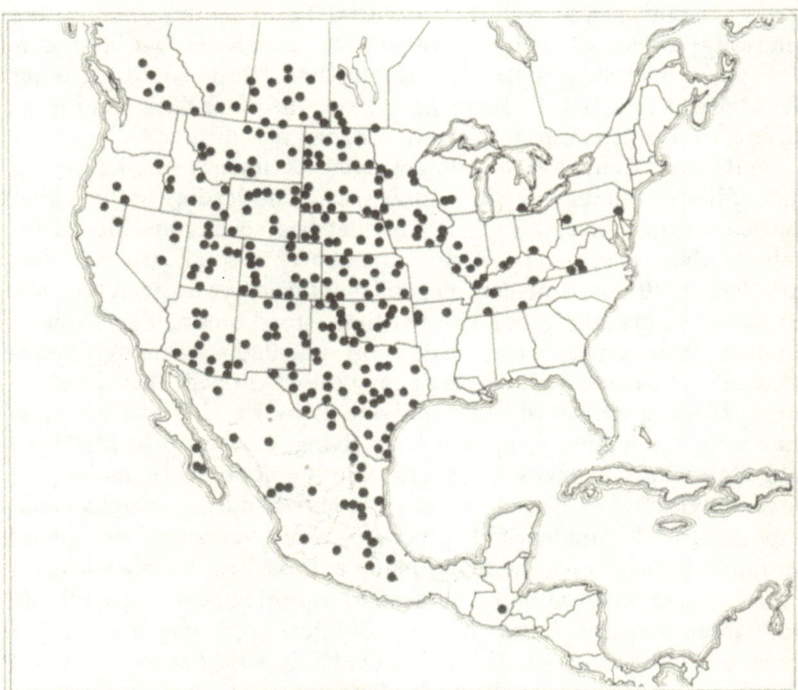

Trachyrhachys kiowa

bare or sparse areas. On prairie sites, it was found in rocky or bare spots or along roadways." In southern Arizona Ball et al. (1942), under *T. mexicana,* note: "At times one of the most common and destructive range grasshoppers in the southern grassland. Most abundant in areas of sparse grass." In northern Arizona, under *T. kiowa,* they say: "A very important range grasshopper, most common in sparse stands of grama grass." Scoggan and Brusven (1972) write: "In Idaho *T. kiowa* is most closely associated with sagebrush, but extends into both the wheatgrass-bluegrass and shadscale communities." Mulkern et al. (1969) report this species as a grass feeder preferring *Bouteloua gracilis.* In southern Arizona Otte and Joern (1975) found this species feeding on a wide variety of plants. In northern Dakota Hubbell (1922a) notes: "extremely abundant on the western plains, but somewhat less numerous in the east. It was fairly common in grassy fields and pastures, roadsides, cultivated fields, and on stony and sandy flats . . . On the western side of the state it was the most abundant of the bare-ground Oedipodinae in all of the localities where collecting was done. Its favorite habitat appears to be the dry uplands, where a considerable amount of bare soil is exposed among the scanty vegetation of grasses and xerophytic plants."

LIFE CYCLE. Egg-overwintering. In Idaho adults appear in late July. In eastern Kansas adults were found from June through September, in Arizona from June to November.

REFERENCES. Blatchley 1920, Hubbell 1922, Ball et al. 1942, Alexander and Hilliard 1969, Otte 1970, Scoggan and Brusven 1972, Smith 1981.

Trachyrhachys aspera Scudder Pl. 20

DISTRIBUTION. Wyoming south to central Mexico.

TAXONOMY. As one goes south in Mexico, specimens become smaller, the hindwings more opaque, the pronotum smoother, and the pronotal disk more pointed posteriorly. Bruner applied the name *inconspicua* to these specimens.

RECOGNITION. Fig. 61*i, j.* Basal area of hindwings pale greenish yellow or yellow, somewhat opaque, outer dark band located at periphery of wing. In the United States this species is rather compact in shape and has a relatively large head. In Mexico it is smaller and the hindwings are more opaque. It differs from *kiowa* as follows: (1) lower edge of lateral lobe not as slanted; (2) median pronotal carina on metazona distinctly higher; (3) hind femora broader and with a marked transition to the narrower part; (4) dark bands on upper face of hind femora more slanted. As in *kiowa,* hind tibiae are blue or blue-gray, but they become black at distal end. Body length to end of forewings 16–23 mm in males, 21–27 mm in females.

HABITAT. In west Texas Tinkham (1948) notes: "This species is fairly common in the Davis Range. In the Chinati Mountains it was encountered only on the second highest peak of that range, Blue Mountain, and at the considerable altitude of 6,700 feet."

LIFE CYCLE. In Colorado and New Mexico adults were collected in August and September, in Texas from August to October, in Mexico from July to November.

REFERENCES. Tinkham 1948.

Trachyrhachys coronata Scudder Pl. 20

DISTRIBUTION. Southern Wyoming and western Nebraska south to Durango, Mexico. Four specimens in the Philadelphia Academy of Natural Sciences are labeled: "College Station, Brazos County, Texas, Aug. 13, 1921, C. D. Duncan, Collector." Because these are the only specimens form a region that has been intensively collected, it seems probable that they were mislabeled.

RECOGNITION. Fig. 61*k, l, r.* Hindwings opaque yellow or opaque orange in basal area and with a dark, well-defined outer band. Forewings with at least seven rounded dark spots. Hind femora with three inner dark bands. Dorsal margin of hind femora slightly indented at anterior dark band. Hind tibiae yellowish to greenish blue. This species is superficially similar to *Xanthippus corallipes* or *Pardalophora haldemani,* but those two species have yellow to red hind tibiae and a different median pronotal crest. Body length to end of forewings 22–25 mm in males, 26–32 mm in females.

HABITAT. Ball et al. (1942) report that the species is rare and found in short grass at the lower edge of the Transition Zone near Springerville, Arizona. Hebard (1929) notes that in Colorado the species is common on "high mesas near the mountain front and rare on a hill out on the plains."

REFERENCES. Hebard 1929, Ball et al. 1942.

Trachyrhachys funeralis Strohecker Pl. 20

DISTRIBUTION. This species is presently known only from the vicinities of Apatzingan and Capirio, Michoacán, Mexico. Thirty-nine specimens are in the Philadelphia Academy of Natural Sciences collection. Two specimens are in the Michigan Museum of Zoology.

RECOGNITION. Hindwings largely black, but with an interrupted pale band in the distal third; extreme base of hindwings often bluish. Forewings sometimes quite blackish, so pale bands are obscured. Inner face of hind femora largely black. Body length to end of forewings 24–30 mm in males, 32–39 mm in females.

HABITAT. Not recorded in the literature.

LIFE CYCLE. All adults of this species were collected in August.

REFERENCES. Strohecker 1945.

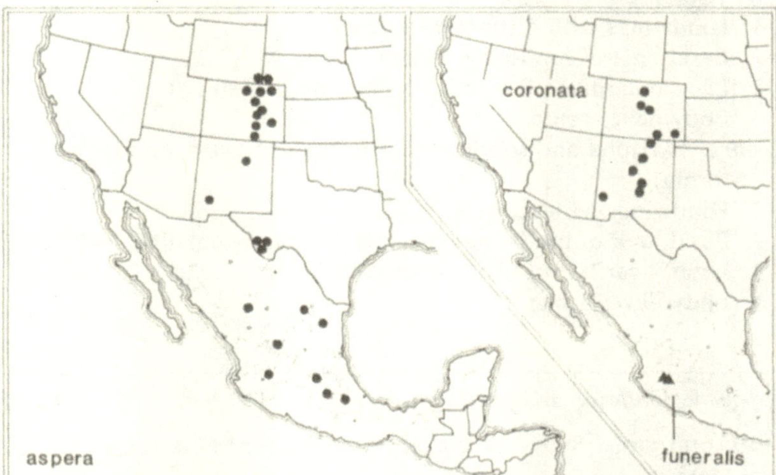

Trachyrhachys aspera, T. coronata, and *T. funeralis*

Genus HIPPOPEDON Saussure

TAXONOMY. The genus *Rehnita* Hebard is a synonym of *Hippopedon.* Hebard never compared *Rehnita* with *Hippopedon* perhaps because *H. saltator,* the type of *Hippopedon,* is from southern Mexico, while *gracilipes* and *capito* are from the United States and northern Mexico.

DISTRIBUTION. Southwestern United States to southern Mexico.

RECOGNITION. Fig. 61*s–v.* Most easily confused with *Trachyrhachys,* but differs as follows: FDE and FDI are distinct and form between them contrastingly pale lateral carinae; fastigium narrower; hind femora narrower and with an oblique dark band in medial area;

middle lobe of pronotal crest very low and uncrestlike or obsolete; hindwings with or without an outer dark band.

REFERENCES. Saussure 1861.

• Identification of Hippopedon Species

saltator (central and southern Mexico) Fig. 61*s, t*

1. Hindwings without dark band
2. Basal area of hindwings colorless or slightly pink
3. Lateral carinae of prozona not broken by first sulcus
4. Dorsum of body, head, and pronotum often green

capito (Colorado to central Mexico) Fig. 61*u, v*

1. Hindwings with distinct dark band
2. Basal area of hindwings yellow
3. Lateral carinae of prozona broken by first sulcus
4. Body never green

gracilipes (Arizona and southern California to Sonora and Baja California)

1. Hindwings without dark band
2. Basal area of hindwings variable from transparent to yellow
3. Lateral carinae of prozona broken by first sulcus
4. Body never green

Hippopedon saltator Saussure — Pl. 20

DISTRIBUTION. States of Oaxaca, Michoacán, Guerrero, and Jalisco, Mexico.

RECOGNITION. Fig. 61*s, t.* Body often with much green, especially on dorsal field of forewings, head, and pronotum. Hindwings without a dark band; basal area transparent or slightly pink. Lateral pronotal carinae not broken by the first sulcus. Body length to end of forewings 18–22 mm in males, 23–28 mm in females.

HABITAT. I collected the species on green mountain slopes with short grass in Oaxaca, Mexico.

LIFE CYCLE. Adults have been collected in April, July, August, and November.

REFERENCES. Saussure 1861.

Hippodedon gracilipes (Caudell) — Pl. 20

DISTRIBUTION. Sonoran Desert from southern Arizona and southern California to the tip of Baja California.

Hippopedon gracilipes and *H. saltator*

RECOGNITION. This species lacks a dark band on the hindwings; basal area usually yellow, but transparent or nearly transparent in some specimens. Head slightly smaller proportionately than in the largest specimens of *capito* from Texas. Aside from these features the two species are largely indistinguishable. Body length to end of forewings 18–23 mm in males, 23–30 mm in females.

HABITAT. Open, arid regions. Ball et al. (1942) record the species from "bare, dry, and often rocky areas in the Lower and Upper Sonoran Zones" of Arizona.

LIFE CYCLE. Ball et al. (1942) record adults from May to September in Arizona, but they are most abundant in July. In Baja California adults have been collected in January, April, July, September, October, and November.

REFERENCES. Caudell 1905, Hebard 1935c, Ball et al. 1942.

Hippodedon capito (Stål) Pl. 20

DISTRIBUTION. Colorado south to central Mexico.

RECOGNITION. Fig. 61*u, v*. This species differs from *saltator* and *gracilipes* principally in the distinct outer dark band on the hindwings. *H. gracilipes* and *capito* both have yellow basal area. In *saltator* the hindwings are transparent or sightly pink at the base. Body length to end of forewings 20–27 mm in males, 24–33 mm in females.

HABITAT. This species apparently usually inhabits slopes of hills and mountains and broken country. Tinkham (1948) reports: "This rare and graceful hopper was taken at about 5,000 feet on gently sloping hill tops in the Chinati Mountains." It has also been collected in the hill country of central Texas.

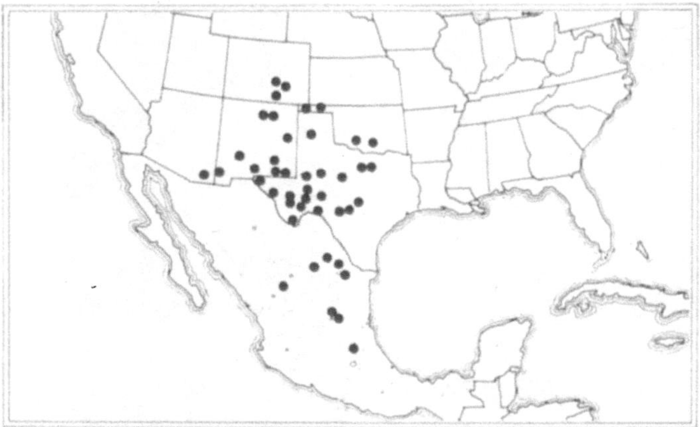

Hippopedon capito

LIFE CYCLE. Adults have been collected from July to October.
REFERENCES. Hebard 1935c, Tinkham 1948.

Genus METATOR McNeill

DISTRIBUTION. Great Plains to California and Oregon.

RECOGNITION. Fig. 64*a, b, d, f.* Very similar to *Trachyrhachys* in appearance but with spoon-shaped cerci. Frontal ridge very shallowly grooved or not grooved at dorsal extremity. Forewings tend to be more spotted than banded. Hind tibiae yellow or blue. Medial area of hind femora without crossbands.

REFERENCES. McNeill 1901.

Metator pardalinus (Saussure) Pl. 21

DISTRIBUTION. Widespread through the Great Plains and mountain states and the prairie provinces of Canada.

RECOGNITION. Fig. 64*a.* Hindwings always with a strong outer dark band; basal area varies from yellow to orange, deep orange, or rosy red. Hind tibiae blue. This species differs from *nevadensis* as follows: dark band on hindwings distinctly broader; forewings with numerous black spots (spots are arranged in three major dark bands in *nevadensis*); dorsal field of forewings with two very clearly defined, narrow, posteriorly converging pale streaks (in *nevadensis* either the entire dor-

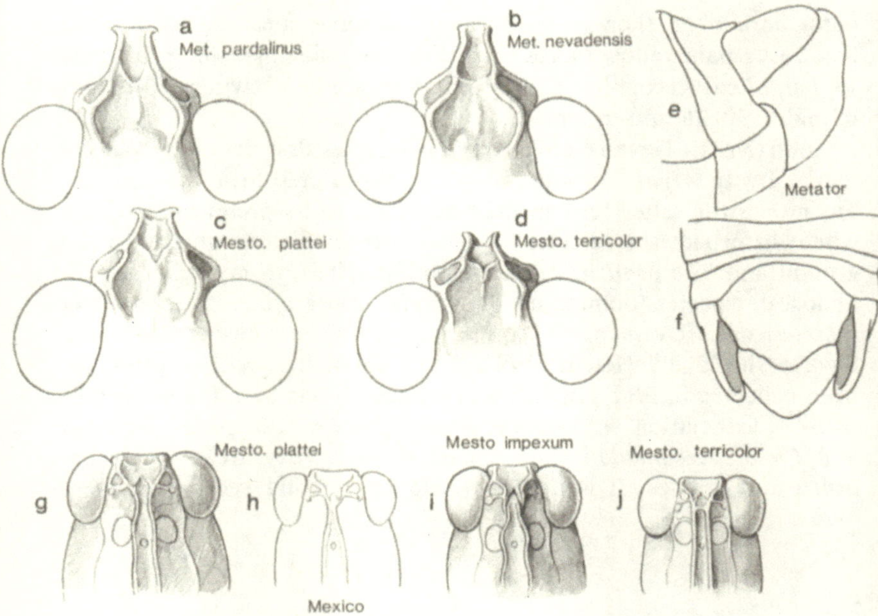

Fig. 64. *Metator* and *Mestobregma.*

sal field is pale, or the pale stripes are broader or less clearly defined). Body length to end of forewings 26–38 mm in males, 32–45 mm in females.

HABITAT. In Idaho, according to Hewitt and Barr (1967): "This is primarily a grassland species and is found both in mountain meadows and short grass prairies." The species inhabits prairies all across Nevada. Ball et al. (1942) consider the species "an important range grasshopper which is sometimes injurious to wheat and other grain crops." In Arizona it inhabits the grasslands of the Upper Sonoran and Transition zones.

LIFE CYCLE. In Arizona adults have been recorded from June 20 to September 1; in more northern states from July to September.

REFERENCES. Ball et al. 1942, Hewitt and Barr 1967.

Metator nevadensis **(Bruner)** Pl. 21

DISTRIBUTION. Northern California, Nevada, and Idaho.

RECOGNITION. Fig. 64*b*. Hindwings usually with a less distinct outer dark band than in *pardalinus,* but sometimes just as distinct; basal area pale yellow or rosy red. For other distinguishing features, see *pardalinus* Recognition. Body length to end of forewings 22–28 mm in males, 30–36 mm in females.

HABITAT. In Nevada LaRivers (1948) notes that *nevadensis* "is associated with *Arphia pseudonietana* in well-watered meadow bottoms, but unlike the latter seldom calls attention to its presence by flying when disturbed; the instinct of *nevadensis* is to drop quietly to the ground and lose itself in dense vegetation. [It is] in my experience a Shadscale species [dominated by *Atriplex confertifolia* and *Artemisia spinescens,* LaRivers p. 657] probably with some overlap into the Sagebrush Zone." Hewitt and Barr (1967) write: "This species has been collected in Idaho in July and August. It has been found on hillsides in association with *Trimerotropis gracilis, Cratypedes neglectus* and *Circotettix undulatus.* It appears to prefer dry areas with sagebrush and grasses. It is a grass feeder, preferring species of *Agropyron.*"

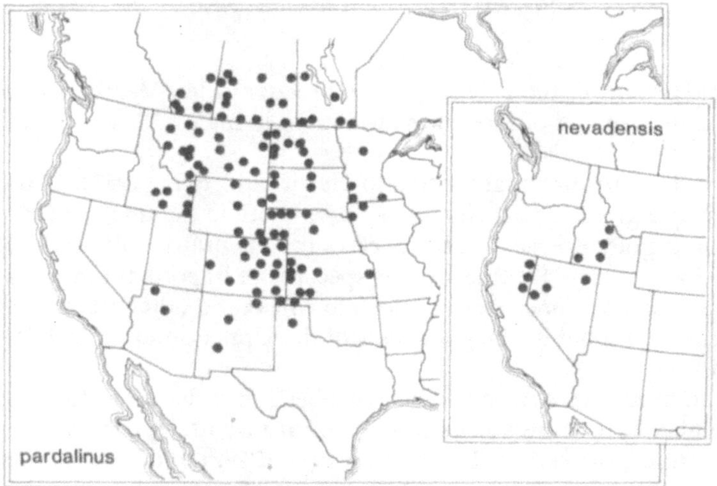

Metator pardalinus and *M. nevadensis*

LIFE CYCLE. Adults from July to September.

REFERENCES. LaRivers 1948, Hewitt and Barr 1967, Strohecker, Middlekauff, and Rentz 1968.

Genus MESTOBREGMA Scudder

DISTRIBUTION. Southeastern United States and northern Mexico, but not in California or Baja California.

RECOGNITION. Fig. 64*c, d, g–j.* Cerci flattened and spoon-shaped.

Mestobregma plattei (Thomas) Pl. 21

TAXONOMY. Rehn (1919) recognized three subspecies of *M. plattei: "plattei plattei* which is the form of the Great Plains region, south to southern Colorado; *plattei corrugata* (Scudder), ranging from northern New Mexico southward; and *plattei rubripenne* (Bruner) of central and southern Arizona. In our detailed projected study of the genus we will discuss the relationship, synonymy and variation, as well as detailed distribution, of these forms." Rehn unfortunately never did provide such a detailed study. Ball et al. (1942) mention only one difference between *p. rubripenne* and *p. corrugata:* the former has the posterior margin of the pronotum "rounded," while the latter has it "slightly produced." I have examined hundreds of specimens of *plattei* and find no instances where several characters vary concordantly; consequently I have not used the subspecies names. Perhaps Rehn retained the names in deference to his predecessors who had described them as distinct species.

DISTRIBUTION. Extreme northern Mexico, north to Montana and Dakotas.

RECOGNITION. Fig. 64*c, g, h.* Basal area of hindwings pale yellow, yellow, orange, or rosy red. The two most characteristic features of this species are the dark band on the lateral lobes and the dark stripes on the face slanting from the lower front of the eyes to the frontal ridge. This latter feature is, however, quite often faint or obsolete. Lateral field of forewings with two dark bands, one above base of hind femora and another above middle of femora. Hind tibiae yellow or pale gray-blue. Unlike the bands in *terricolor* (but as in *impexum*), these bands extend across R vein. *M. plattei* is confusingly similar to *impexum* from northwestern Arizona, Utah, Nevada, and California. It differs from that species as follows: in southwestern states the frontal ridge is not strongly constricted just above the level of the antennal sockets, in central Mexico it is rather strongly constricted; face more angulate at junction of fastigium with frontal ridge; dark bands on forewings do not extend onto dorsal field; inner face of hind femora has a single pale crossband at narrowest part. For differences between *plattei* and *terri-*

color, see *terricolor* Recognition. Body length to end of forewings 22–32 mm in males, 27–40 mm in females.

HABITAT. In southern Arizona Ball et al. (1942) report: "A moderately common grasshopper which is sometimes the most abundant of this subfamily on the range. Most common in rocky soil with sparse grass in the desert grassland." Tinkham (1948) notes that in west Texas: "This species is at home in tall sparse grass on rocky hillsides and hilltops from 5,000 to 6,000 feet elevation." I found the species inhabiting open sand patches near Great Sand Dune National Monument in Colorado (Otte, 1970).

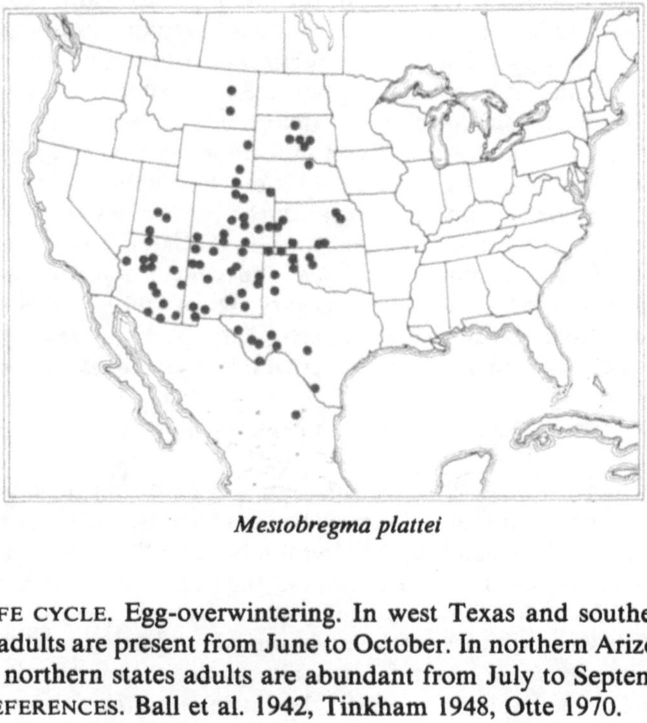

Mestobregma plattei

LIFE CYCLE. Egg-overwintering. In west Texas and southern Arizona adults are present from June to October. In northern Arizona and more northern states adults are abundant from July to September.

REFERENCES. Ball et al. 1942, Tinkham 1948, Otte 1970.

Mestobregma impexum Rehn Pl. 21

DISTRIBUTION. Northern Arizona, Utah, Nevada, and California.

RECOGNITION. Fig. 64*i*. Hindwings yellow or rosy red in basal area, dark outer band less distinct than in *plattei.* Inner face of hind femora usually black with two pale bands, but in specimens from Kane,

Wayne, and Emery counties in Utah it is entirely pale. *M. impexum* differs from *plattei* as follows: frontal ridge distinctly constricted just above the level of the antennae; dark bands on forewings sometimes extend indistinctly onto the dorsal field. Hind tibiae yellowish to pale gray-blue. Body length to end of forewings 22–28 mm in males.

HABITAT. Collected by Rehn in foothills from "sage brush covered ridges at 5,000 feet and on relatively bare slopes with scattered sage and yellow flowered bushes, at 4,900 to 5,000 feet elevation" (Rehn 1919: 242). I collected the species on open gravelly ground in desert grassland near Bicknell, Utah.

BEHAVIOR. I observed courtship behavior in specimens from Utah. These I incorrectly identified as *plattei corrugata* (Otte 1970). In four out of seven approaches, males stridulated one to three times. Sound was sometimes produced on both the up- and the downstroke, yielding two closely spaced pulses on each stroke. I did not notice femur-tipping in these approaches. *M. plattei* from Colorado never stridulated during courtship.

LIFE CYCLE. Adults from early July to September in Utah and Nevada, July and August in California.

REFERENCES. Rehn 1919, Otte 1970.

Mestobregma terricolor Rehn — Pl. 21

DISTRIBUTION. Arizona, New Mexico, west Texas, Colorado, and northern Mexico.

RECOGNITION. Fig. 64*j*. Basal area of hindwings rosy red with a brownish outer dark band. Body the color of dry, gray-brown earth. Lateral lobes with dark bands, fading noticeably dorsally. This species differs from *plattei* and *impexum* as follows: face never has a black band between eyes and frontal ridge; sides of frontal ridge are nearly parallel; median carinae of pronotum between the two sulci are lower than immediately adjacent lateral projections; dark band on lateral field of forewings does not cross R vein. Inner face of hind femora largely black, but with a pale band across narrowest part. Body length to end of forewings 23–28 mm in males, 30–36 mm in females.

HABITAT. Ball et al. (1942) report the species from "bare adobe flats" from the Willcox Plain to Bright Angel Point, Grand Canyon. In west Texas Tinkham (1948) writes: "This very rare oedipodine was taken in the adobe hill region which appears to be its native habitat; its clay coloration blends perfectly with its terricolorous nature." Rehn

(1919) writes: "At Pecos the species occurred on the bare spots of an adobe flat, where it was fairly numerous, but very shy and in scattered colonies. At Sierra Blanca [Texas] the species was also taken on bare adobe, while on the rocky hills at the same place its relative *Mestobregma plattei corrugata* occurred."

LIFE CYCLE. Probably egg-overwintering. Adults have been collected in July, August, and September.

REFERENCES. Rehn 1919, Ball et al. 1942, Tinkham 1948.

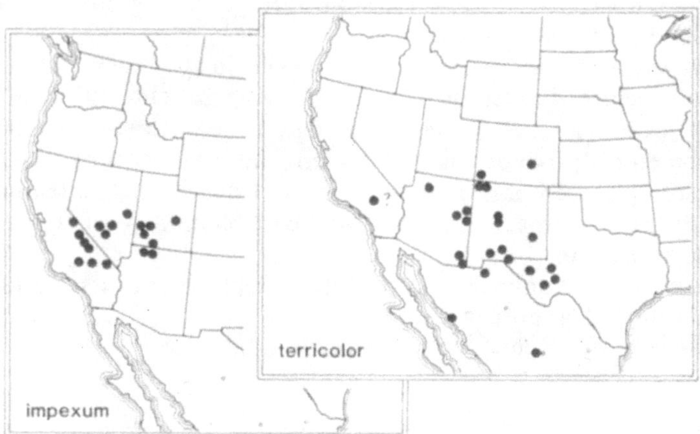

Mestobregma impexum and *M. terricolor*

Genus TREPIDULUS McNeill

DISTRIBUTION. Extreme southwestern United States and Baja California.

RECOGNITION. Fig. 65. Small slender species with markings most similar to those of *Mestobregma*. Differs from *Mestobregma* in having a short frontal ridge, in lacking a dark band on the hindwings, and in having very narrow forewings.

REFERENCES. McNeill 1901, Rehn 1904.

• Identification of Trepidulus Species

rosaceus (widespread over southwestern deserts) Fig. 65 *a, b, c*

1. Basal area of hindwings usually faintly rosy red or pink, occasionally colorless

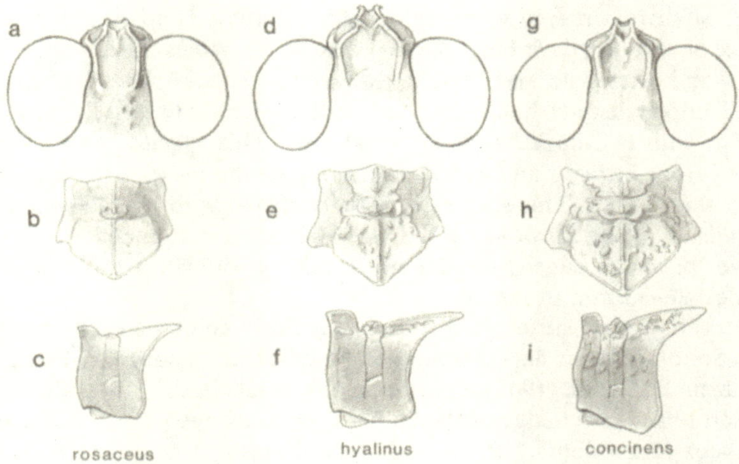

Fig. 65. *Trepidulus* species. Comparison of head, fastigium, and pronotum.

2. Ridges on pronotal disk low, rounded; metazona without distinct ridges and protuberances
3. Dorsal field of forewings usually unbanded, occasionally with very faint bands

hyalinus (California north of Los Angeles, mainly Central Valley) Fig. 65*d, e, f*

1. Hindwings always colorless
2. Pronotal ridges distinct, sharp; metazona usually with longitudinal ridges (similar to *concinens*)
3. Dorsal field of forewings unbanded

concinens (Baja California, Mexico) Fig. 65*g, h, i*

1. Hindwings clear
2. Ridges of pronotal disk distinct, sharp; metazona with protuberances and ridges (similar to *hyalinus*)
3. Dorsal field of forewings often banded; bands less distinct than on lateral field

---

***Trepidulus rosaceus* (Scudder)** Pl. 21

DISTRIBUTION. Sonoran Desert: Arizona, southern Utah, southern Nevada, southern California, and northern Baja California.

RECOGNITION. Fig. 65*a–c*. Basal area of hindwings usually faintly pinkish or rosy red; in some regions it is colorless. Hindwings lack any trace of an outer dark band. Dorsal field of forewings usually speckled gray and often with an irregular scattering of small dark spots, assembled into indistinct bands in some individuals. Lateral field of forewings with two dark bands in costal area. This species can be separated from *hyalinus* and *concinens* mainly on the basis of the pronotal disk sculpturing. This species may be confused with *Derotmema delicatulum,* but in *Derotmema* the lateral lobes are rounded along the lower posterior edge. Body length to end of forewings 18–25 mm in males, 26–33 mm in females.

HABITAT. In California this species is fairly common on the southern Sonoran desert flats (Strohecker, Middlekauff, and Rentz 1968). In Arizona Ball et al. (1942) report it on "bare flats in the Lower Sonoran desert to the lower edge of the Upper Sonoran Zone [from] Yuma and Willcox to Fredonia." Rehn and Hebard (1908) found it to be common in desert growth at the edge of Tucson.

LIFE CYCLE. Egg-overwintering. Adults from June to October.

REFERENCES. Ball et al. 1942, Strohecker, Middlekauff, and Rentz 1968.

Trepidulus hyalinus (Scudder) Pl. 21

DISTRIBUTION. Central Valley of California, mostly along the western side, and mountains northwest of Santa Barbara.

RECOGNITION. Fig. 65*d–f.* Hindwings entirely colorless. Dorsal field of forewings often straw-yellow, sometimes pale gray. Inner face of hind femora variable, sometimes almost entirely pale. Body length to end of forewings 21–24 mm in males, 27–31 mm in females.

HABITAT. Strohecker, Middlekauff, and Rentz (1968) report this species from dry grassy areas. Like *rosaceus,* it probably runs about on bare patches of ground.

LIFE CYCLE. Probably egg-overwintering. Adults from July to September.

REFERENCES. Strohecker, Middlekauff, and Rentz 1968.

Trepidulus concinens Otte, n. sp. Pl. 21

DISTRIBUTION. Most specimens of this species were collected in Baja California (see Specimens).

RECOGNITION. Fig. 65*g–i.* Hindwings entirely colorless. Bands usually cross entire forewing, although they are much less pronounced on dorsal field (*hyalinus* lacks bands on the dorsal field). Pronotal disk with distinct ridges and protuberances (relatively smooth in *rosaceus*). Inner face of hind femora has two distinct dark bands in proximal two-thirds. Body length to end of forewings 18–22 mm in males, 23–29 mm in females.

HABITAT. Desert and desert grassland.

LIFE CYCLE. All thirty-five adults of this species were collected between June 21 and August 6.

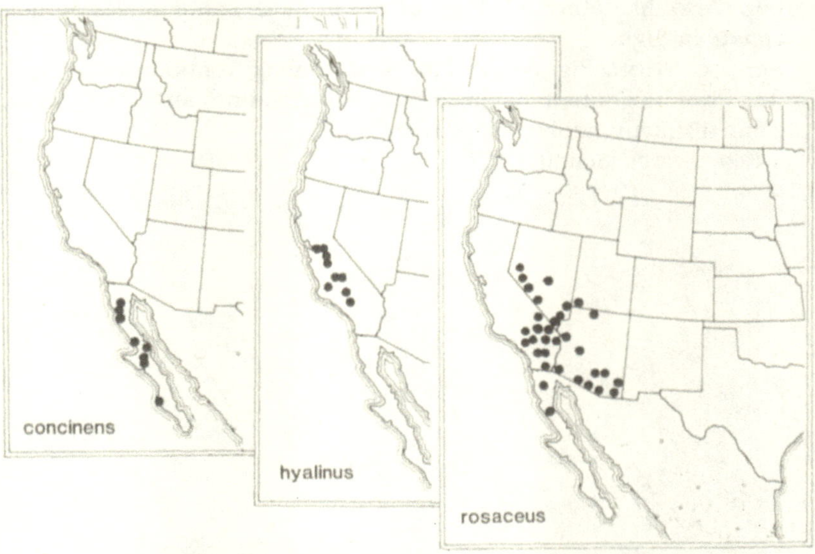

Trepidulus rosaceus, T. hyalinus, and *T. concinens*

SPECIMENS. Thirty-four specimens of this species are deposited in California Academy of Sciences (CAS) and Academy of Natural Sciences of Philadelphia (ANSP). Specimen labels bear the following collecting data: BAJA CALIFORNIA, 0.5 km south of Rosario, on Highway 1 (CAS); 10 miles south of Punta Prieta (CAS); San Domingo (CAS). 5 km from main road on road to Sierra San Pedro Martir National Park (CAS and ANSP); km 34 marker on road to Sierra San Pedro Martir National Park, CAS; 8 km south of San Quintin (CAS); 26 km north of Colonet (CAS and ANSP). 1.8 km west of Guerrero Negro on road to Estero de San Jose (ANSP); 24 km along dirt road to Punta Abreojos off Highway 1 (CAS and ANSP).

Genus DEROTMEMA Scudder

DISTRIBUTION. Western United States, southwestern Canada, and northwestern Mexico.

RELATIONSHIP. Fig. 62. Hebard placed this genus in the Group Psinidiae in his synonymic checklist (unpublished). The genus differs from other members of this group in several features: *Derotomema* has rounded lateral lobes, and the epiphalli are more like those in *Trimerotropis.* However, the narrower forewings and the head and pronotal sculpturing are more like the Psinidia group than the Sphingonotus group. Also, like other members of the Psinidia group, males do not crepitate in flight.

RECOGNITION. Fig. 66*a–n*. This genus differs from other members of the Psinidia group in having a rounded lateral pronotal lobe and in lacking distinctly bilobed epiphallic lophi.

REFERENCES. Scudder 1876.

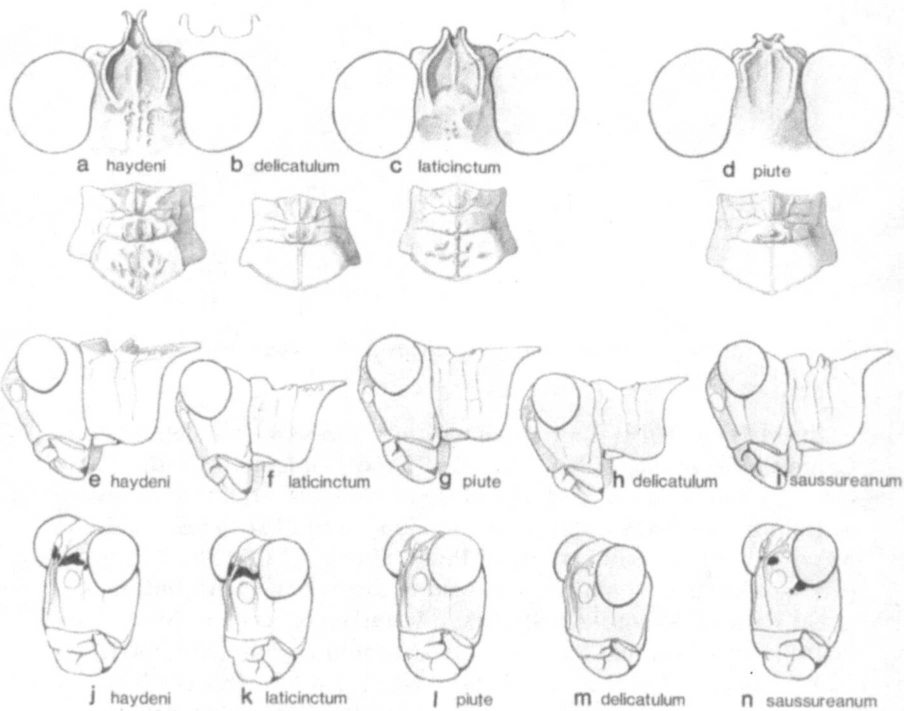

Fig. 66. *Derotmema* species. Comparison of fastigium, pronotum, and head.

• Identification of Derotmema Species

haydeni (Great Plains to Arizona and Nevada) Fig. 66*a, e, j*

1. Posterior margin of pronotal disk distinctly angulate
2. Disk of pronotum has distinct protuberances and ridges on metazona
3. Basal area of hindwings yellow or reddish; outer dark band extends to front margin.
4. Frontal ridge located well in front of eyes

laticinctum (widespread from Texas to California) Fig. 66*c, f, k*

1. Posterior margin of pronotal disk rounded
2. Disk of pronotum variable: smooth to strongly tuberculate
3. Basal area of hindwings yellow; outer dark band varies from indistinct to distinct and reaches anterior margin
4. Frontal ridge located well in front of eyes

delicatulum (Arizona, Utah, southern California, Baja California) Fig. 66*b, h, m*

1. Posterior margin of pronotal disk rounded
2. Disk of pronotum smooth or with rounded projections
3. Basal area of hindwings pale yellow; outer dark band does not reach front margin
4. Frontal ridge very near front of eyes

saussureanum (extreme southern California and Baja California) Fig. 66*i, n*

1. Posterior margin of pronotal disk angulate
2. Disk of pronotum with pronounced toothlike projections
3. Basal area of hindwings yellow to yellow-green or pale blue
4. Front ridge located well in front of eyes

piute (western Nevada) Fig. 66*d, g, l*

1. Posterior margin of pronotum slightly rounded
2. Disk of pronotum smooth, rounded
3. Basal area of hindwings very pale yellow; outer dark band indistinct or obsolete
4. Frontal ridge very near front of eyes

Derotmema haydeni (Thomas) Pl. 22

TAXONOMY. This species was divided into four subspecies by Rehn and Hebard. The name *haydeni haydeni* was given to specimens with more brightly and contrastingly colored hindwings from the Great Plains west to Arizona. In these specimens the frontal ridge is generally cut more deeply. The name *haydeni mesembrinum* was applied to specimens said to have more bulbous eyes. The distinction between *h. haydeni* and *h. mesembrinum* is not definite, however, as Rehn himself (1919: 233) notes: "The Mexican specimens are typical of [h. *mesembrinum*], as are all the paratypic individuals. The Sierra Blanca specimens are practically typical, occasional individuals [are more similar to] *h. haydenii.* The Marathon series is virtually intermediate between [*h. haydeni*] and [*mesembrinum*]." The name *h. rileyanum* has been applied to specimens from a wide area, including parts of Wyoming, New Mexico, Colorado, Utah, Nevada, and Idaho. These generally have pale yellow hindwings with a less distinct dark band; they also usually have the metazona shorter than the prozona plus the mesozona, and the frontal ridge is usually less constricted. Were it not for the existence of numerous intermediates between *h. rileyanum* and *h. haydeni* and the presence of discordant variation in the above-mentioned differences, one could apply the names without too much difficulty. As it is, one name will suffice.

D. laticinctum has been incorrectly considered a subspecies of *haydeni* by some authors (Ball et al. 1942). Hebard (1935: 292) caused this error by stating, incorrectly, that Rehn (1919) had reduced the name to racial status. Rehn (1919) does not mention *laticinctum.*

DISTRIBUTION. Western United States (except California and Washington) and northern Mexico.

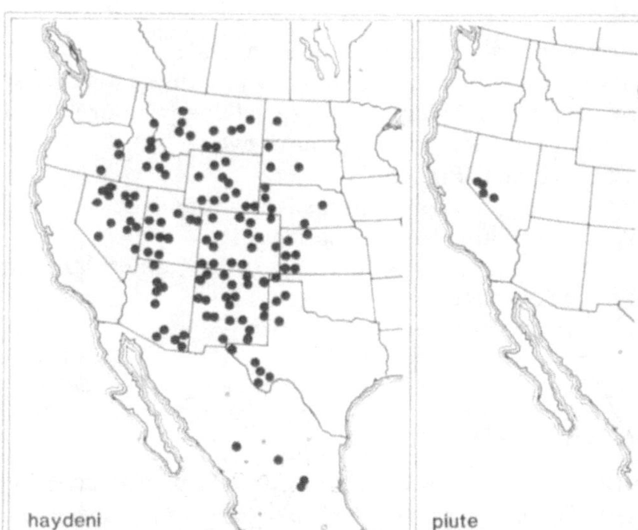

Derotmema haydeni and *D. piute*

RECOGNITION. Fig. 66*a, c, j*. Basal area of hindwings yellow or reddish. The proportion of red wings is much higher in the eastern part of the range and in Arizona and is most common in populations having a darker peripheral band. The outer dark band is wider and darker in the east and in parts of Arizona and New Mexico, and narrower and fainter in Utah, Nevada, Idaho, and parts of Colorado and New Mexico. This species overlaps broadly with *laticinctum;* it differs from that species in having a pronotal disk that is distinctly angulate along the posterior margin and that has strong protuberances and ridges. Body length to end of forewings 18–24 mm in males, 22–30 mm in females.

HABITAT. Grasslands, usually with patches of bare ground. Tinkham (1948) reports the species from "bare adobe patches on the Marfa prairies. At Presidio it was taken only on an adobe flat with clumps of mesquite scattered about." Rehn and Hebard (1908) collected the species on "soil in an open field [and] in an open place heavily overgrown with low vegetation" in Arizona. In Nebraska Hauke (1953) reports it on gravelly and sandy ground.

LIFE CYCLE. Adults from June to September.

REFERENCES. Ball et al. 1942, Tinkham 1948.

Derotmema delicatulum Scudder Pl. 22

DISTRIBUTION. Sonoran Desert of southwestern United States and Baja California.

RECOGNITION. Fig. 66*b, h, m*. Basal area of hindwings very faint yellow; outer dark band does not reach anterior margin. Posterior margin of pronotal disk is rounded, as in *laticinctum;* differs from that species in having a pronotal disk without strong ridges and toothlike projections and in lacking a dark facial band above the antennae. Frontal ridge nearly flush with front of eyes (side view). This species can be confused with *Trepidulus rosaceus,* but it differs as follows: posterior margin of pronotal disk more rounded (more angulate in *T. rosaceus*); lower posterior margin of lateral lobes more broadly rounded; dark markings on the lateral lobe different in shape; dorsal field of forewings with at least two darker crossbands; color of hindwings different.

Body length to end of forewings 18–24 mm in males, 23–28 in females.

HABITAT. This species is restricted to very dry deserts on hard, bare soil with sparse vegetation. Strohecker, Middlekauff, and Rentz (1968) report it from the "extreme desert of the Lower Sonoran Zone." Rehn and Hebard (1908) report: "The habitat of this form is the Mojave and Yuma deserts . . . specimens were collected during a train stop in a most arid and desolate location."

LIFE CYCLE. Adults from July to September.

REFERENCES. Strohecker, Middlekauff, and Rentz 1968.

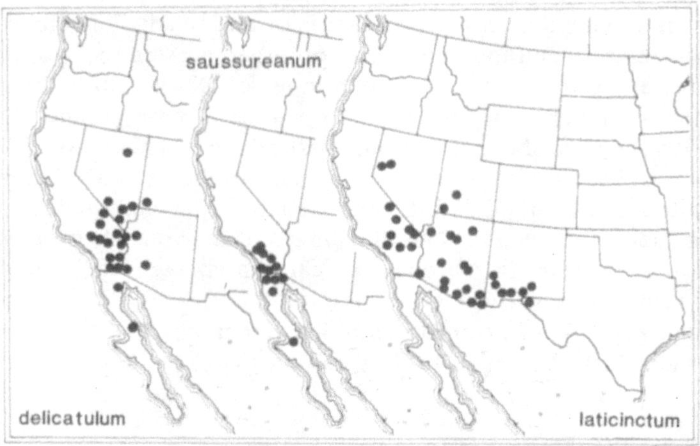

Derotmema delicatulum, D. saussureanum and *D. laticinctum*

Derotmema laticinctum **Scudder** Pl. 22

TAXONOMY. This name has appeared as a subspecies of *haydeni* in the literature, apparently because of an incorrect statement by Hebard (1935: 292) that Rehn (1919) had reduced the name to racial status. *D. haydeni* and *laticinctum* are quite distinct in a number of features.

DISTRIBUTION. Chihuahuan and Sonoran deserts, from west Texas to southern California and Nevada.

RECOGNITION. Fig. 66*c, f, k.* Basal area of hindwings pale yellow; outer dark band extends to front margin, but sometimes this band is faint. This species can be distinguished as follows: unlike *haydeni* and *saussureanum* the posterior margin of the pronotal disk is rounded (*delicatulum* is similar). It differs from *delicatulum* in having a very rough pronotal disk with toothlike projections on the mesozona and in having a more continuous dark band on the hindwings. Body length to end of forewings 17–22 mm in males, 23–30 mm in females.

HABITAT. Deserts, often on the lower slopes of hills. Rehn and Hebard (1909) write: "[At Aden, New Mexico] the species was found chiefly in grass prairie land, but in other situations as well, such as on cinders along the railroad track where it was inactive until the sun warmed the ground. In the irrigated tract near El Paso the species was fairly common on the barer spots, usually in colonies, while at the foot of the mesa it was not common on pieces of bare ground." At El Paso, Texas, it was collected on sandy tracts.

LIFE CYCLE. Adults from July to December.

REFERENCES. Rehn and Hebard 1908, 1909, Ball et al. (1942), Strohecker, Middlekauff, and Rentz 1968.

Derotmema piute Rehn Pl. 22

DISTRIBUTION. Western Nevada.

RECOGNITION. Fig. 66*d, g, l*. Basal area of hindwings very pale yellow; outer dark band indistinct or obsolete. Spots on forewings are gathered into bands, so the dorsal field has from one to three crossbands. Frontal ridge barely extends forward of the eyes. Pronotal disk smooth and with rounded bumps; posterior margin slightly rounded. Body length to end of forewings 19–21 mm in males, 22–27 mm in females.

HABITAT. Desert flats. Rehn (1919) does not discuss the habitat.

LIFE CYCLE. All adults in the Philadelphia Academy of Natural Sciences collection were collected in September.

REFERENCES. Rehn 1919.

Derotmema saussureanum Scudder Pl. 22

DISTRIBUTION. Extreme southern California and Baja California.

RECOGNITION. Fig. 66*i, n*. Basal area of hindwings pale yellow, pale greenish yellow, or pale blue; outer dark band variable in intensity, usually reaching anterior margin but sometimes slightly interrupted near the front. Differs from *laticinctum* and *delicatulum* in that the posterior margin of the pronotal disk is angulate. Differs from *delicatulum* in having a rough pronotal disk with toothlike projections. Body length to end of forewings 19–24 mm in males, 23–28 mm in females.

HABITAT. Desert flats. Rehn and Hebard (1909) found this species "on dry spots in and near the Arroyo Seco" at Pasadena, California.

LIFE CYCLE. Adults from June to September.

REFERENCES. Rehn and Hebard 1909, Strohecker, Middlekauff, and Rentz 1968.

Appendixes

References

Taxonomic Index

APPENDIX 1

Explanation of Characters

This appendix lists the characters used to derive the various phylogenies given in the text. In the figures a black dot is placed on the branch that possesses the presumed derived (apomorphic) condition, but both derived and ancestral (plesiomorphic) conditions are described here. In many cases it could not be determined whether a condition is derived or ancestral; in such cases the dot was arbitrarily placed on one of the branches. The letters following each character state indicate that the state is presumed to be derived (d) or ancestral (a), respectively. In most cases out-group comparisons were used to determine which characters are synapomorphous and which are symplesiomorphous.

Arphia Genus Group: Fig. 11

1d, point of aedeagus slender and serrated, without an enlarged lobe (Fig. 63*i–m*); **1a,** aedeagus thicker, unserrated, less pointed, with a larger basal lobe (Fig. 63*a–f*). Out-group comparison.

2d, hindwings without a distinctly defined outer dark band; **2a,** with such a band (secondarily lost in *secreta*). Out-group comparison.

3d, anterior lobe of hindwings with enlarged windowlike cells (secondarily lost in *Shotwellia?*); **3a,** without such cells. Out-group comparison.

4d, male-male behavior: femur-shaking with wing-striking (behavior of *Tomonotus* and *Leuronotina* not studied) (Otte 1970); **4a,** silent shaking or shaking with substrate-striking.

5d, dorsum of forewings has a pale median band in some individuals (*Arphia, Lactista*); **5a,** dorsum of forewings without a pale band. Out-group comparison.

6d, forewings without crossbands or numerous large spots; **6a,** with crossbands or with scattered dark markings (indistinct in some *Lactista* and some *Tomonotus*).

7d, median pronotal carina indistinctly cut or not cut by principal sulcus; **7a,** distinctly or deeply cut. Out-group comparison.

8d, solitary males do not crepitate; **8a,** solitary males crepitate. Out-group comparison.

9d, dark band on hindwings located near periphery and with a long dark spur directed toward base; **9a,** not as above. Out-group comparison.

10d, median pronotal crest very high; **10a,** crest not high. Out-group comparison.

11d, antennae ensiform; **11a,** antennae filiform. Out-group comparison.

12d, hind femur narrows suddenly on lower margin; **12a,** narrows suddenly on upper margin. Out-group comparison.

Genus *Arphia:* Fig. 12

1d, rostrum angulate and lateral foveolae rectangulate; **1a,** rostrum rounded, foveolae triangulate. Out-group comparison.

2d, fastigium pointed anteriorly; **2a,** fastigium rectangulate. Out-group comparison.

3, eastern distribution; *behrensi* has western distribution.

4d, median ridge of pronotum moderately to strongly arched; **4a,** not arched or slightly arched. Out-group comparison.

5d, dark band on hindwings rather broad; **5a,** band narrower. Out-group comparison.

6, geographical and seasonal relations between *xanthoptera* and *granulata* suggest that the latter is derived from the former.

7, a common stem for *conspersa, pulchripennis,* and *pecos* is based on their almost indistinguishable phenotypes. No synapomorphic characters could be found.

8d, pronotal crest not cristate; **8a,** crest cristate. Out-group comparison.

9, *pecos* and *pulchripennis* are both late summer and fall species.

10, a close relationship between *nietana* and *pseudonietana* is suggested by their very similar (almost indistinguishable) appearance in Durango.

11, common stem for *saussureana* and *pseudonietana* is suggested by a similar overall appearance and by their geographical relations. In the recent literature they have been considered synonyms.

12d, hindwings red; **12a,** hindwings yellow or orange. Out-group comparison.

***Leuronotina, Tomonotus,* and *Lactista:* Fig. 18**

1d, fastigium much longer than wide; **1a,** wider than long. Out-group comparison.

2, hindwings orange in *ritensis* and *philorites;* rosy red, pink, or transparent in *orizabae.* Apomorphic and plesiomorphic conditions uncertain.

3d, lower posterior margin of lateral lobes with a prominent projection or tooth; **3a,** lateral lobe rounded. Out-group comparison.

4d, body foreshortened, head held closely against pronotum; **4a,** body more elongate. Out-group comparison.

5, morphological and geographical relations do not suggest clearly which species should be clustered; the more southern distribution of *stramineus* suggests that it is most closely related to *pellepidus.*

6, *gibbosus* is probably an offshoot of the widespread *aztecus;* likewise, *micrus* and *humilis* are probably derived from the more widespread *aztecus.*

***Chortophaga, Chimarocephala, Encoptolophus,* and *Shotwellia:* Fig. 24**

1d, distance between front of eyes and front of head relatively long; **1a,** distance relatively short. Out-group comparison.

2, a common stem for *Chortophaga* and *Chimarocephala* is also based on an overall resemblance in shape and color.

3d, fastigium elongate and deep; **3a,** fastigium triangulate and flat. out-group comparison.

4d, frontal ridge of head deeply grooved; **4a,** frontal ridge grooved shallowly at most. Out-group comparison.

5d, distinct dark markings on sides of forewings and upper portion of hind femora; **5a,** forewings and hind femora not distinctly banded. Out-group comparison.

6d, frontal ridge not concave at level of antennae; **6a,** concave. Out-group comparison.

7d, hind tibiae bluish; **7a,** hind tibiae dirty yellow to brown. Out-group comparison.

8d, hindwings without enlarged cells in anterior section; **8a,** with enlarged cells. Out-group comparison.

9d, solitary males do not crepitate (*subgracilis* and *pallidus; robustus* and *californicus* are unstudied); **9a,** solitary males crepitate (*sordidus* and *costalis; fuliginosus* and *otomitus* remain unstudied). Out-group comparison.

10, a common stem for *costalis, fuliginosus,* and *otomitus* is also suggested by their geographical distributions.

11d, upper face of hind femora with an oblique dark band; **11a,** upper face with a triangular band or no band. Out-group comparison.

12d, fastigium deeply excavated and narrow; **12a,** fastigium shallow and broad. Out-group comparison.

13d, lateral carinae of pronotum continuous; **13a,** lateral carinae interrupted. Out-group comparison.

14d, fastigium deeply excavated; **14a,** fastigium not deeply excavated. Out-group comparison.

15d, body straw-yellow; **15a,** body brownish. Out-group comparison.

16d, low median pronotal crest and very poorly defined lateral carinae; **16a,** pronotal crest and lateral carinae more distinct. Out-group comparison.

17d, fastigium narrow and shallow; **17a,** fastigium wider and often deeper. Out-group comparison.

18d, median pronotal ridge obsolete; **18a,** median ridge present, often crestlike. Out-group comparison.

Hippiscus Genus Group: Fig. 32

1d, median pronotal carina very low, obsolete or nearly so; **1a,** median pronotal carina distinctly raised. Out-group comparison.

2d, forewings spotted; **2a,** forewings banded. Out-group comparison.

3d, hind tibiae blue; **3a,** hind tibiae all red or red and black. Out-group comparison.

4d, folded forewings without converging pale stripes on dorsum; **4a,** with converging pale stripes. Out-group comparison.

5d, forewings with dark crossbands of equal width; **5a,** forewing crossbands less distinct and of unequal width. Out-group comparison.

6d, head relatively small; **6a,** head relatively large. Out-group comparison.

7d, intercalary vein absent; **7a,** intercalary vein present. Out-group comparison.

8d, spring adults; **8a,** fall adults. Out-group comparison.

9d, forehead broadly rounded, fastigium largely convex; **9a,** forehead more angulate, fastigium concave. Out-group comparison.

10d, epiphallus not bilobate; **10a,** epiphallus bilobate.

11d, median crest of pronotum cut by two sulci; **11a,** cut by one sulcus. Out-group comparison.

12d, forewings uniform in color or with small, indistinctly defined spots; **12a,** forewings with numerous large, well-defined spots.

***Xanthippus, Pardalophora, Cratypedes, Agymnastus, Sticthippus:* Fig. 33**

1d, Median pronotal crest cut by one sulcus; **1a,** cut by two sulci. Out-group comparison.

2d, forewings without numerous, distinctly defined dark markings; **2a,** with numerous distinctly defined dark markings. Out-group comparison.

3d, sexual dimorphism very strong, with large flightless females; **3a,** sexual dimorphism normal, with flying females. Out-group comparison.

4, a common stem for *neglectus* and *lateritius* is based on their overall great similarity—only the widening lateral lobes can reliably be used to distinguish *neglectus.*

5, the position of *californicus* is uncertain; its spotted forewings suggest a closer relationship to *Xanthippus* than to *Cratypedes,* but in appearance it seems intermediate between the two.

6, a common stem for *corallipes, olancha, brooksi,* and *aquilonius* is based on their overall great resemblance and their geographical relations; the latter three species are hypothesized to be derived from the widely distributed *corallipes.*

7, the position of *P. haldemani* is uncertain; occasional specimens have the pronotal crest cut twice as in *Xanthippus.* In overall appearance they appear to be more similar to *X. corallipes* than to other *Pardalophora* species.

8d, forewing area between veins Cu_1 and Cu_2 largely dark; **8a,** forewing area has several spots. Out-group comparison.

9d, inner face of hind femora bluish near base; **9a,** inner face of hind femora yellow, orange, or red near base. Out-group comparison. The geographical distributions of these two species also suggest that they share a common stem.

10d, inner face of hind femora mostly blue, with an orange band at the narrowest part; **10a,** inner face banded with black and yellow, sometimes with orange, but never with an orange band at narrowest part.

***Leprus, Hadrotettix,* and *Heliastus:* Fig. 36**

1, a common stem for *magnificus* and *scotodes* is based on the less distinct forewing banding in these species, but it remains uncertain whether this feature is derived.

2d, anterior crossband of forewings indistinct, especially on dorsum; **2a,** band distinct. Out-group comparison.

3d, lower posterior corner of lateral lobes produced; **3a,** lobes rounded. Out-group comparison.

4d, body dorso-ventrally flattened; **4a,** body shape normal.

5d, a common stem for *elephas* and *wheeleri* is based on their greater overall resemblance, but which of the features is synapomorphous is uncertain. The similarities are listed under 3, 4, and 5 in the Identification of *Leprus* Species.

6d, pronotum rather strongly sculptured, prozona with definite shoulders; **6a,** pronotum smooth, rounding gradually onto lateral lobes. Out-group comparison.

7d, lower inner marginal area of hind femora deep orange and without a black band; **7a,** lower inner marginal area yellowish, with at least one black band. Out-group comparison.

Bryodema and Sphingonotus Groups: Fig. 38

1d, hindwings with an expanded fan and with greatly thickened anal veins and reinforcing network of veins (the somewhat reduced wing of *Circotettix maculatus* may be the result of a secondary reduction); **1a,** hindwing fan not expanded. Out-group comparison.

2d, forewings very broad; **2a,** forewings intermediate in width. Out-group comparison.

3d, hindwings lack an outer dark band, or dark band has poorly defined margins; **3a,** outer dark band well defined (secondarily lost in a few species). Out-group comparison.

4, the branching pattern within Eurasian species of the Bryodema group has not been analyzed; on the basis of their geographic proximity it is likely that they share a common stem that does not include *Circotettix.*

5d, pronotal crest cut by one sulcus (except in *E. phenax, S. campestris, S. bunites,* and some *S. equale*); **5a,** pronotal crest cut twice. Within-group trend.

6d, hindwing coloration as in Pl. 12, *Eximacris;* **6a,** hindwing coloration variable, not as above. Out-group comparison.

7d, crossbands on forewings distinctly marked on lateral surfaces of closed wing, fading completely on dorsum; **7a,** crossbands on forewings cross the dorsum (but a reduction of the dorsal band occurs in several *Trimerotropis* species, for example, *Cincta* and *Occulens* groups). Out-group comparison.

8d, pronotal ridge usually cristate; **8a,** pronotal crest usually low or forming a broad ridge. Out-group comparison.

9d, anterior transverse sulcus of pronotum often not nearly so deep as posterior sulcus, sometimes does not cut median crest; **9a,** anterior transverse sulcus about as deep as posterior one. Out-group comparison.

10, a common stem for *Microtes* and *Conozoa* is based on their geographic proximity.

***Spharagemon, Dissosteira, Eximacris:* Fig. 39**

1d, hindwings smoky brown without distinct bands; **1a,** hindwings banded. Out-group comparison.

2d, inner and outer faces of hind femora have a dark band covering the narrowest part; **2a,** narrowest part of femora not banded. Out-group comparison.

3d, body green and yellow, forewings unbanded and unspotted and reddish brown; **3a,** body brownish or yellowish, never greenish; forewings sometimes banded or spotted and reddish brown. Out-group comparison.

4d, hindwings not yellow on fan; **4a,** hindwings yellow on fan. Out-group comparison.

5, pronotal carina high and bilobed; whether this feature is synapomorphous is problematical, for it occurs in some species of *Spharagemon* and *Eximacris.*

6d, forewings at least sometimes spotted (usually unicolorous in *carolina*); **6a,** forewings banded. Ex-group comparison.

7d, basal area of hindwings black; **7a,** basal area not black, plesiomorphic condition probably yellowish. Out-group comparison.

8, the phylogenetic position of *Dissosteira pictipennis* is problematical. On the basis of pronotal shape and external body coloration it appears to be more similar to *D. spurcata* than to any other species.

9, hind tibiae partly or entirely orange or red (except in *bunites,* which has blue tibiae); possibly this character is not apomorphic.

10d, hind tibiae red and black; **10a,** tibiae orange or red (blue tibiae in *bunites* probably apomorphic within the *equale* subgroup). Out-group comparison.

11d, metazona of pronotum very pale in relation to prozona in at least some individuals; **11a,** metazona never pale. Out-group comparison.

12d, median carina of pronotum forms high, narrow ridge; **12a,** median carina low. Out-group comparison.

13d, body finely speckled; **13a,** body not finely speckled.

14d, inhabits sandy ground; **14d,** inhabits firm ground. Out-group comparison.

15d, pronotal crest at least occasionally cut by two sulci; **15a,** pronotal crest always cut by one sulcus. Out-group comparison.

16d, pronotal crest nearly always cut by two sulci; **16a,** nearly always cut by one sulcus. Out-group comparison.

17d, inhabits high mountain meadows; **17a,** inhabits lower-altitude grasslands. Out-group comparison.

18d, pronotal crest high; **18a,** pronotal crest low. I am uncertain about which character is apomorphic.

19, a common stem for *bolli* and *crepitans* is suggested by their distributions—*crepitans* evidently speciated from *bolli* when the sea inundated northern Florida.

20d, inhabits lichen-covered ground in the eastern and northeastern forests, **20a,** inhabits open grassy patches at lower elevations. Out-group comparison.

***Conozoa* and *Microtes:* Fig. 44**

1d, crossbands on forewings do not reach dorsal horizontal surface; **1a,** crossbands reach horizontal surface and usually cross this plane (do not cross it in *T. thalassica, T. koebelei, T. lauta, T. cincta,* and some *T. fontana*). Out-group comparison.

2d, small, wide-bodied species; **2a,** medium length and narrow-bodied. Out-group comparison.

3d, body noticeably hirsute; **3a,** not noticeably hairy. Out-group comparison.

4d, inhabits sand; **4a,** inhabits firm ground. Out-group comparison.

5d, lower posterior margin of pronotum with a tooth or projection; **5a,** without a tooth or projection. Out-group comparison.

6d, pronotal carinae strongly bilobate on prozona, anterior sulcus always deep; **6a,** anterior sulcus often much shallower than posterior one, and prozona often scarcely bilobate. Out-group comparison with *Microtes* and *Mioscirtus.*

7d, central dark crossband on forewings usually does not reach anterior margin; **7a,** central band touches anterior margin. Out-group comparison.

8, a common stem for *carinata, clementina,* and *nicola* is based mainly on the fact that in all three species the anterior pronotal sulcus either does not cut the median carina or does so only shallowly.

9d, lower posterior margin of lateral lobes lacks a tooth or projection; **9a,** has a tooth. Out-group comparison within *Conozoa.*

10, a common stem for *clementina* and *nicola* is also suggested by the fact that both are from the islands south and southwest of Los Angeles. Of course, their progenitors may have reached the two islands independently from the mainland.

11, the branching sequence in *Microtes* is difficult to estimate—*occidentalis* and *helferi* have similar tibial spurs and wing lengths; *occidentalis* and *pogonata* have similar forewing banding; *pogonata* and *helferi* have similar pronotal crests (see Identification of *Microtes* Species).

***Trimerotropis:* Fig. 48**

1, how the five or six major groupings of *Trimerotropis* are related to one another remains uncertain, for there are few superficial characters that can be used to clarify these relationships.

2, a common stem for the *maritima* and *latifasciata* groups is highly tentative. It is based principally on the fact that these two groups possess yellow, orange, or red hind tibiae (*arenacea* sometimes has gray or bluish tibiae). In relation to other groups, it seems likely that this feature is synapomorphic.

3, I cannot identify any synapomorphic characters that unite the species on this stem.

4d, lower posterior part of lateral lobes produced (less evident in *maritima*); **4a,** lateral lobes not produced, with or without a small tooth. Out-group comparison.

5d, lateral lobes very long and produced along lower posterior margin, body dorso-ventrally flattened; **5a,** lateral lobes only moderately produced.

6, a common stem for *arenacea* and *whitei* is tentative and is based on their nearly identical appearance. Indeed, there is some question as to whether they have reached specific status.

7d, hind tibiae pale yellow to white; **7a,** hind tibiae orange or red. Out-group comparison.

8, it is uncertain if *californica* shares an exclusive stem with *melanoptera-latifasciata* or with *pacifica-titusi.* The arrangement depicted is preferred because *californica* and *latifasciata* are easily confused.

9d, abdomen reddish on dorsum; **9a,** yellowish. Out-group comparison.

10, a common stem for *tolteca* and *pistrinaria* is postulated because these two species sometimes appear to be almost indistinguishable.

11d, pronotal disk rounds gradually onto lateral lobes on prozona; **11a,** prozona with shoulders. Out-group comparison.

12d, lateral lobe has an arching black mark; **12d,** lacks such a black mark. Out-group comparison.

13d, prozona often black medially, metazona nearly white; **13a,** not so marked. Out-group comparison.

14d, dark crossbands on forewings are indistinct on dorsum of folded wings or do not cross the dorsal surface; **14a,** dark crossbands more distinctly indicated on dorsum (occasionally similar in *fontana*). Out-group comparison.

15d, dorsum of forewings darkly pigmented where it disappears under the pronotum (not visible in all specimens); **15a,** forewings not so darkened. Out-group comparison.

16, common stem for these three species is not based upon any identifiable apomorphic characters, but rather on the fact that they are virtually indistinguishable and on their geographical distributions; all three appear to be derived from a single species that ranged along the coastal ranges of central California.

17d, crossbands on forewings distinct only on lateral field, not crossing dorsal field; **17a,** crossbands on forewings present on dorsal field (except in *lauta,* which lacks crossbands altogether). Out-group comparison.

18d, dark band runs between the eyes and crosses the face below the antennae; **18a,** face lacks such a dark band. Out-group comparison.

19d, body contrastingly marked; **19a,** body color not so contrastingly marked. Out-group comparison.

20, inhabiting high-elevation grasslands, often in association with open pine woodlands; other species in this group are variable—*thalassica* also inhabits such areas, but *lauta* is known only from desert habitats.

21d, hind tibiae yellow to brownish; **21a,** variable, blueish, greenish, or yellowish. The plesiomorphic condition is uncertain; the majority of *Trimerotropis* species have bluish tibiae.

22, head of *salina* and *schaefferi* (measured through eyes) relatively broad (Fig. 53) and body coloration of western *salina* and *schaefferi* almost indistinguishable (Fig. 52).

23, the branching sequence involving these species is not known; *saxatilis* becomes increasingly similar to *pallidipennis* from east to west and may not represent a distinct species; the geographically separated *huroniana* is much smaller than normal *pallidipennis* and inhabits sand dunes; its reproductive behavior, however, is like that of *pallidipennis* (Otte 1970).

24d, poorly defined crossbands on forewings (bands most distinct in *pseudofasciata, cyaneipennis, fratercula,* and *bernardi*); **24a,** forewing crossbands distinct (secondarily lost in sand-inhabiting groups such as the *maritima* group). Out-group comparison. This character does not in itself adequately unite the species arising from this stem. See **25.**

25, a common stem for the *cyaneipennis* and *suffusa* groups was first suspected by the discovery that *fratercula* (usually yellow-winged) and *cyaneipennis* (usually blue-winged) are difficult to distinguish in New Mexico, where greenish-winged specimens of both species occur.

26d, hindwings bluish or greenish blue in basal area (blue to yellowish in *sparsa*); **25a,** hindwings yellow to transparent (with a greenish tinge in New Mexico *fratercula*). Out-group comparison.

27d, outer dark band on forewings indistinct to obsolete; **27a,** outer dark band distinct. Out-group comparison.

28, I was unable to discern a possible branching sequence within this cluster of species. Possibly *arizonensis-leucophaea* branch off first. A very close relationship between *cyaneipennis* is likely, and the two may not be specifically distinct (see *cyaneipennis* Taxonomy). It is my guess that the geographically restricted *inyo, bernardi,* and *fratercula* are all derived from the widely distributed *cyaneipennis.* The resemblance between *bernardi* and *fratercula* is very close, but their geographic separation suggests they are independent branches of *cyaneipennis.*

29d, male hindwings with distinctly thickened anal veins; **29a,** male anal veins slightly thickened or not noticeably thickened. Out-group comparison.

30, a common stem for *cyaneipennis* and *fratercula* is suggested by their morphological convergence in the area of contact; the degree of similarity suggests that they may not be distinct species.

31, how *gracilis* is related to other *Trimerotropis* species is uncertain; it resembles *sparsa* more than any other species.

32d, sloping pronotal desk on prozona; **32a,** prozona with slight shoulders. Ex-group comparison.

33d, fastigium very narrow; **33a,** fastigium wider. Out-group comparison.

34d, lower posterior margin of pronotum with tooth; **34a,** without tooth. Out-group comparison.

35, *tolteca* and *modesta* were long considered to be conspecific; they remain difficult to distinguish; their preference for mountain habitats may be symplesiomorphous.

Circotettix: **Fig. 59**

1d, basal area of hindwings black or transparent; **1a,** basal area yellowish, bluish, or greenish. Out-group comparison.

2d, hindwing veins 6A, 7A, and 8A thickened; **2a,** veins 6A, 7A, and 8A not thickened. This feature suggests a common stem for *carlinianus* and *stenometopus,* but the lack of thickened veins in *maculatus* is probably apomorphic.

3d, fan of hindwings blue-green or blue; **3a,** fan yellow. Out-group comparison.

4d, vein Cu_1 on hindwings about equidistant between MP and 1A; **4a,** vein Cu_1 closer to MP than to 1A. Out-group comparison.

5d, vein $2A_a$ of hindwings: anterior branch absent or very short; **5a,** anterior branch of $2A_a$ long. Out-group comparison.

6, there is no direct evidence of a common stem for *crotalum* and *rabula.* The distribution of *crotalum* suggests that it is more closely related to *rabula* than to *shastanus.*

7, the only species with black in the basal area are *carlinianus* and *maculatus.* This feature is probably apomorphic and suggests a common stem.

Psinidia Genus Group: Fig. 62

1, on the basis of geographic proximity it is not unreasonable to expect the North American genera to represent a monophyletic group.

2d, antennae ensiform; **2a,** antennae filiform. Out-group comparison.

3d, frontal ridge of head concave in profile; **3a,** frontal ridge straight or convex in profile. Out-group comparison.

4d, forewings lack crossbands; **4a,** forewings have crossbands. Out-group comparison.

5d, dorsal field of forewings often has two posteriorly converging pale bands; **5a,** forewings lack such bands, or bands are indistinct. Out-group comparison.

6d, forewings often have more than three blotches or bands on lateral field; **6a,** forewings with at most three blotches or bands. Out-group comparison.

7, the branching sequence involving *Hippopedon, Trachyrhachys,* and *Metator* is difficult to determine. No synapomorphous characters can be identified.

8d, cerci spoon-shaped; **8a,** cerci finger-shaped. Out-group comparison.

9d, lateral pronotal carinae form discrete pale lines, bordered on both sides by very dark pigmentation; **9a,** lateral carinae may be pale but do not form thin pale streaks with sharply defined margins. Out-group comparison.

10d, pronotal region between first and second sulcus has three lobes or knobs, with the lateral knobs often as high as or higher than the median knob; **10a,** this region has a prominent median lobe and very low lateral ones. Out-group comparison.

11d, anterior median lobe of prozona squarer and higher than the posterior lobe; **11a,** anterior posterior lobes both rounded, anterior one not much higher than posterior one. Out-group comparison.

12d, lower posterior margin of lateral lobe broadly rounded; **12a,** lateral lobe produced and angulate. Out-group comparison.

13d, lateral lobe has a sloping or arching dark band; **13a,** lateral lobe has a small white spot surrounded on three sides by black. Out-group comparison.

14d, hindwings pink; **14a,** hindwings yellow. Out-group comparison.

15d, lateral pronotal carinae continuous in central part; **15a,** lateral carinae interrupted. Out-group comparison.

16, a common stem for *capito* and *gracilipes* is based on both their geographic proximity and their almost indistinguishable morphologies.

17, I can find no synapomorphic characters on which to base a branching sequence.

18d, forewings have numerous ovoid spots on lateral surface; **18a,** forewings have several bands. Out-group comparison.

19d, fastigium strongly ridged; **19a,** fastigium not strongly ridged. Out-group comparison.

20d, hindwings mostly black; **20a,** hindwings banded or mostly transparent. Out-group comparison.

21, common stem is based on a very close overall similarity among these species; I can find no synapomorphic characters to establish monophyly.

22d, head large relative to pronotum; **22a,** head size normal. Out-group comparison.

23d, central dark band on upper face of hind femora strongly oblique; **23a,** dark band triangular or square. Out-group comparison.

25d, hindwings lack an outer dark band; **25a,** hindwings have an outer dark band. Out-group comparison.

26, I cannot identify any synapomorphic characters; a common stem here is based on the near indistinguishability of *impexum* and *plattei.*

27, I cannot identify any synapomorphic characters on which to base a branching pattern.

28d, face lacks black markings; **28a,** face has black markings. Which character is apomorphic is uncertain. The presence of dark faces in *Mestobregma* and *Trachyrhachys* suggests that this feature is plesiomorphous.

29, I am unable to identify synapomorphic characters that will establish the branching pattern.

APPENDIX 2

Signal Repertoire of Oedipodinae

Most of the information in the following table of courtship and male-male signals in Oedipodinae is summarized from Otte (1970). Other unpublished information is also included. The species misidentified in Otte (1970), with the correct name given in parentheses, are: *Leprus robustus* (*wheeleri*); *Heliastus benjamini* (*aztecus*); *Spharagemon humile* (*equale*); *Trimerotropis agrestis* (*barnumi*); *Trimerotropis koebelei* (*inconspicua*); *Trimerotropis calignosa* (*occidentalis*); *Trimerotropis caerulipes* (*occidentalis*); *Trimerotropis albescens* (*bifasciata*); *Trimerotropis bilobata* (*Conozoa texana*); *Circotettix thalassinus* (*undulatus*).

T, femur-tipping; OS, O-stridulation; VS, V-stridulation; FSw, femur-shaking, femora strike wings; FSg, femur-shaking, tibiae strike ground; FS, femur-shaking without striking wings or ground; TS, T-stridulation; Tws, tipping with wing striking; AS, alternating stridulation.

	Male Signals	
Species	Male–female courtship interactions	Male–male interactions
Tropidolophus formosus	OS	T, FS
Arphia sulphurea	T, OS, VS	T, FSw
simplex	T, OS, VS	T, FSw
conspersa	T, OS, VS	T, FSw
pseudonietana	T, OS, VS	T, FSw
xanthoptera	T, OS	T, FSw
granulata	T, OS	T, FSw
pecos	T, FSw	T, FWw?
Chortophaga viridifasciata	OS	T, FS
australior	T	T, FS
Encoptolophus sordidus	T, VS, TS	T, FSg
costalis	T, OS, VS	T, FSg
subgracilis	T, OS	T, FSg
Lactista gibbosus	T?, OS, FSw (sometimes)	T, FSw
elota	T, OS	T, FSw
punctatus	FSw	T, FSw

Species	Male Signals	
	Male–female courtship interactions	Male–male interactions
Lactista aztecus	OS	T, FSw
Pardalophora apiculata	OS, VS	T, FS
phoenicoptera	OS	T, FS
saussurei	OS	T, FS
haldemani	FSg	T, FSg
Hippiscus ocelote	OS	T, FS
Xanthippus corallipes	T, OS, VS	?
montanus	T, OS	T, FSg
Leprus wheeleri	T, OS	T, FSg
Camnula pellucida	OS	T, FS
Hadrotettix trifasciatus	T, OS	T, FSg
Heliastus aztecus	T, FSg	T
sumichrasti	FSw	T, FSw
Spharagemon marmorata	T, OS, alt. vib.[1]	T, FSg
collare	OS	T, FSg
cristatum	T, OS, FSg	T, FSg
bolli	T, OS, FSg	T, FSg
crepitans	T, FSg	T
saxatile	OS, VS	T, FSg
equale	VS, FSg	T, FSg
campestris	T, OS, VS	T, FSg
Dissosteira carolina	T, Tws, AS	T, FSg
pictipennis	T, OS, TS	T, FSg
Trimerotropis		
pallidipennis	T, OS	T, FSg
diversellus (California)	T, OS(2),[2] FSw or FS	T, FSg?
huroniana	T, OS	T, FSg
albescens	T, OS	?
californica	T, FSg	T, FSg (some populations)
maritima	T, OS, FSg	T, FSg
v. verruculata	OS	T, FSg
v. suffusa	OS	T, FSg, sometimes FSw
gracilis	T, OS	T, FSg, sometimes FSw
barnumi	T, OS	T, FSG
fratercula	OS	T, FSG, sometimes FSw

Species	Male Signals	
	Male–female courtship interactions	Male–male interactions
Trimerptropis cincta	T, OS, FS	T, FS
inconspicua	T, OS	T, FSg
occidentalis	T, OS	T, FSg
melanoptera	OS	T, FSg
thalassica	OS	T, FSg
fontana	OS, FS	T
cyaneipennis	T, OS	FSw (Kaibab Plateau)
bifasciata	OS	T, FSg (So. Calif.)
pseudofasciata	OS, FS (So. Calif.)	T, FSg (So. Calif.)
Conozoa rebellis	OS	T, FSg
texana	OS	T, FSg
carinata	T, OS, FSg	T
sulcifrons	T, FSg	T
Circotettix rabula	T, OS	T, FSg
undulatus	OS	T, FSg, sometimes FSw
shastanus	?	T, FSg, sometimes FSw
Psinidia fenestralis	T, OS	T, FSg
Trachyrhachys kiowa	T, OS (Michigan)	T, FSg
Mestobregma impexum	OS	T, FSg
plattei	T, OS	T, FSg
Derotmema haydeni	T, OS	T, FSg
saussureana	T?, OS	

1. Alternating vibration
2. Two kinds.

APPENDIX 3

Ephiphalli and Aedeagi of Palaearctic and Nearctic Species of Oedipodinae

Note: the figures are not all drawn to the same scale.

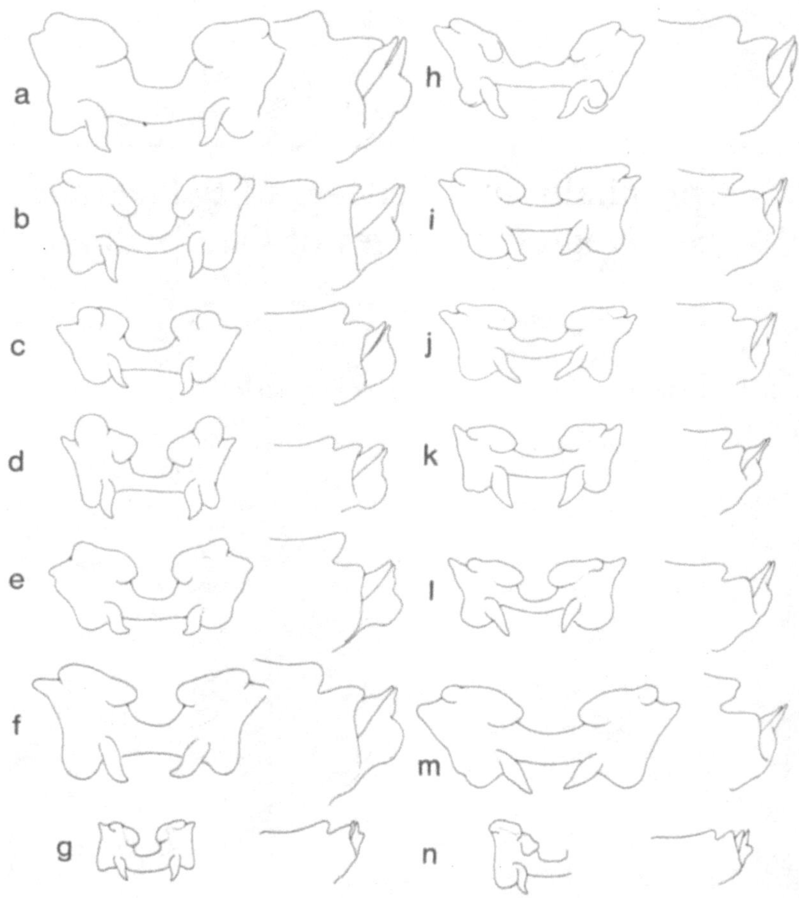

Fig. 67. *a, Arphia sulphurea; b, A. nietana; c, Lactista aztecus; d, Lactista elota; e, Leuronotina ritensis; f, Tomonotus ferruginosus; g, Oedipoda fuscocincta* (Sardinia); *h, Machaerocera mexicana; i, Chortophaga viridifasciata; j, Chimarocephala pacifica; k, Encoptolophus subgracilis; l, E. costalis; m, Shotwellia isleta; n, Celes variabilis.*

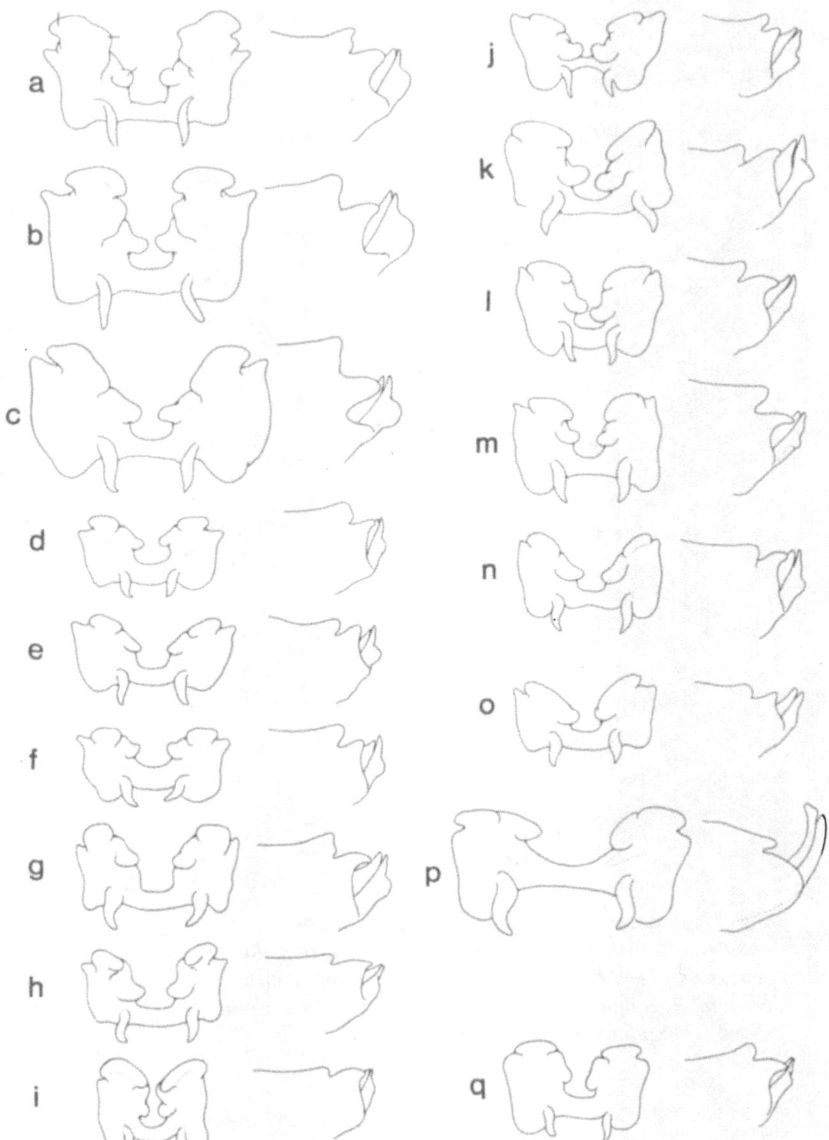

Fig. 68. *a, Heliastus infrequens; b, H. sumichrasti; c, H. aztecus; d, Hadrotettix scotodes; e, Had. magnificus; f, Had. trifasciatus; g, Leprus wheeleri; h, Hyalorrhipis calcarata; i, Hyal. calusi; j, Hippiscus ocelote; k, Pardalophora apiculata; l,* Xanthip*pus corallipes; m, X. aquilonius; n, Agymnastus venerabilis; o, Sticthippus californicus; p, Camnula pellucida; q, Leptoternis gracilis.*

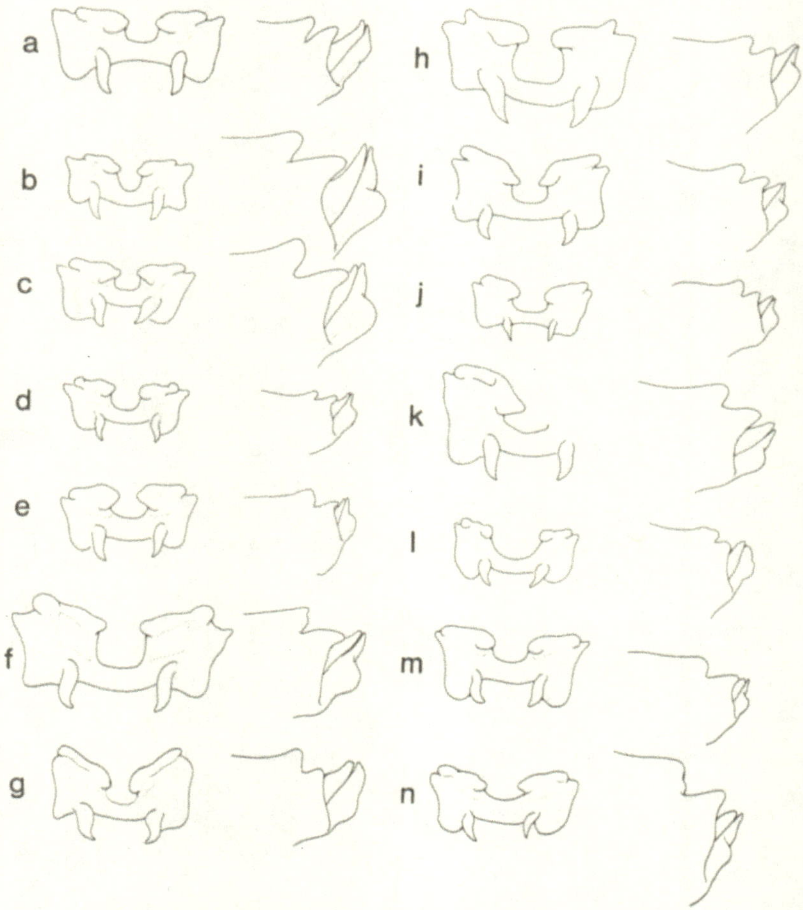

Fig. 69. *a, Dissosteira longipennis; b, D. carolina; c, Spharagemon equale; d, S. collare; e, S. cristatum; f, S. marmorata picta; g, Mioscirtus wagneri; h, Aiolopus oxianus; i, Crinita hirtipes; j, Ptetica cristata; k, Sphingoderus carinatus; l, Pyrgodera armata; m, Bryodema tuberculata; n, B. semenovi.*

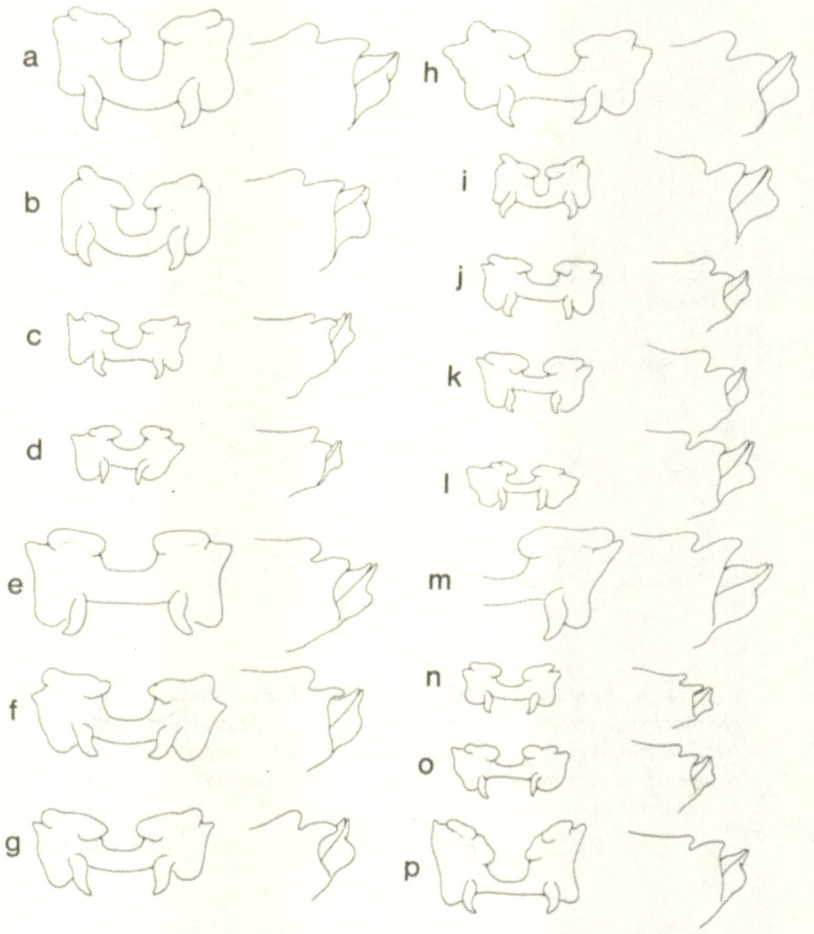

Fig. 70. *a, Circotettix undulatus; b, C. maculatus; c, C. carlinianus; d, C. rabula; e, Conozoa sulcifrons; f, Microtes occidentalis; f, Conozoa texana; h, Trimerotropis inconspicua; i, T. barnumi; j, T. latifasciata; k, T. pistrinaria; l, T. v. verruculata; m, T. agrestis; n, Anconia integra; o, A. hebardi; p, Sphingonotus haitensis.*

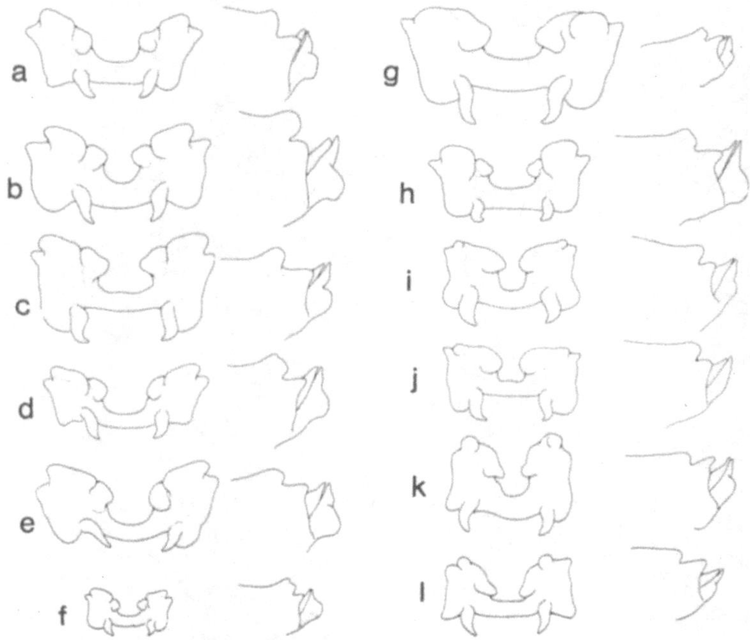

Fig. 71. *a, Psinidia fenestralis; b, Trachyrhachys kiowa; c, Trachyrhachys funeralis; d, Hippopedon capito; c, Mestobregma plattei; f, Mestobregma terricolor; g, Metator pardalinus; h, Trepidulus rosaceus; i, Derotmema haydeni; j, Derotmema laticinctum; k, Acrotylus insubricus; l, Acrotylus longipes.*

APPENDIX 4

Taxonomic Changes Made in This Volume

This list includes all new synonyms and generic reassignments adopted in this work (see Appendixes 5 and 6 for generic and species nomenclature).

Aerochoreutes Rehn, new synonym of **Circotettix** Scudder
Anconia caeruleipennis Bruner, new synonym of **Anconia integra** Scudder
Anconia grisea Bruner, new synonym of **Anconia integra** Scudder
Arphia aurantipennis Bruner, new synonym of **Arphia nietana** (Saussure)
Arphia pallidipennis Bruner, new synonym of **Arphia conspersa** Scudder
Chimarocephala behrensi Saussure, new synonym of **Chimarocephala pacifica** (Thomas)
Chortophaga meridionalis Bruner, new synonym of **Chortophaga viridifasciata** (DeGeer)
Circotettix coconino Rehn, new synonym of **Circotettix rabula** Rehn and Hebard
Circotettix splendidus Rehn and Hebard, new synonym of **Circotettix shastanus** Bruner
Circotettix stenometopus (Strohecker and Buxton), new combination for *Aerochoreutes stenometopus*
Conozoa clementina (Rentz and Weissman), new combination for *Scirtetica clementina*
Conozoa hyalina (McNeill), new combination for *Trimerotropis hyalina*
Conozoa nicola (Rentz and Weissman), new combination for *Microtes nicola*
Conozoa rebellis (Saussure), new combination for *Trimerotropis rebellis*
Conozoa texana (Bruner), new combination for *Trimerotropis texana*
Conozoa wallula (Scudder), new synonym of **Conozoa sulcifrons** (Scudder)
Derotmema haydenii mesembrinum Rehn, new synonym of **Derotmema haydeni** (Thomas)
Derotmema rileyanum Saussure, new synonym of **Derotmema haydeni** (Thomas)
Eximacris Otte, new genus
Eximacris phenax Otte, new species
Eximacris superbum (Hebard), new combination for *Spharagemon superbum*
Hadrotettix magnificus (Rehn), new combination for *Trimerotropis magnifica*

Hadrotettix scotodes Otte, new species
Heliastus cirrhoides Otte, new species
Heliastus costaricensis Rehn, new synonynm of **Heliastus sumichrasti** (Saussure)
Heliastus guatemalae Saussure, new synonym of **Heliastus sumichrasti** (Saussure)
Heliastus infrequens Otte, new species
Heliastus rubellus Otte, new species
Heliastus venezuelae Saussure, new synonym of **Heliastus sumichrasti** (Saussure)
Hippiscus australis Bruner, new synonym of **Trimerotropis tolteca** (Saussure)
Hippiscus rugosus (Scudder), new synonym of **Hippiscus ocelote** (Saussure)
Hippopedon capito (Stål), new combination for *Rehnita capito*
Hippopedon gracilipes (Caudell), new combination for *Rehnita gracilipes*
Lactista aztecus (Saussure), new combination for *Platylactista aztecus*
Lactista elota Otte, new species
Lactista micrus (Hebard), new combination for *Platylactista micrus*
Lactista oslari Caudell, new synonym of **Lactista gibbosus** Saussure
Leprus cyaneus Cockerell, new synonym of **Leprus intermedius** Saussure
Leprus glaucipennis Scudder, new synonym of **Leprus wheeleri** (Thomas)
Leprus interior Bruner, new synonym of **Leprus intermedius** Saussure
Leprus robustus Hebard, new synonym of **Leprus wheeleri** (Thomas)
Leuronotina obesa Otte, new species
Leuronotina philorites Otte, new species
Leuronotina ritensis (Rehn), new combination for *Scirtetica ritensis*
Machaerocera magna Bruner, new synonym of **Machaerocera mexicana** Saussure
Microtes helferi (Strohecker), new combination for *Trimerotropis helferi*
Microtes pogonata (Strohecker), new combination for *Trimerotropis pogonata*
Platylactista Hebard, new synonym of **Lactista** Saussure
Rehnita Hebard, new synonym of **Hippopedon** Saussure
Scirtetica Saussure, new synonym of **Spharagemon** Scudder
Spharagemon bunites Otte, new species
Spharagemon campestris (McNeill), new combination for *Trimerotropis campestris*
Spharagemon humile Morse, new synonym of **Spharagemon equale** (Say)
Spharagemon marmorata (Harris), new combination for *Scirtetica marmorata*
Sphingonotus haitensis cubensis Saussure, new status
Trachyrhachys inconspicua Bruner, new synonym of **Trachyrhachys aspera** Scudder
Trachyrhachys mexicana Saussure, new synonym of **Trachyrhachys kiowa** (Thomas)
Trachyrhachys occidentalis Bruner, new synonym of **Conozoa sulcifrons** (Scudder)
Trepidulus concinens Otte, new species
Trimerotropis acta Hebard, new synonym of **Trimerotropis maritima** (Harris)

Trimerotropis agrestis gracewileyae Tinkham, new synonym of **Trimerotropis barnumi** Tinkham

Trimerotropis barnumi Tinkham, new status for *Trimerotropis agrestis barnumi*

Trimerotropis bilobata Rehn and Hebard, new synonym of **Conozoa texana** (Bruner)

Trimerotropis citrina Scudder, new synonym of **Trimerotropis maritima** (Harris)

Trimerotropis citrina neomexicana Tinkham, new synonym of **Trimerotropis maritima** (Harris)

Trimerotropis cristata McNeill, new synonym of **Conozoa texana** (Bruner)

Trimerotropis huasteca (Saussure), *nomen dubium*

Trimerotropis laticincta Saussure, new synonym of **Trimerotropis latifasciata** Scudder

Trimerotropis modesta Bruner, elevated from subspecies of **T. tolteca**

Trimerotropois monticola Saussure, *nomen dubium*

Trimerotropis munda (Caudell), new synonym of **Trimerotropis pacifica** Bruner

Trimerotropis salina Thomas, elevated from subspecies of **T. pallidipennis**

Trimerotropis santamonica Rentz and Weissman, new synonym of **Trimerotropis thalassica** Bruner

Trimerotropois strenua McNeill, new synonynm of **Trimerotropis californica** Bruner

Trimerotropis townsendi Bruner, new synonym of **Trimerotropis tolteca** (Saussure)

Xanthippus affrictus Scudder, new synonym of **Xanthippus corallipes** (Haldeman)

Xanthippus calthulus Saussure, new synonym of **Cratypedes lateritius** (Saussure)

Xanthippus croceus Scudder, new synonym of **Cratypedes neglectus** (Thomas)

Xanthippus griseus Scudder, new synonym of **Cratypedes lateritius** (Saussure)

Xanthippus sierra (Rhen), new synonym of **Cratypedes lateritius** (Saussure)

Xanthippus validus Scudder, new synonym of **Cratypedes lateritius** (Saussure)

Nomina Dubia

Trimerotropis monticola Saussure 1884: 170. The type of this species, a female from Tezuitlan, Puebla, Mexico, has evidently been replaced by a male of *Trimerotropis melanoptera.* The specimen bears two labels: "Tesuitlan" and "monticola Ss Type." No other specimens bear similar labels.

Trimerotropis monticola var. *huasteca* Saussure 1888: 63. Holotype ♂, Lake Metztitlan, Hidalgo, Mexico. GM. *Pseudotrimerotropis monticola,* Kirby 1910. *Trimerotropis monticola,* Caudell 1911. *Trimerotropis huasteca,* Hebard 1932: 260. I could not find this specimen in the Geneva Museum.

APPENDIX 5

The Genera of Oedipodinae

Each genus is listed in bold type, and its synonyms in italics, along with the reference to the original description of the genus and the name of the type species.

Agymnastus Scudder 1897. Type species: *Agymnastus ingens* (Scudder), by monotypy.

Anconia Scudder 1876. Type species: *Anconia integra* Scudder, by monotypy.

Arphia Stål 1873. Type species: *Arphia sulphurea* (Fabricius), subsequent designation by Rehn 1904b.

Camnula Stål 1873. Type species: *Camnula tricarinata* Stål (synonym of *C. pellucida*), by monotypy.

Chimarocephala Scudder 1876. Type species: *Chimarocephala pacifica* (Thomas), by monotypy.

Chortophaga Saussare 1884. Type species: *Chortophaga viridifasciata* (De Geer), by monotypy.

Tragocephala Harris 1841. Name preoccupied by *Tragocephala* Dupont.

Circotettix Scudder 1876. Type species: *Circotettix undulatus* (Thomas), subsequent designation by Kirby 1910.

Aerochoreutes Rehn 1921. Type species: *Circotettix carlinianus* (Thomas), by original designation. NEW SYNONYM

Conozoa Saussure 1884. Type species: *Conozoa behrensi* Saussure (synonym of *C. sulcifrons*), subsequent designation by Kirby 1910.

Agonozoa McNeill 1901. Type species: *Agonozoa texana* (Bruner), subsequent designation by Caudell 1911.

Cratypedes Thomas 1876. Type species: *Cratypedes neglectus* (Thomas), subsequent designation by Kirby 1910.

Derotmema Scudder 1876. Type species: *Derotmema cupidineum* Scudder (synonym of *Derotmema haydeni*) by monotypy.

Dissosteira Scudder 1876. Type species: *Dissosteira carolina* (Linn.), subsequent designation by Kirby 1910.

Encoptolophus Scudder 1875. Type species: *Encoptolophus sordidus* (Burmeister), subsequent designation by Rehn 1904b.

Eximacris Otte, NEW GENUS. Type species: *Eximacris superbum* (Hebard) (previously under *Spharagemon*), here designated.

Hadrotettix Scudder 1876. Type species: *Hadrotettix trifasciatus* (Say), by monotypy.

Heliastus Saussure 1884. Type species: *Heliastus sumichrasti* (Saussure), subsequent designation by Rehn 1904b.

Hippiscus Saussure 1861. Type species: *Hippiscus ocelote* Saussure, by monotypy.

Hippopedon Saussure 1861. Type species: *Hippopedon saltator* Saussure, by monotypy.

Rehnita Hebard 1935. Type species: *Rehnita gracilipes* (Caudell), by original designation. NEW SYNONYM.

Lactista Saussure 1884. Type species; *Lactista punctatus* (Stål), subsequent designation by Rehn 1904b.

Platylactista Hebard 1932. Type species: *Platylactista azteca* (Saussure), by monotypy. NEW SYNONYM.

Leprus Saussure 1861. Type species: *Leprus elephas* Saussure, by monotypy.

Leuronotina Hebard 1932. Type species: *Leuronotina orizabae* (Saussure), by monotypy.

Mestobregma Scudder 1876. Type species: *Mestobregma plattei* (Thomas), by monotypy.

Metator McNeill 1901. Type species: *Metator pardalinus* (Saussure), by original designation.

Microtes Scudder 1900. Type species: *Microtes nubila* Scudder (synonym of *Microtes occidentalis*) by monotypy.

Pardalophora Saussure 1884. Type species: *Pardalophora phoenicoptera* (Burmeister), subsequent designation by Kirby 1910.

Psinidia Stål 1873. Type species: *Psinidia fenestralis* (Serville), subsequent designation by Kirby 1910.

Shotwellia Gurney 1940. Type species: *Shotwellia isleta* Gurney, by original designation.

Spharagemon Scudder 1875. Type species: *Spharagemon equale* (Say), subsequent designation by Rehn 1904.

Scirtetica Saussure 1884. Type species: *Scirtetica marmorata* (Harris), subsequent designation by Kirby 1910.

Sphingonotus Fieber 1852. Type species: *Sphingonotus caerulans* (Linn.), subsequent designation by Kirby 1910.

Sticthippus Scudder 1892. *Sticthippus californicus* Scudder, subsequent designation by Kirby 1910.

Tomonotus Saussure 1861. Type species: *Tomonotus mexicanus* Saussure, subsequent designation by Kirby 1910.

Trachyrhachys Scudder 1876. Type species: *Trachyrhachys aspera* Scudder, subsequent designation by Kirby 1910.

Trepidulus McNeill 1901. Type species: *Trepidulus rosaceus* (Scudder), by original designation.

Aereopteryx Caudell 1901. Type species: *Aereopteryx penelopae,* by monotypy. Synonymized by Rehn 1904.

Trimerotropis Stål 1873. Type species: *Trimerotropis maritima* (Harris), subsequent designation by Rehn 1904.

Pseudotrimerotropis Rehn 1902. Type species: *Pseudotrimerotropis cincta* (Thomas), subsequent designation by Kirby 1910.

Tropidolophus Thomas 1873. Type species: *Tropidolophus formosus* (Say), by monotypy.

Xanthippus Saussure 1884. Type species: *Xanthippus corallipes* (Haldeman), subsequent designation by Kirby 1910.

APPENDIX 6

The Species of Oedipodinae

This index lists in alphabetical order all species of Oedipodinae (boldface) and their synonyms (italics). Literature references, type localities, and all generic reassignments are included. The present locations of the types are given by the following abbreviations:

ANSP Academy of Natural Sciences of Philadelphia, Philadelphia
BM British Museum (Natural History), London
CAS California Academy of Sciences, San Francisco
GM Muséum d'Histoire Naturelle, Geneva
UMMZ University of Michigan Museum of Zoology
USNM United States National Museum, Washington D.C.
VM Naturhistorisches Museum, Vienna

Agymnastus ingens (Scudder)

Leprus ingens 1877: 32. Holotype ♀, Sausalito, California, June 7. ANSP. *Xanthippus?,* Saussure 1884. *Agymnastus,* Scudder 1897: 75.

Hippiscus (*Xanthippus*) *stigmosus* Scudder 1892: 363. Lectotype ♂, here designated, Ukiah, California. ANSP. *Xanthippus,* Kirby 1910. *Hippiscus,* Caudell 1911. *Agymnastus ingens,* Hebard 1928.

Hippiscus (*Xanthippus*) *aurilegulus* Scudder 1892: 362. Lectotype ♂, California, 79. ANSP. *Xanthippus,* Kirby 1910. *Hippiscus,* Caudell 1911. *Agymnastus ingens,* Hebard 1928: 239.

Xanthippus vitellinus Saussure 1884: 90. Nevada. Location of type not known. *Hippiscus* (*Xanthippus*), Scudder 1892. *Xanthippus,* Kirby 1910. *Hippiscus,* Caudell 1911. *Agymnastus ingens,* Strohecker, Middlekauff, and Rentz 1968: 85.

Agymnastus venerabilis Rentz 1978: 83. Holotype ♂, Dardanelle, Tuolumne County, California, July 19, 1961. CAS.

Anconia hebardi Rehn 1919: 250. Holotype ♀, El Paso, Texas, 3,650 feet, July 10, 1907. ANSP.

Anconia integra Scudder 1876: 515. Lectotype ♀, Mojave Desert, California. Scudder's type no. 1876. ANSP.

Anconia grisea Bruner 1906: 186. Holotype ♂, Antelope Valley, California. ANSP. NEW SYNONYM.

Anconia caeruleipennis Bruner 1906: 186. Holotype ♀, Hawthorne, Nevada. June. ANSP. NEW SYNONYM.

Arphia behrensi Saussure 1884: 72. Lectotype ♂, here designated, Napa Soda Spring, California, May 30, 1970, 602–35. GM. *Arphia conspersa behrensi,* Strohecker, Middlekauff, and Rentz 1968.

Arphia conspersa Scudder 1875: 514. Lectotype ♀, April 26, Dallas, Texas, Boll's no. 84, Scudder's type 1875, type no. 15895. ANSP.

Arphia arcta Scudder 1876: 263. Type ♂, Denver, Colorado, July 10. ANSP. *Arphia frigida,* Walker 1910: 293. Synonymized by Hebard 1931: 148.

Arphia frigida Scudder 1875: 344. Lectotype ♂, type no. 15896, designated by Hebard 1937: 363, Yukon River, Alaska. ANSP. Synonymized by Hebard 1931: 386

Arphia teporata Scudder 1876: 508. Lectotype ♂, designated by Hebard 1929: 333n., southern Colorado, USNM.

Arphia acta, Caudell 1903: 785. *Arphia frigida,* Walker 1910: 293. *Arphia conspersa,* Hebard 1931: 386.

Arphia infernalis Saussure 1884: 70. Lectotype ♂, here designated, Fort McKinley, Wyoming. GM. Synonymized by Hebard 1931: 148.

Arphia canora Rehn 1904: 564. Lectotype ♂, designated by Rehn and Hebard 1912: 99, Salt Lake City, July 15, 1902. ANSP. Synonymized by Hebard 1937: 363.

Arphia aberrans Bruner 1905: 129. Lectotype ♀, designated by Rehn and Hebard 1912: 126, Huachuca Mountains, Arizona. ANSP. NEW SYNONYM, but see discussion under genus.

Arphia pallidipennis Bruner 1905: 131. Lectotype ♀, designated by Rehn and Hebard 1912: 114, Eslava, D.F., Mexico. ANSP. NEW SYNONYM.

Arphia granulata Saussure 1884: 67. Lectotype ♂, here designated, Georgia. GM.

Arphia nietana (Saussure)

Tomonotus nietanus Saussure 1861: 70. Type ♂, Cordova t.c, Mexico, GM. *Arphia nietana,* Saussure 1884: 70.

Arphia imperfecta Bruner 1905: 126. Type ♀, Tlalpam, Mexico, November 1887. ANSP. Synonymized by Hebard 1932: 244.

Arphia truculenta Rehn 1904: 522. Holotype ♂, Guadalajara, Jalisco, Mexico, September 18, 1903. ANSP. NEW SYNONYM.

Arphia aurantipennis Bruner 1905: 125. Lectotype ♂, here designated, Amula, Guerrero, Mexico, 6,000 ft., August. ANSP. NEW SYNONYM.

Arphia novissima Otte, NEW SPECIES. Holotype ♂, Alamos, Sonora, Mexico, August 12, 1960 (D. C. Rentz). ANSP.

Arphia pecos Otte, NEW SPECIES. Holotype ♂, Pecos River at Palmdale, Val Verde County, Texas, July 13, 1970. ANSP.

Arphia pseudonietana (Thomas)

Tomonotus pseudonietanus Thomas 1870: 82. Holotype ♂, near Cañon City, Colorado, in mountain canyon, USNM. *Arphia pseudonietana,* Caudell 1903: 786. *Arphia pseudonietana pseudonietana,* Hebard 1935: 286.

Oedipoda tenebrosa Scudder 1872: 251. 2♀, Nebraska City; ♂, Platte, Nebraska. ANSP. *Tomonotus,* Thomas 1873: 107. *Arphia,* Scudder 1875: 344. *Tomonotus,* Thomas 1876: 63. *Arphia,* Saussure 1884: 68. *Arphia pseudonietana,* Caudell 1903: 786. *Arphia pseudonietana pseudonietana,* Hebard 1935: 286.

Arphia sanguinaria Stål 1873. Type ♂, Insula Vancouver Americae borealis occidentalis. SM. *Tomonotus tenebrosus,* Thomas 1875: 874. *Arphia pseudonietana,* Caudell 1903: 786.

Arphia ovaticeps Saussure 1888: 165. Type ♀, no locality data on specimen, just 602 and 33 and Arphia ovaticeps Sauss. GM. Synonymized by Hebard 1935: 286.

Tomonotus theresiae Brunner de Wattenwyl 1895: 277. Type ♀, Fort Yates, North Dakota. VM. Synonymized by Hebard 1925: 68.

Arphia calida Bruner 1905: 127. Lectotype ♂, designated by Rehn and Hebard 1912: 114, Zacatecas, Mexico, November 1887. ANSP. Synonymized by Hebard 1932: 244.

Arphia crassa Bruner 1905: 128. Lectotype ♂, designated by Rehn and Hebard 1912: 114, southern Arizona. ANSP. *Arphia pseudonietana crassa,* Hebard 1935: 286. NEW SYNONYM.

Arphia pulchripennis Bruner 1905: 131. Holotype ♂, vicinity of Los Angeles, California. ANSP. *Arphia conspersa,* Strohecker, Middlekauff, and Rentz, 1968.

Arphia ramona Rehn 1902: 142. Lectotype ♂, designated by Rehn and Hebard 1912, San Diego, California, April 1901. ANSP. *Arphia conspersa ramona,* Strohecker, Middlekauff, and Rentz 1968.

Arphia koebelei Bruner 1905: 133. Lectoptype ♂, designated by Rehn and Hebard 1912: 114, San Mateo County, California. ANSP. *Arphia conspersa behrensi,* Strohecker, Middlekauff, and Rentz 1968.

Arphia robertsi Otte, NEW SPECIES. Holotype ♂, Zitácuaro, Michoacán, Mexico, March 13, 1938 (M.J.D. White). ANSP.

Arphia saussureana Bruner 1890: 63. Lectotype ♂, designated by Rehn and Hebard 1912: 64, San Francisco, California, November 1887. ANSP. *Arphia pseudonietana,* Strohecker, Middlekauff, and Rentz, 1968.

Arphia secreta Otte, NEW SPECIES. Holotype ♂, Cuatrocienegas, Coahila, Mexico, May 20, 1975 (R. Rice). ANSP.

Arphia simplex Scudder 1875: 514. Type ♀, Dallas, Texas, Boll's no. 83, type no. 15890. ANSP.

Arphia luteola Scudder 1875: 515. Cotypes 2♂, Dallas, Texas (Boll). ANSP. Synonymized by Hebard 1925: 71.

Arphia decepta Bruner 1905: 132. Lectotype ♀, designated by Rehn and Hebard 1912: 114, Orizaba, Mexico, November. ANSP. Synonymized by Hebard 1925: 71.

Arphia sulphurea (Fabricius)

Gryllus sulphureus Fabricius 1781: 369. North America, location of types not known. *Acrydium sulphureum,* Olivier 1791: 227. *Oedipoda,* Burmeister 1838: 643. *Locusta,* Harris 1841: 143. *Tomonotus,* Saussure 1861: 21. *Oedipoda,* Scudder 1862: 470. *Locusta,* Harris 1862: 177.

Oedipoda, Walker 1870: 721. *Arphia sulphurea,* Stål 1873: 119. *Tomonotus,* Thomas 1873: 105. *Oedipoda* (*Arphia*), Provancher 1883: 39. *Arphia,* Saussure 1884: 71.

Arphia townsendi Bruner 1905: 122. Lectotype ♂, designated by Rehn and Hebard 1912: 114, Colonia Gracia, Chihuahua, Mexico, ANSP.

Arphia xanthoptera (Burmeister)

Oedipoda xanthoptera Burmeister 1838: 643. Sex and location of type not known, based on specimens from "Carolina."

Oedipoda carinata Scudder 1869: 306. Type ♀ lost, locality unknown. *Tomonotus carinatus,* Thomas 1873: 106. *Tomonotus sulphureus carinatus,* Thomas 1876: 63. *Tomonotus carinatus,* Putnam 1876: 265. *Arphia carinata,* Saussure 1884: 67. *Arphia xanthoptera carinata,* Blatchley 1891: 78. *Arphia carinata,* Lugger 1898: 139. *Arphia xanthoptera carinata,* Blatchley 1920: 250. *Arphia xanthoptera,* Hebard 1931: 149.

Arphia crepusculum Saussure 1884: 67. Lectotype ♂, Dallas County, Iowa. GM. Synonymized by Hebard 1931: 149.

Arphia cristata Bruner. *Nomen nudum.* Hebard 1931: 149.

Camnula pellucida (Scudder)

Oedipoda pellucida Scudder 1862: 472. Type ♂, Massachusetts. ANSP. *Camnula pellucida,* Scudder 1874.

Oedipoda atrox Scudder 1872: 253. 1♂, 7♀, California. ANSP. *Camnula atrox,* Scudder 1874: 378. Synonymized by Saussure 1884: 81.

Oedipoda tricarinata Stål 1873: 120. Types ♂, ♀, Vancouver Island, Canada. SM. Synonymized by Saussure 1884: 81.

Stenobothrus obiona Thomas 1871a: 279. Types lost, "first seen on the Arkansas in the vicinity of Canon City." *Tragocephala,* Thomas 1873b: 100. *Chimarocephala,* Bruner 1883: 56. *Camnula pellucida* var. *obiona,* Bruner 1888: 164. *Camnula pellucida,* Scudder 1900: 32.

Chimarocephala elongata Rentz 1977: 146. Holotype ♂, Wilbur Springs, Colusa County, California, June 27, 1961. CAS.

Chimarocephala pacifica (Thomas)

Tragocephala pacifica Thomas 1873: 101. Lectotype ♂, designated by Hebard 1927, California. USNM. *Chimarocephala,* Scudder 1875: 484. *Chimarocephala,* Caudell 1906: 123. *Chimarocephala pacifica obtusa,* Caudell 1912: 160.

Chimarocephala behrensi Saussure 1873: 75. Lectotype ♂, here designated, California. GM. NEW SYNONYM.

Tragocephala brevipennis Scudder 1875: 483. Types 3♀, California. ANSP. *Chimarocephala,* Scudder 1876: 508. *Chortophaga,* Saussure 1884: 73. *Chimarocephala pacifica,* Strohecker, Middlekauff, and Rentz 1968: 93.

Chimarocephala pacifica incisa Caudell 1906: 124. Lectotype ♂, designated by Caudell and Hebard 1912, California. USNM. *Chimarocephala incisa,* Rehn and Hebard 1910: 435. *C. pacifica incisa* Hebard 1931: 119. NEW SYNONYM.

Chortophaga australior Rehn and Hebard 1911: 589. Lectotype ♀ lost. Neotype ♂, here designated, Key West, Florida, March 15–16, 1910. ANSP. *Chortophaga viridifasciata australior,* Blatchley 1920: 257.

Chortophaga cubensis (Scudder)

Tragocephala cubensis Scudder 1875a. Lectotype ♂, here designated, firmina, Cuba. ANSP. *Chimarocephala,* Scudder 1876: 508. *Chortophaga,* Saussure 1884: 74.

Chortophaga mendocino Rentz 1977: 142. Holotype ♂, Mendocino, Mendocino County, California, March 27, 1963. CAS.

Chortophaga viridifasciata (DeGeer)

Acridium viridifasciatum DeGeer 1773. Type ♀, Pennsylvania, SM. *Locusta viridifasciata* Harris 1835: 56. *Tragocephala,* Scudder 1862: 461. *Chimarocephala,* Scudder 1876: 508. *Chortophaga,* Saussure 1884: 73.

Acridium hemipterum Beauvois. Synonymized by Thomas 1873: 103.

Acridium marginatum Olivier. Synonymized by Thomas 1873: 103.

Gryllus virginianus Fabricius. Synonymized by Thomas 1873: 103.

Gryllus chrysomelas Gmelin. Synonymized by Thomas 1873: 103.

Tomonotus zimmermanni Saussure. Synonymized by Saussure 1884: 73.

Locusta (*Tragocephala*) *infuscata* Harris 1841. No type. Synonymized by Scudder 1875: 80.

Locusta radiata Harris 1835. Synonymized by Thomas 1873: 104.

Chortophaga meridionalis Bruner 1905: 136. Lectotype ♀, designated by Rehn and Hebard 1912: 114, Costa Rica. ANSP. NEW SYNONYM.

Circotettix carlinianus (Thomas)

Oedipoda carliniana Thomas 1870: 81. Lectotype ♂, designated by Hebard 1927, eastern Colorado. USNM. *Oedipoda carlingiana,* Thomas 1871, *Circotettix carlingiana,* Saussure 1884. *Circotettix carlinianus,* Scudder 1900. *Aerochoreutes carlinianus carlinianus,* Rehn 1921.

Aerochoreutes carlinianus strepitus Rehn 1921: 174. Holotype ♂, Prince Royal Canyon, Star Peak Range, Humboldt County, Nevada, 4,500–5,000 feet, September 16, 1919. ANSP. NEW SYNONYM.

Circotettix crotalum Rehn 1921: 191. Holotype ♂, Lee Canyon, Spring Mountains, Clark County, Nevada, 8,000–8,500 feet, August 19, 1919. ANSP.

Circotettix maculatus Scudder 1880: 26. Lectotype ♂, here designated, Sierra Nevada, July 17, 1872. Type no. 15866. ANSP. *Aerochoreutes maculatus,* Strohecker, Middlekauff, and Rentz 1968.

Circotettix rabula Rehn and Hebard 1906: 393. Lectotype ♀, designated by Rehn ansd Hebard 1912, summit of hills at head of Mammoth Hot Springs, Yellowstone National Park, Wyoming, 7,000 feet, August 8, 1904. ANSP. *Circotettix rabula rabula,* Rehn 1921.

Circotettix nigrafasciata Beamer 1917: 123. Lectotype ♂, designated by Rehn 1921: 181n., Trego County, Kansas, July 12, 1912. Snow Entomological Museum, University of Kansas. *Circotettix rabula nigrafasciata,* Rehn 1921: 181.

Circotettix rabula altior Rehn 1921: 183. Holotype ♂, Cloudcroft, Sacramento Mountains, Otero County, New Mexico, 8,600–8,700 feet, June 12, 1902. ANSP. NEW SYNONYM.

Circotettix coconino Rehn 1921: 187. Holotype ♂, Bill Williams Mountain, Coconino County, Arizona, September 14, 1917. ANSP. NEW SYNONYM.

Circotettix shastanus Bruner 1889: 76. Holotype ♂, Hazel Creek, Shasta County, California, July 1885. USNM.

Circotettix splendidus Rehn and Hebard 1909: 456. Lectotype ♂, designated by Rehn and Hebard 1912, Mount Lowe, San Gabriel Mountains, California, 5,300 feet, July 8, 1907. ANSP. NEW SYNONYM.

Circotettix stenometopus (Strohecker and Buxton) NEW COMBINATION

Aerochoreutes stenometopus Strohecker and Buxton 1963: 261. Holotype ♂, Telephone Campgrounds, Glenn County, California. California Department of Agriculture.

Circotettix undulatus (Thomas)

Oedipoda undulata Thomas 1872: 460. Lectotype ♀, designated by Rehn 1921, Colorado. USNM. *Circotettix,* Scudder 1876.

Circotettix thalassinus Saussure 1884: 177. Holotype ♂, Nevada, VM. *Circotettix undulatus thalassinus,* Strohecker, Middlekauff, and Rentz 1968. NEW SYNONYM.

Circotettix lobatus Saussure 1888: 65. Type ♀. Hebard 1928: 263, states in a footnote: "Saussure's type of *lobatus* . . . came originally from the Bruner collection and is labelled correctly 'Salmon City, Idaho, August 1883' (as is other material of *undulatus* of the same series before us) but bears also the label, erroneous as to locality, 'type, Colorado.' GM.

Circotettix lapidicolus Bruner 1890: 75. Lectotype ♂, designated by Rehn and Hebard 1912, Salmon City, Idaho, July 1883. ANSP. Synonymized by Rehn 1921: 178.

Conozoa carinata Rehn 1907: 38. Holotype ♂, Carr Canyon, Huachuca Mountains, Cochise County, Arizona, August 1905. ANSP.

Conozoa clementina (Rentz and Weissman) NEW COMBINATION

Scirtetica clementina Rentz and Weissman 1981: 74. Holotype ♂, naval barracks area, San Clemente Island, Los Angeles County, California. August 8, 1969. CAS.

Conozoa hyalina (McNeill) NEW COMBINATION

Trimerotropis hyalina McNeill 1901: 405. Holotype ♀ from California lost from USNM.

Trimerotropis gratiosa Scudder 1900: 330. Lectotype ♂, designated by Rentz and Weissman 1980, Ceres, Stanislaus County, California. ANSP. *Trimerotropis hyalina,* Strohecker, Middlekauff, and Rentz 1968.

Conozoa nicola (Rentz and Weissman) NEW COMBINATION

Microtes nicola Rentz and Weissman 1981: 72. Holotype ♂, Thousand Springs, Celery Canyon, San Nicolas Island, Ventura County, California, July 16, 1969. CAS.

Conozoa rebellis (Saussure) NEW COMBINATION

Conozoa rebellis Saussure 1888: 61. The types, 5♂ 1♀ in the Brunner collection, VM, are from California. *Trimerotropis,* McNeill 1900, 1901. *Conozoa,* Kirby 1910. *Trimerotropis,* Strohecker, Middlekauff, and Rentz 1968.

Conozoa albolineata Bruner 1889: 66. The types, from Los Angeles, California, were destroyed. *Trimerotropis,* McNeill 1900. Synonymized by Hebard 1932: 259.

Conozoa sulcifrons (Scudder)

Psinidia sulcifrons Scudder 1876b: 293. Holotype ♂, near Mohave River, San Bernardino County, California, July. ANSP. *Conozoa sulcifrons,* Saussure 1884: 166. *Conozoa sulcifrons sulcifrons,* Hebard 1929: 351.

Psinidia wallula Scudder 1880: 27. Lectotype ♂, Wallula, Washington, designated by Gurney and Strohecker 1959: 79. ANSP. *Conozoa wallula,* Saussure 1884: 166. NEW SYNONYM.

Conozoa behrensi Saussure 1884: 165. Described from California. Synonymized by Thompson and Buxton 1964: 18, under *Conozoa wallula.*

Conozoa acuminata Scudder 1902: 32. Holotype ♂, Mesilla, New Mexico, June 27, 1897. ANSP. *Conozoa sulcifrons acuminata,* Rehn 1929.

Trachyrhachys occidentalis Bruner 1905: 174. Lectotype ♂, designated by Rehn and Hebard 1912, hilltops southwest of San Francisco, California. ANSP. NEW SYNONYM.

Conozoa texana (Bruner) NEW COMBINATION

Conozoa texana Bruner 1889: 65. The types from El Paso, Texas, are lost (Rentz and Weissman 1980). Neotype ♂, here designated, El Paso, Texas, July 11, 1907, 3,650 feet. ANSP. Trimerotropis texana, McNeill 1900.

Trimerotropis cristata McNeill 1901: 408. Types from San Julio, Baja California, probably lost or destroyed; not in ANSP as indicated by Rentz and Weissman. NEW SYNONYM.

Trimerotropis bilobata Rehn and Hebard 1906: 382. Lectotype ♂, designated by Rehn and Hebard 1912, Antlers, Mesa County, Colorado, 5,310 feet, August 5, 1904. ANSP. *Conozoa,* Kirby, 1910. *Trimerotropis,* Hebard 1929: 351. NEW SYNONYM.

Cratypedes lateritius (Saussure)

Xanthippus lateritius Saussure 1884: 92. Type ♀, Nevada, GM. *Hippiscus* (*Xanthippus*), Scudder 1892. *Xanthippus,* Kirby 1910. *Hippiscus,* Caudell 1911. *Xanthippus,* Strohecker, Middlekauff, and Rentz 1968: 130.

Xanthippus calthulus Saussure 1884: 93. Lectotype ♂, here designated, Nevada. GM. *Hippiscus* (*Xanthippus*), Scudder 1892. *Xanthippus,* Kirby 1910: 210. *Hippiscus,* Caudell 1911. *Xanthippus,* Strohecker, Middlekauff, and Rentz 1968: 129. NEW SYNONYM.

Hippiscus (*Xanthippus*) *griseus* Scudder 1892: 362. Types 3♀, Palmer's sorting no. 1209, no locality labels on specimens, assumed to be Mokiak Pass, Utah, April 20–30. ANSP. *Xanthippus,* Kirby 1910. NEW SYNONYM.

Hippiscus sierra Rehn 1907b: 350. Holotype ♀, Mt. Tyndall, 12,000 ft, Tulare County, California. ANSP. *Xanthippus,* Kirby 1910. *Hippiscus,* Caudell 1911. *Xanthippus,* Strohecker, Middlekauff, and Rentz 1968. NEW SYNONYM.

Hippiscus validus Scudder 1901: 91. Lectotype ♀, here designated, Blaine County, Idaho, July 23. ANSP. *Xanthippus,* Kirby 1910. *Hippiscus,* Caudell 1911. NEW SYNONYM.

Cratypedes neglectus (Thomas)

Oedipoda neglecta Thomas 1870: 81. Holotype ♀, New Mexico. USNM. *Hippiscus,* Scudder 1876. *Arphia,* Thomas 1876. *Xanthippus,* Saussure 1884. *Hippiscus* (*Xanthippus*), Scudder 1892. *Cratypedes,* Kirby 1910. *Hippiscus,* Caudell 1911. *Xanthippus,* Buckell 1924. *Cratypedes neglectus neglectus,* Hebard 1928: 238.

Cratypedes putnami Thomas 1876a: 257. Types 4♀, Empire, Colorado, 1♀, Middle Park, Colorado. USNM. *Hippiscus* (*Xanthippus*), Scudder 1892.

Hippiscus (*Xanthippus*) *croceus* Scudder 1901: 91. Lectotype ♀, here designated, Blaine County, Idaho, July 23. ANSP. *Cratypedes?,* Kirby 1910: 205. *Xanthippus,* Kirby 1910: 210. *Hippiscus croceus,* Caudell 1911: 162. NEW SYNONYM.

Hippiscus (*Xanthippus*) *glaucipes* Scudder 1901: 90. Lectotype ♂, here designated, Nev. F.—Cl'ds Rest, Y.V., California, August 12, 1897. ANSP. *Xanthippus,* Kirby 1910. *Hippiscus,* Caudell 1911. *Cratypedes neglectus glaucipes,* Hebard 1928: 238.

Hippiscus (*Xanthippus*) *obscurus* Scudder 1892: 359. Lectotype ♀, designated by Hebard, British Columbia. ANSP. *Xanthippus,* Kirby 1910. *Hippiscus,* Kirby 1910. *Hippiscus,* Caudell 1911. *Xanthippus neglectus,* Rehn and Hebard in Buckell 1924: 10.

Derotmema delicatulum Scudder

Derotmema (*Derotmema*) *delicatulum* Scudder 1900: 390. Lectotype ♂, designated by Rehn and Hebard 1912: 96, Lancaster, California, July 13, 1897. ANSP.

Derotmema haydeni (Thomas)

Oedipoda haydenii Thomas 1872: 460. Types lost. Neotype ♂, here designated, Pineridge, South Dakota, September 2, 1924. ANSP. *Derotmema haydeni,* Bruner 1897: 132. *Derotmema* (*Derotmema*) *haydeni,* Scudder 1900: 392. *Derotmema haydenii haydenii,* Rehn 1919: 230.

Derotmema brunnerianum Saussure 1884: 155. Holotype ♀, Colorado. GM. Synonymized by Scudder 1900: 392.

Derotmema cupidineum Scudder 1876: 513: Lectotype ♂, here designated, northern New Mexico, Scudder's Type, 1876. ANSP. *Derotmema haydenii haydenii,* Rehn 1919: 230.

Derotmema haydenii var. *flavum* Scudder and Cockerell 1902: 32. No types. Var. *flavum* referred to yellow-winged specimens. *Derotmema haydenii haydenii,* Rehn 1919: 230.

Derotmema rileyanum Saussure 1884: 156. Type ♂, Amer. Sept. GM. *Derotmema (Tmetodera) rileyanum,* Scudder 1900: 392. *Tmetodera rileyanum,* Kirby 1910: 244. *Derotmema haydenii rileyanum,* Rehn 1919: 230. NEW SYNONYM.

Derotmema haydenii mesembrinum Rehn 1919: 231. Holotype ♂, Doubl Windmill (or Twin Mills), Brewster County, Texas, 2,725 feet, September 3, 1912, ANSP. NEW SYNONYM.

Derotmema laticinctum Scudder

Derotmema (*Derotmema*) *laticinctum* Scudder 1900: 389. Lectotype ♂, designated by Rehn and Hebard 1912: 96, Mesilla, New Mexico, June 30, 1897. ANSP. *Derotmema haydenii laticinctum,* Hebard 1935: 292. (Hebard states incorrectly that Rehn 1919 placed this species as a subspecies of *D. haydenii*).

Derotmema piute Rehn 1919: 233. Holotype ♀, Mason, Lyon County, Nevada, 4,500 feet, September 5, 1910. ANSP.

Derotmema saussureanum Scudder 1900: 391. Lectotype ♂, designated by Rehn and Hebard 1912: 96, San Bernardino, California, July 16, 1897. ANSP.

Dissosteira carolina (Linnaeus)

Gryllus (*Locusta*) *carolinus* Catesby 1743. This is the earliest reference to *carolinus* and predates Linnaeus 1758. However, the species is created to Linnaeus. Types nonexistent. Neotype ♂, here designated, Charleston County, South Carolina, September 23, 1917. ANSP. *Acridium carolinum,* DeGeer 1773: 491. *Gryllus carolinus,* Fabricius 1775: 291. *Acrydium carolinum,* Oliver 1791: 225. *Acridium carolinianum,* Beauvois 1805: 147. *Gryllus carolinus,* Thunberg 1815: 239. *Locusta carolina,* Harris 1833: 583. *Acridium carolinianum,* Hahn 1835. *Oedipoda carolina,* Serville 1839. *Oedipoda carolina,* Walker 1870: 729. *Oedipoda carolina,* Stål 1873: 129. *Dissosteira carolina,* Scudder 1876.

Dissosteira longipennis (Thomas)

Oedipoda longipennis Thomas 1872. Lectotype ♂, selected by Rehn and Hebard 1912: 5, Kansas. USNM. *Dissosteira longipennis,* Scudder 1876b. *Dissosteira* (*Dissosteira*) *longipennis,* Saussure 1884: 137. *Dissosteira longipennis,* Bruner 1885: 133.

Oedipoda nebrascensis Bruner 1876: 133. Type ♀, West Point, Nebraska. *Dissosteira nebrascensis,* Bruner 1876. Synonymized by Bruner 1885b: 133.

Dissosteira pictipennis Bruner 1905. Lectotype ♀, designated by Rehn and Hebard 1912: 115, California. ANSP.

Oedipoda venusta Stål 1861. Combination preoccupied, Fieber 1853. See Rehn 1945: 132.

Trimerotropis porrecta McNeill 1901: 409. Type ♂, California (Koebele), evidently lost. Synonymized by Rentz and Weissman 1980: 247, without explanation.

Dissosteira spurcata Saussure

Dissosteira (*Dissosteira*) *spurcata* Saussure 1884: 137. Type ♀, California. GM.

Oedipoda obliterata Thomas 1880. Combination preoccupied by Germar (in Burmeister 1838).

Encoptolophus californicus (Bruner)

Encoptolophus californicus Bruner 1905: 142. Lectotype ♂, designated by Rehn and Hebard 1912: 115, Los Angeles, California. ANSP. *Chimarocephala,* Strohecker, Middlekauff, and Rentz 1968: 92.

Encoptolophus costalis (Scudder)

Oedipoda costalis Scudder 1862: 473. Type ♂, Texas. ANSP. *Tragocephala costalis,* Stål 1873: 119. *Encoptolophus,* Scudder, 1875b: 78. *Encoptolophus sordidus costalis,* Hebard 1934b: 104.

Encoptolophus coloradensis Bruner 1905: 140. Lectotype ♂, designated by Rehn and Hebard 1912: 123, Fort Collins, Colorado (Bruner). ANSP. *Encoptolophus costalis,* Hebard 1925c: 74. *E. sordidus costalis,* Hebard 1934b: 104.

Encoptolophus montanus Bruner 1905: 140. Lectotype ♂, designated by Rehn and Hebard 1912: 115, Bozeman, Montana. ANSP. *Encoptolophus parvus,* Caudell 1908: 74. *Encoptolophus costalis,* Hebard 1925c: 74. *Encoptolophus sordidus costalis,* Hebard 1934b: 104.

Encoptolophus parvus Scudder 1875b: 79. Types 2♂, Dallas, Texas. ANSP. *Encoptolophus costalis,* Hebard 1925c: 74. *E. sordidus costalis,* Hebard 1934b: 104.

Encoptolophus fuliginosus Bruner 1905: 141. Lectotype ♂, designated by Rehn and Hebard 1912: 115, Colonia Garcia, Chihuahua, Mexico. ANSP.

Encoptolophus otomitus (Saussure)

Tomonotus otomitus Saussure 1861: 322. Type ♂, Mexico, location of type not known. *Chimarocephala otomita,* Saussure 1884: 75. *Encoptolophus otomitus,* Hebard 1932: 246.

Encoptolophus herbaceus Bruner 1905: 142. Lectotype ♂, designated by Rehn and Hebard 1912: 115, Chapultepec, Federal District, Mexico. ANSP. Synonymized by Hebard 1932: 246.

Encoptolophus pallidus Bruner 1893a: 266. Lectotype ♂, designated by Rehn and Hebard 1912: 126, Panamint Valley, California. USNM. *Encoptolophus pallidus pallidus,* Hebard 1935: 287.

Encoptolophus robustus Rehn and Hebard 1910: 435. Lectotype ♀, designated by Rehn and Hebard 1912: 106, Los Angeles River, Los Angeles, California, ANSP.

Encoptolophus sordidus (Burmeister)

Oedipoda sordida Burmeister 1838: 643. Type ♂, Pennsylvania (Zimmerman). Type location unknown. *Acridium* (*Oedipoda*) *sordidum,* DeHaan 1842: 143. *Tragocephala sordida,* Stål 1873: 119. *Encoptolophus sordidus,* Scudder 1875b: 78. *E. sordidus sordidus,* Hebard 1934b: 104.

Locusta nebulosa Harris 1841. *Oedipoda nebulosa,* Erichson. Synonymized by Scudder 1875b: 479.

Locusta periscelidis Say. Synonymized by Scudder 1875b: 479.

Encoptolophus subgracilis Caudell 1903a: 163. Types 2♂, 1♀, Phoenix, Arizona, October 27, 1902. USNM. *Encoptolophus pallidus subgracilis,* Hebard 1935: 287.

Encoptolophus texensis Bruner 1905: 142. Lectotype ♂, designated by Rehn and Hebard 1912: 115, Carrizo Springs, Texas, October 1884. ANSP. *Encoptolophus subgracilis,* Hebard 1925a: 276. *E. subgracilis texansis,* Hebard 1938: 16.

Eximacris phenax Otte, NEW SPECIES. Holotype ♂, Big Cedar, Oklahoma, June 24, 1937 (Standish-Kaiser). ANSP.

Eximacris superbum (Hebard) NEW COMBINATION.

Spharagemon superbum Hebard 1937: 366. Holotype ♂, Katharine, Willacy County, Texas, August 8, 1912. ANSP.

Hadrotettix magnificus (Rehn) NEW COMBINATION

Trimerotropis magnifica Rehn 1907: 42. Holotype ♂, Carr Canyon, Huachuca Mountains, Cochise County, Arizona. ANSP.

Hadrotettix scotodes Otte, NEW SPECIES. Holotype ♂, Durango, Mexico, on Highway 40, 137 miles east of junction with Highway 15, 7,000 feet, July 29, 1964. ANSP.

Hadrotettix trifasciatus (Say)

Gryllus trifasciatus Say 1828. Types lost. Neotype ♂, here designated, Prescott, Arizona, July 18, 1917. ANSP.

Heliastus aztecus (Saussure)

Oedipoda azteca Saussure 1884: 214. Holotype ♂, bearing the following labels: "Meztill t.t./Oedipoda azteca Ss. type/Tomon. azteca Sauss/Oedipoda azteca Sauss. Holotypus C.S.C. [Carlos Carbonell] 1966." GM.

Heliastus benjamini Caudell 1905: 474. Lectotype ♀, selected by Caudell and Hebard 1912, Hauchuca Mountains, Arizona, July 16, 1903. USNM.

Heliastus cirrhoides Otte, NEW SPECIES. Holotype ♂, Huajuapan, Oaxaca, Mexico, 1,597 meters, September 13, 1948 (H. O. Wagner). UMMZ.

Heliastus infrequens Otte, NEW SPECIES. Holotype ♂, 26 miles north of La Mira, on road from Playa Azul to Uruapan, Michoacán, Mexico, November 8, 1981. (Otte). ANSP.

Heliastus rubellus Otte, NEW SPECIES. Holotype ♂, 12 miles south of Tzitzio, Michoacán, Mexico, September 23, 1959 (Cantrall and Cohn). UMMZ.

Heliastus subroseus Caudell.

Heliastus sumichrasti subrosea Caudell 1904: 951. Lectotype ♂, Galveston, Texas, April 1903. USNM. *Heliastus subroseus,* Hebard 1937: 368.

Heliastus sumichrasti (Saussure)

Oedipoda sumichrasti Saussure 1861: 324. "Mexico Calida." Location of types not known. Neotype ♂, here designated, Tampico, Tamaulipas, Mexico, October 22, 1966 (Otte). ANSP. *Heliastus sumichrasti,* Saussure 1884.

Heliastus venezuelae Saussure 1884: 213. America meridionalis septentrionalior; Venezuela, Colombia, Panama. Location of type not known; not in GM or VM. NEW SYNONYM.

Heliastus guatemalae Saussure 1888: 91. Guatemala. Location of type not known; not in GM or VM. NEW SYNONYM.

Heliastus costaricensis Rehn 1905: 402. Holotype ♀, Turrialba, Costa Rica (Schild and Burgdorf). USNM. NEW SYNONYM.

Hippiscus ocelote (Saussure)

Oedipoda (*Hippiscus*) *ocelote* Saussure 1861: 398. Holotype ♀, Cordova, Mexico. GM. *Hippiscus,* Thomas 1873: 200. *Hippiscus,* Kirby 1910: 204.

Oedipoda rugosa Scudder 1862: 469. Type ♂, Massachusetts. ANSP. *Hippiscus rugosus,* Scudder 1874: 377. NEW SYNONYM.

Oedipoda transtrigata Walker 1870: 733. Types 2♂, Oaxaca, Mexico. BM. *Hippiscus,* Kirby 1910. Synonymized by Uvarov 1925: 275.

Hippiscus compactus Scudder 1892: 288. Types ♂, ♀, Carolina and Maryland. ANSP. *Hippiscus rugosus,* Rehn and Hebard 1916: 181.

Hippiscus suturalis Scudder 1892: 301. Type ♂, Moline, Rock Island County, Illinois. ANSP. *Hippiscus rugosus,* Rehn and Hebard 1916: 181.

Hippiscus variegatus Scudder 1892: 268. Types series 3♂, 2♀, Pennsylvania and Georgia. ANSP. *Hippiscus rugosus,* Blatchley 1903: 272.

Hippiscus citrinus Scudder 1901: 88. Type ♂, Alabama. ANSP. *Hippiscus rugosus,* Rehn and Hebard 1916: 181.

Hippiscus immaculatus Morse 1906: 119. Holotype ♂, Clarendon, Texas. ANSP. *Hippiscus rugosus,* Rehn and Hebard 1916: 181.

Hippopedon capito (Stål) NEW COMBINATION

Psinidia capito Stål 1873: 133. Types ♂, ♀, Texas. SM. *Psinidia* (*Trachyrhachis*), Saussure 1884. *Trachyrhachis,* Saussure 1888. *Mestobregma,* Scudder 1897. *Trachyrhachis,* Bruner 1905. *Mestobregma,* Rehn and Hebard 1906. *Trachyrhachys,* Kirby 1910. *Mestobregma,* Hebard 1929. *Rehnita capito,* Hebard 1935: 57.

Conozoa melleola Scudder 1902: 34. Types 2♂, La Cueva, Organ Mountains, New Mexico, 5,300 feet, September 23. ANSP. *Trepidulus,* Caudell 1911. *Mestobregma,* Hebard 1929. Synonymized by Hebard 1935: 57.

Hippopedon gracilipes (Caudell) NEW COMBINATION

Mestobregma gracilipes Caudell 1905: 472. Lectotype ♀, designated by Caudell and Hebard 1912: 160, Huachuca Mountains, Arizona, July 25, 1903. USNM. *Trepidulus,* Bruner 1905, Kirby 1910. *Rehnita gracilipes,* Hebard 1935: 57.

Hippopedon saltator Saussure 1861: 23. Type ♀. Mexico. GM.

Lactista aztecus (Saussure) NEW COMBINATION

Oedipoda azteca Saussure 1861: 397. Holotype ♂, Meztill, Mexico, type bears label "holotypus CSC 1966." GM. *Tomonotus,* Saussure 1884. *Platylactista,* Hebard 1932: 254.

Lactista boscanus Rehn 1903: 721. Type ♀, Ysleta, El Passo County, Texas. ANSP. Synonymized by Bruner 1905: 169.

Lactista elota Otte, NEW SPECIES. Holotype ♂, 33 miles north of Elota, Sinaloa, Mexico, October 17, 1966, Otte, ANSP.

Lactista gibbosus Saussure 1884: 143. Type ♂, California. GM.

Arphia hesperophila Rehn 1902: 143. Type ♂, San Diego, California, April 4, 1901. ANSP. Synonymized by Bruner 1905: 168.

Lactista oslari Caudell 1905: 468. Holotype ♂, Nogales, Arizona, July 3, 1903. USNM. NEW SYNONYM.

Lactista humilis Hebard 1932: 252. Holotype ♂, Cuernavaca, Morelos, Mexico, January 4, 1899. ANSP.

Lactista micrus (Hebard) NEW COMBINATION

Platylactista micrus Hebard 1932: 255. Holotype ♀, Ocotlán, Oaxaca, Mexico, March 20, 1922. ANSP.

Lactista pellepidus Saussure 1884: 144. Lectotype ♂, here designated, Valladolid, Yucatán. Labeled "Typus" by C. S. Carbonell, 1966. GM.

Lactista punctatus (Stål)

Oedipoda punctata Stål 1873: 130. Type ♀, Mexico. SM. *Lactista punctatus,* Saussure 1884.

Lactista stramineus (Erichson)

Oedipoda straminea Erichson 1848: 582. Type lost? British Guiana. *Trimerotropis pallidipennis,* Saussure 1884, Kirby 1910, and Caudell 1911. *Lactista stramineus,* Rehn 1940: 401.

Lactista pulchripennis Saussure 1884: 144. Types, 3♂, Colombia. GM. Synonymized by Rehn 1940: 401.

Lactista australis Bruner 1913: 457. Types: ♂ and ♀, Bonda, Colombia, June. ANSP. *Lactista pulchripennis,* Hebard 1923: 217. *Lactista stramineus,* Rehn 1940: 401.

Leprus elephas Saussure

Oedipoda (*Leprus*) *elephas* 1861: 398. Holotype ♀, Mexico. Labeled "Holotypus" by Carbonell. GM *Leprus elephas,* Walker 1870, IV: 795.

Leprus corpulentus Saussure 1884: 96. Holotype ♂, Mexico. Labeled "Holotypus" by Carbonell. GM. Synonymized by Scudder 1900: 75.

Leprus intermedius Saussure 1884: 95. Holotype ♀, Santa Rosa Island, California. VM.

Leprus cyaneus Cockerell 1902: 306. Holotype ♂, Raton, New Mexico, July 26. USNM. NEW SYNONYM.

Leprus interior Bruner 1905: 160. Lectotype ♂, designated by Rehn and Hebard 1912: 115, Salt Lake City, Utah. ANSP. NEW SYNONYM.

Leprus wheeleri (Thomas)

Oedipoda wheeleri Thomas 1875: 879. Holotype ♀, taken on Wheeler expedition [probably Colorado]. USNM. *Leprus wheeleri,* Saussure 1888: 169.

Leprus glaucipennis Scudder 1900: 76. Lectotype ♀, here designated, San

Luis Potosí, Mexico. ANSP. *Leprus intermedius* Strohecker, Middlekauff, and Rentz 1968: 110. NEW SYNONYM.

Leprus robustus Hebard 1935a: 121. Holotype ♀, San Bernardino Ranch, Cochise County, Arizona, 3,900–3,950 feet, September 24, 1922. ANSP. NEW SYNONYM.

Leuronotina obesa Otte, NEW SPECIES. Holotype ♂, Sierra Fria, 40 miles south of Rincon de Romus, Aguascalientes, Mexico, 8,200 ft, March 9, 1953 (I.J. Cantrall). UMMZ.

Leuronotina orizabae (Saussure)

Oedipoda orizabae Saussure 1861. Type ♂, Tescutlan, Mexico. GM. *Tomonotus,* Saussure 1884. *Leuronotina,* Hebard 1932: 256.

Leuronotina philorites Otte, NEW SPECIES. Holotype ♂, 142 miles southwest of Durango, Mexico, on Rte. 40, August 26, 1963 (Grant and Howard). ANSP.

Leuronotina ritensis (Rehn) NEW COMBINATION

Scirtetica ritensis Rehn 1910. Holotype ♀, Santa Rita Mountains, Pima County, Arizona, 5,000–8,000 feet, July 1907. ANSP.

Machaerocera mexicana Saussure 1859: 391. Types ♂, ♀, Mexico calida. GM.

Oxycoryphus tibialis Walker 1870: 787. Types ♂, ♀, Oaxaca, Mexico. BM. Hebard 1925 synonym.

Machaerocera sumichrasti Thomas 1874. 70. Type ♂, Mexico. Location of type not known. Hebard 1925 synonym.

Machaerocera obscura Bruner 1904: 50. Lectotype ♂, here designated, cotype, Rio Papagaio, Guerrero, Mexico, October. ANSP. Hebard 1925 synonym.

Machaerocera pacifica Bruner 1904: 51. Lectotype ♂, selected by Rehn and Hebard 1912, Tepic, Mexico. ANSP. Hebard 1925 synonym.

Machaerocera magna Bruner 1904: 51. Holotype ♀, Durango or Sinaloa, Mexico. BM. NEW SYNONYM.

Mestobregma impexum Rehn 1919: 240. Holotype ♂, Milford, Beaver County, Utah, 4,900–5,000 feet, September 5, 1909. ANSP.

Mestobregma plattei (Thomas)

Oedipoda plattei Thomas 1873: 124. Type lost, neotype ♂, here designated, Pine Mountain (slopes), Davis Mountains, Texas, August 29, 1912, 6,400–7,000 feet. ANSP. *Psinidia,* Bruner 1897: 132. *Trachyrhachis,* Saussure 1888: 170. *Mestobregma,* Scudder 1897: 75. *Trachyrhachis,* Bruner 1905: 260. *Mestobregma,* Rehn and Hebard 1906: 380. *Trachyrhachys,* Kirby 1910: 248. *Mestobregma plattei plattei,* Rehn 1919: 238.

Conozoa corrugata Scudder and Cockerell 1902: 33. Holotype ♀, Fillmore Canyon, Organ Mountains, New Mexico, August 27. ANSP. *Mestobregma plattei corrugata,* Rehn 1919: 238.

Trachyrhachis rubripennis Bruner 1905: 177. Types ♂, ♀, Oracle, Arizona. USNM. *Mestobregma rubripenne,* Rehn and Hebard 1906: 380. *Mestobregma plattei rubripenne,* Rehn 1919: 238.

Mestobregma terricolor Rehn 1919: 242. Holotype ♂, Pecos, Reeves County, Texas, 2,596 feet, September 12, 1912. ANSP.

Metator nevadensis (Bruner)

Trachyrhachis nevadensis Bruner 1905: 275. Holotype ♀, Esmeralda, Nevada, June 1881. ANSP. *Metator nevadensis,* Hebard 1925:86.

Metator pardalinus (Saussure)

Psinidia (*Trachyrhachis*) *pardalina* Saussure 1884: 161. Lectotype, here designated, Colorado. VM. *Trachyrhachis,* Saussure 1888 *Mestobregma,* Scudder 1897. *Metator pardalina,* McNeill 1901. *Metator pardalinus,* Rehn and Hebard 1906.

Psinidia (*Trachyrhachis*) *maculosa* Saussure 1884: 162. Holotype ♂, Fort McLeod, British Columbia. GM. *Trachyrhachis,* Saussure 1888. *Mestobregma,* Scudder 1897. *Trachyrhachis,* Bruner 1905. Synonymized by Rehn and Hebard 1906: 381.

Microtes helferi (Strohecker) NEW COMBINATION

Trimerotropis helferi Strohecker 1960: 32. Holotype ♂, Cleone, Mendocino County, California. USNM.

Microtes occidentalis (Bruner)

Scirtetica occidentalis Bruner 1893: 267. Holotype ♀, Argus Mountains, California, May 1981 (Koebele, Death Valley Expedition). Type no. 1071. USNM.

Microtes nubila Scudder 1900: 330. Lectotype ♂, designated by Gurney and Strohecker 1959: 77, Monterey, California, July 16. Synonymized by Caudell 1915.

Dissosteira planipennis Bruner 1905: 164. Lectotype ♀, designated by Rehn and Hebard, 1912, California. ANSP. Synonymized by Gurney and Strohecker 1959: 77.

Microtes pogonata (Strohecker) NEW COMBINATION

Trimerotropis pogonata Strohecker 1963: 157. Holotype ♂, Grover City, ♂, Grover City, San Luis Obispo County, California, CAS.

Pardalophora apiculata (Harris)

Locusta apiculata Harris 1835: 576. Type lost. Neotype here designated, Saint Anthony Park, Minnesota, June 7, 1894. ANSP. *Hippiscus,* Rehn and Hebard 1911: 630. *Pardalophora,* Rehn and Hebard 1916: 180. *Hippiscus,* Blatchley 1920: 263. *Pardalophora,* Hebard 1925b: 76.

Acridium tuberculatum Palisot de Beauvois 1805. This is the first name applied to this species. Palisot believed his species to be identical to *Gryllus tuberculatus* Fabricius, an Old World species. The species was renamed by Harris 1835. See Rehn and Hebard 1910: 630.

Oedipoda obliterata Burmeister 1838: 643. Type ♂, Halle Museum. *Oedipoda phoenicoptera* Scudder 1862: 468, *Hippiscus* (*Pardalophora*) *tuberculatus,* Saussure 1884: 87.

Locusta corallina Harris 1833. Type lost. *Oedipoda phoenicoptera,* Scudder 1862: 469. *Hippiscus tuberculatus,* Saussure 1884: 87.

Pardalophora haldemani (Scudder)

Oedipoda haldemanii Scudder 1872: 251. Lectotype ♂, here designated, Platte River, Nebraska. ANSP. *Hippiscus,* Scudder 1876: 264. *Pardalophora,* Kirby 1910: 206.

Hippiscus (*Xanthippus*) *tigrinus* Scudder 1892: 334. Lectotype ♀, designated by Rehn and Hebard 1925: 77. West Point, Nebraska, July 1887 (L. Bruner). Type no. 977. ANSP. *Xanthippus tigrinus,* Kirby 1910: 208. *Hippiscus tigrinus,* Caudell 1911: 162. *Pardalophora haldemanii,* Hebard 1925: 77.

Hippiscus nanus Saussure 1884: 86. Type ♂, Illinois. GM. *Hippiscus haldemanii,* Scudder 1892: 286.

Pardalophora phoenicoptera (Burmeister)

Oedipoda phoenicoptera Burmeister 1838: 643. Type ♀, South Carolina. Martin Luther Universität, Halle, DDR. *Hippiscus,* Scudder 1874. *Pardalophora,* Kirby 1910: 206. *Hippiscus,* Caudell 1911: 162. *Pardalophora,* Rehn and Hebard 1916: 180. *Hippiscus,* Blatchley 1920: 263. *Pardalophora,* Hebard 1928: 237.

Oedipoda discoidea Serville 1839: 724. Lectotype ♀, here designated, Georgia. PM *Hippiscus discoideus,* Stål 1873. *Hippiscus phoenicopterus,* Saussure 1884. *Pardalophora phoenicoptera,* Kirby 1910.

Hippiscus texanus Scudder 1892: 286. Lectotype ♂, here designated, Dallas, Texas. ANSP. *Pardalophora texana,* Kirby 1910. *Pardalophora phoenicoptera,* Rehn and Hebard 1916: 180.

Pardalophora saussurei (Scudder)

Hippiscus saussurei Scudder 1892: 302. Lectotype ♀, here designated, Dallas, Texas. ANSP. *Pardalophora,* Kirby 1910. *Hippiscus,* Caudell 1911. *Pardalophora,* Rehn and Hebard 1916: 180.

Psinidia amplicornis Caudell

Psinidia sulcifrons var. *amplicornis* Caudell 1903. 791. Types ♂, ♀, Victoria, Texas, June 1902. USNM. *Conozoa amplicornis,* Hart 1906. *Psinidia amplicornis,* Rehn 1919: 247.

Psinidia fenestralis (Serville)

Oedipoda fenestralis Serville 1838: 726. Type ♂, North America. Latreille Collection. *Psinidia,* Stål 1873. *Psinidia,* Saussure 1884. *Psinidia fenestralis fenestralis,* Rehn 1919: 247.

Psinidia fenestralis frater, Rehn 1919: 247. Holotype ♀, Katherine, Willacy County, Texas, August 8, 1912. ANSP. NEW SYNONYM.

Locusta eucerata Harris 1833: 583. According to Scudder 1900, *eucerata* remained undescribed until Harris 1841. Types nonexistent, Massachusetts. *Oedipoda,* Uhler 1862, *Psinidia* (*Psinidia*), Saussure 1884. Synonymized by Scudder 1900: 40.

Shotwellia isleta Gurney 1940: 3. Holotype ♂, Bernalillo County, New Mexico, August 24, 1938. Type no. 53956. USNM.

Spharagemon bolli Scudder 1875a: 468. Lectotype ♂, here designated, Dallas, Texas, Boll's no. 93. ANSP. *Dissosteira* (*Spharagemon*) *bolli,* Saussure 1884: 140. *Spharagemon bolli bolli,* Hebard 1937: 363.

Spharagemon inornatum Morse 1895: 291. Type ♀, Hot Springs, New Mexico, 7,000 feet. ANSP. *Spharagemon bolli inornatum,* Hebard 1937: 363. NEW SYNONYM.

Spharagemon balteatum Scudder 1875a: 468. Lectotype ♀, here designated, Norway, Maine. ANSP. *Dissosteira* (*Spharagemon*) *bolli* variety *balteatum,* Saussure 1884: 140. *Sphoragemon balteatum,* Bruner 1885b: 132. *Sphoragemon bolli,* Morse 1894: 228.

Sphoragemon robustum Morse 1895: 291. Type ♀, Coahila, Mexico. ANSP. Synonymized by Hebard 1937: 365.

Spharagemon bunites Otte, NEW SPECIES. Holotype ♂, Glacier Point Road, Yosemite National Park, Mariposa County, California, August 7, 1940. ANSP.

Spharagemon campestris (McNeill) NEW COMBINATION

Trimerotropis campestris McNeill 1901:423. Type lost. Described from Pine Bluffs, Wyoming. Neotype ♂, here designated, Pine Bluffs, Wyoming. ANSP.

Trimerotropis longicornis E. M. Walker 1902: 4. Holotype ♂, Vernon, British Columbia, September 10, 1897. Royal Ontario Museum. *Trimerotropis campestris,* Hebard 1925: 89. Removed from synonymy by Vickery 1979.

Spharagemon collare (Scudder)

Oedipoda collaris Scudder 1872: 250. Lectotype ♂, Platte, Nebraska. ANSP. *Spharagemon collare,* Scudder 1875e. *Dissosteira (Spharagemon) collare,* Saussure 1884. *Spharagemon collare,* Morse 1894. *Spharagemon collare collare,* Morse 1895.

Oedipoda wyomingiana Thomas 1872. Lectotype ♂, designated by Hebard 1927: 5, eastern Wyoming. USNM. *Spharagemon wyomingianum,* Scudder 1875e. *Dissosteira (Spharagemon) collaris* var. *wyomingiana,* Saussure 1884. *Spharagemon collare wyomingianum,* Morse 1895. *Spharagemon wyomingianum,* Scudder 1900. *Spharagemon collare,* Kirby 1910. *Spharagemon wyomingianum,* Blatchley 1920. *Spharagemon collare,* Hebard 1925b.

Oedipoda belfragii Stål 1873: 129. Type ♀, Illinois. SM. *Spharagemon belfragii,* Kirby 1910. *Spharagemon aequale,* Caudell 1911. Synonymized by Hebard 1929: 346.

Spharagemon collare pallidum Morse 1895: 299. Lectotype ♀, designated by Morse and Hebard 1915, White River, Colorado. ANSP. *Spharagemon collare,* Kirby 1910.

Spharagemon aequale scudderi Morse 1894: 225. Lectotype ♀, designated by Morse and Hebard 1915, Sherborn, Mass. ANSP. *Spharagemon collare scudderi,* Morse 1895. *Spharagemon collare,* Kirby 1910. *Spharagemon wyomingianum,* Blatchley 1920. *Spharagemon collare,* Hebard 1925b: 83.

Spharagemon oculatum Morse 1894: 232. Lectotype ♀, designated by Morse and Hebard 1915, Marshall County, Indiana. ANSP. *Spharagemon collare wyomingianum,* Morse 1895. (See *Oedipoda wyomingiana* for history).

Spharagemon collare angustipenne Morse 1895: 298. Lectotype ♀, designated by Morse and Hebard 1915, Salt Lake Valley, Utah, 4,300 feet. ANSP. *Spharagemon angustipenne,* Caudell 1903b. *Spharagemon collare,* Kirby 1910.

Oedipoda utahensis Thomas 1875: 883. Type lost (Hebard 1927). *Dissosteira utahensis,* Saussure 1888. *Spharagemon collare utahense,* Morse 1895. *Spharagemon collare,* Kirby 1910.

Spharagemon crepitans (Saussure)

Dissosteira (*Spharagemon*) *bolli* var. *crepitans* Saussure 1884: 140, types 2♀, Georgia. GM. *Spharagemon crepitans,* Morse 1895.

Spharagemon cristatum Scudder 1875e: 470. Lectotype ♂, here designated, Belfrage, Texas. ANSP. *Dissosteira* (*Spharagemon*) *cristata,* Saussure 1884. *Spharagemon cristatum,* Morse 1895. *Spharagemon collare cristatum,* Hebard 1932: 249. *Spharagemon collare,* Hebard 1937: 366.

Spharagemon equale (Say)

Gryllus equalis Say 1825: 307. Neotype ♂, here designated, 20 miles west of Colby, Sherman County, Kansas, July 2, 1966. ANSP. *Locusta aequalis,* Harris 1835: 56. *Oedipoda aequalis,* Uhler 1862. *Trimerotropis aequale,* Scudder 1874. *Spharagemon aequale,* Scudder 1875e. *Dissosteira* (*Spharagemon*) *aequalis,* Saussure 1884. *Spharagemon aequale,* Bruner 1885b. *Spharagemon aequale,* Morse 1895. *Spharagemon aequale,* Hebard 1937: 364.

Dissosteira (*Spharagemon*) *texensis* Saussure. 1884: 140. *Spharagemon texense,* Scudder 1900. Synonymized by Caudell 1911.

Spharagemon humile Morse 1895: 292. Lectotype ♂, designated by Morse and Hebard 1915, Garden of the Gods, Colorado. Type no. 16152. ANSP. *Trimerotropis humile,* Hebard 1937: 363. *Spharagemon humile,* Rentz and Weissman 1980. NEW SYNONYM.

Spharagemon marmorata marmorata (Harris) NEW COMBINATION

Locusta marmorata Harris 1841: 145. Types lost. Neotype ♂, here designated, Provincetown, Massachusetts, September 5, 1891. ANSP. *Oedipoda marmorata,* Uhler 1862. *Dissosteira* (*Scirtetica*) *marmorata,* Saussure 1884. *Psinidia marmorata,* Davis 1889. *Scirtetica marmorata,* Bruner 1890. *Scirtetica marmorata,* Morse 1897. *Scirtetica marmorata marmorata,* Rehn and Hebard 1912b.

Locusta cerineipennis Harris 1833. *Nomen nudum,* according to Scudder 1900.

Spharagemon marmorata picta (Scudder) NEW COMBINATION

Trimerotropis picta Scudder 1877: 31. Lectotype ♂, here designated, Fort Reed, Florida, April 21, 1876. ANSP. *Dissosteira* (*Scirtetica*) *picta,* Saussure 1884. *Scirtetica picta,* Scudder 1900. *Scirtetica marmorata picta,* Rehn and Hebard 1912b.

Spharagemon saxatile Morse 1894: 229. Lectotype ♀, designated by Morse and Hebard 1915, Blue Hill, Massachusetts, September 20, 1891. ANSP.

Spharagemon saxatile planum Morse 1904b: 13. Lectotype ♀, designated by Morse and Hebard 1915, Wytheville, Virginia, September 4, 1903, 2,300 feet. ANSP. *Spharagemon planum,* Fox 1928. *Spharagemon planum,* Hebard 1937: 365.

Sphingonotus haitensis haitensis (Saussure)

Oedipoda haitensis Saussure 1861: 323. Type from Haiti in GM. *Sphingonotus haitensis haitensis,* NEW STATUS.

Sphingonotus jamaicensis Saussure 1884: 202. Type ♂, Jamaica. GM. Synonymized by Rehn and Hebard 1938: 211.

Sphingonotus haitensis cubensis Saussure

Sphingonotus caerulans cubensis Saussure 1884: 201. Type from Cuba in GM. *Sphingonotus cubensis,* Rehn and Hebard 1938: 214. *Sphingonotus haitensis cubensis,* NEW STATUS.

Sticthippus californicus (Scudder)

Hippiscus (*Sticthippus*) *californicus* Scudder 1892: 317. Lectotype ♂, Gilroy, Santa Clara County, California, ANSP. *Sticthippus,* Kirby 1910. *Hippiscus,* Caudell 1911. *Sticthippus,* Hebard 1928.

Hippiscus (*Sticthippus*) *marmoratus* Scudder 1892: 318. *Sticthippus,* Kirby 1910. *Hippiscus,* Caudell 1911. *Sticthippus,* Hebard 1928.

Tomonotus ferruginosus Bruner (in Caudell 1905: 470). Lectotype ♂, designated by Rehn and Hebard 1912, Fort Grant, Arizona. ANSP.

Tomonotus mexicanus Saussure 1861: 321. Lectotype ♂, here designated, Cuautla, Mexico. GM.

Trachyrhachys aspera Scudder 1876: 512. Holotype ♀, plains of northern New Mexico, eastern slope, October 14–31, 1875. ANSP. *Mestobregma,* Scudder 1897. *Trachyrhachis,* Bruner 1905: 260. *Mestobregma,* Rehn and Hebard 1906. *Trachyrhachys,* Kirby 1910.

Trachyrhachis compacta Bruner 1905: 176. Lectotype ♀, designated by Rehn and Hebard 1912, Silver City, New Mexico, July 1884. ANSP. Synonymized by Hebard 1929: 349.

Trachyrhachis inconspicua Bruner 1905: 176. Lectotype ♂, designated by Rehn and Hebard 1912, Zacatecas, Mexico, November 1887. ANSP. NEW SYNONYM.

Trachyrhachys coronata Scudder 1876: 292. Holotype ♀, northern New Mexico, August–September. ANSP. *Mestobregma,* Scudder 1897. *Trachyrhachis,* Rehn and Hebard 1906.

Psinidia (*Trachyrhachis*) *borealis* Saussure 1884: 164. Types, ♂, ♀, Colorado, Brunner Collection. *Trachyrhachis,* Saussure 1888: 59. *Mestobregma,* Rehn and Hebard 1906. *Trachyrhachys,* Kirby 1910. Synonymized by Hebard 1929: 349.

Trachyrhachys funeralis Strohecker 1945: 214. Holotype ♂, Apatzingan, Michoacán, August 7, 1940. Strohecker Collection. Paratypes in ANSP.

Trachyrhachys kiowa (Thomas)

Oedipoda kiowa Thomas 1872: 461. Lectotype ♀, desgnated by Hebard 1927: 5, Colorado. USNM. *Psinidia,* Thomas 1875. *Mestobregma,* Thomas 1876. *Psinidia* (*Trachyrhachis*), Saussure 1884. *Trachyrhachis,* Saussure 1888. *Mestobregma,* Scudder 1897. *Trachyrhachis,* Bruner 1905. *Mestobregma,* Rehn and Hebard 1906. *Trachyrhachis,* Morse 1907. *Trachyrhachys,* Caudell 1911.

Psinidia fuscifrons Stål 1873: 134. Holotype ♂, Texas. SM. *Psinidia*

(*Trachyrhachis*), Saussure 1888. *Mestobregma*, Scudder 1897. *Trachyrhachis*, Bruner 1905. *Mestobregma*, Rehn and Hebard 1906. *Trachyrhachys kiowa* var. *fuscifrons* Morse 1907. *Trachyrhachys fuscifrons*, Kirby 1910. *Trachyrhachys kiowa fuscifrons*, Hebard 1935.

Psinidia (*Trachyrhachis*) *fuscifrons* var. *texana* Saussure 1884: 163. Type ♀, Dallas, Texas. GM? *Trachyrhachis*, Bruner 1905. *Mestobregma*, Rehn and Hebard 1906. *Trachyrhachys*, Kirby 1910. *Trachyrhachis kiowa fuscifrons*, Hebard 1934: 185.

Pisinidia (*Trachyrhachis*) *mexicana* Saussure 1884: 164. Type ♂, San Bartolo, Mexico. VM. *Trachyrhachis*, Saussure 1888. *Trachyrhachis*, Bruner 1905. *Mestobregma*, Rehn and Hebard 1906. *Trachyrhachys*, Kirby 1910. NEW SYNONYM.

Mestobregma pulchella Bruner 1889: 65. Lectotype ♂, designated by Rehn and Hebard 1912, Glendive, Montana (L. Bruner). ANSP. *Trachyrhachis*, Bruner 1905. *Mestobregma*, Rehn and Hebard 1906. *Trepidulus*, Kirby 1910. *Trachyrhachys*, Caudell 1911. *Mestobregma kiowa kiowa*, Hebard 1925: 84. *Trachyrhachys kiowa kiowa*, Hebard 1919: 348.

Trachyrhachis obliterata Bruner 1905: 175. Lectotype ♀, designated by Hebard 1931: 161, Topeka, Kansas. ANSP. *Mestobregma*, Rehn and Hebard 1906. *Mestobregma kiowa obliterata*, Morse 1907. *Trachyrhachys obliterata*, Kirby 1910. *Trachyrhachys kiowa thomasi* Hebard 1931: 161. *Trachyrhachys kiowa fuscifrons*, Hebard 1937.

Mestobregma thomasi Caudell 1904: 125. New name for *M. cincta* of Bruner, Scudder, etc., but not of Thomas 1870. Type lost. *Trachyrhachys*, Bruner 1905. *Mestobregma*, Rehn and Hebard 1906. *Trachyrhachis kiowa* var. *thomasi*, Morse 1907. *Trachyrhachys thomasi*, Kirby 1910. *Trachyrhachys kiowa* var, *thomasi* Hebard 1931.

Trachyrhachis townsendi Bruner 1905: 176. Lectotype ♀, designated by Rehn and Hebard 1912, Colonia Garcia, Chihuahua, Mexico. ANSP. *Trachyrhachys*, Kirby 1910. *Trachyrhachis mexicana*, Hebard 1932: 258.

Trachyrhachis cinctipennis Bruner 1905: 174. No types, not formally described, only mentioned in a key of the genus. *Trachyrhachys*, Kirby 1910. *Nomen nudum.*

Trepidulus concinens Otte, NEW SPECIES. Holotype ♂, 0.5 km south of Rosarito at 0.24 km north of km 53 road marker on Mexico highway 1, July 10, 1978. (D. B. Weissman and D. Lightfoot). CAS.

Trepidulus hyalinus (Scudder)

Mestobregma hyalinum Scudder 1900: 92. Lectotype ♀, here designated, Kern City, California, August 4, 1897. ANSP. *Trepidulus hyalinus*, Bruner 1905: 185.

Trepidulus rosaceus (Scudder)

Mestobregma rosaceum Scudder 1900: 91. Lectotype ♀, here designated, Yuma, Arizona, July 7, 1897. ANSP. *Trepidulus rosaceus*, McNeill 1901: 398.

Araeopteryx penelope Caudell 1901: 102. Holotype ♀, Prescott, Arizona, September 26, 1900. USNM. Synonymized by Rehn 1904: 567.

Trimerotropis agrestis McNeill 1900: 32. Holotype destroyed, neotype ♂, here designated, La Junta, Otero County, Colorado, July 11, 1940. ANSP.

Trimerotropis agrestis hewitti Tinkham 1964: 31. Holotype ♂, Sand Dune Lake, 8 miles NE of Bruneau, Owyhee County, Idaho, September 4, 1962. California Academy of Sciences.

Trimerotropis albescens McNeill 1901: 418. Holotype ♂, Los Angeles, California, lost (Rentz and Weissman 1980). Neotype ♂, here designated, Los Angeles, California (Coquillett). ANSP.

Trimerotropis arenacea Rehn 1910: 13. Lectotype ♀, designated by Rehn and Hebard 1912, Springfield, Bingham County, Idaho, July 25, 1906. ANSP.

Trimerotropis arizonensis Tinkham 1947: 127. Holotype ♂, Black Mountains, near Boulder Dam, Arizona. Ernest R. Tinkham Collection, Indio, California.

Trimerotropis barnumi Tinkham, NEW STATUS

Trimerotropis agrestis barnumi Tinkham 1960: 51. Holotype ♂, Oak City Dunes, 5 miles WSW of Oak City, Millard County, Utah, 4,600 feet, July 26, 1960. Tinkham Collection, Indio, California.

Trimerotropis agrestis gracewileyae Tinkham 1961: 50. Holotype ♂, Cane Springs Dunes, 19 miles north of Hanksville, Emery County, Utah, August 4, 1958. Tinkham Collection, Indio, California. NEW SYNONYM.

Trimerotropis bernardi Rentz and Weissman 1984. Holotype ♂, Mill Creek camping area, 6,000 feet, San Bernardino Mountains, California, July 25, 1981 (D. Weissman). CAS.

Trimerotropis bifasciata Bruner 1890: 70. Holotype ♂, labeled holotype by Rentz and Weissman 1980: 230, California. USNM. See discussion by Rentz and Weissman 1980.

Trimerotropis californica Bruner 1890: 71. Lectotype ♂, designated by Rehn and Hebard 1912, San Luis Valley, California. ANSP.

Trimerotropis strenua McNeill 1901: 432. Syntypes 1♂, 1♀, Salt Lake Valley, Utah, lost (Rentz and Weissman 1980). *Pseudotrimerotropis,* Kirby 1910. *Trimerotropis,* Caudell 1911. NEW SYNONYM.

Trimerotropis montana McNeill 1900: 32. Holotype ♂, Boise City, Idaho, lost (Rentz and Weissman 1980). Synonymized by Hebard 1929: 355.

Trimerotropis cincta (Thomas)

Oedipoda cincta Thomas 1870: 80. Holotype ♀ from New Mexico lost (Hebard 1927:6). Neotype ♂, here designated, Tesuque Creek, Santa Fe County, New Mexico, July 29, 1919. ANSP. *Trimerotropis,* McNeill 1900. *Pseudotrimerotropis,* Kirby 1910. *Trimerotropis,* Caudell 1911.

Trimerotropois cyaneipennis Bruner 1889: 68. Lectotype ♂, designated by Rehn and Hebard 1912, Ogden Canyon, Lake Bonneville, Salt Lake Valley, Utah, September 1883. ANSP.

Trimerotropis cyanea Scudder 1902: 36. Syntypes from Eagle Creek, White Mountains, Arizona; Dripping Spring, Organ Mountains, Arizona; Filmore Canyon, Organ Mountains, Arizona. ANSP. Synonymized by Rentz and Weissman 1980.

Trimerotropis diversellus Hebard 1928: 255. Holotype ♂, Lower Geyser Basin, Yellowstone National Park, 7200 feet, 14 viii 1922. ANSP.

Trimerotropis fontana Thomas 1876: 225. Lectotype not at ANSP as stated by Rentz and Weissman 1980. Neotype ♂, here selected, Mill Creek, 5,200–5,500 feet, San Bernardino Mountains, California, August 28, 1919. ANSP.

Trimerotropis juliana Scudder 1876: 294. Lectotype ♀, designated by Hebard 1929: 360, Julian, San Diego County, California. ANSP. Synonymized by Hebard 1929: 360.

Trimerotropis caerulipes Scudder 1881: 27. Lectotype ♂, designated by Rentz and Weissman 1980, Portland, Oregon. ANSP. Synonymized by Hebard 1928: 259.

Trimerotropis fallax Saussure 1884: 170. Lectotype ♀, designated by Rentz and Weissman 1980, "California, 60, 35." GM. *Trimerotropis obscura,* Scudder and Cockerell 1902: 362. Synonymized by Rentz and Weissman 1980: 233.

Trimerotropis calignosa McNeill 1901: 417. Holotype ♀, from Los Angeles County, California, reported lost (Rentz and Weissman 1980). *Trimerotropis caerulipes,* Rehn and Hebard 1909: 446. *Trimerotropis fontana,* Hebard 1928: 259.

Trimerotropis tessellata McNeill 1901: 417. Holotype ♀, from Turkey Tanks, Arizona, missing from USNM (Rentz and Weissman 1980). *Trimerotropis caerulipes* var. *tessellata,* Caudell 1908. *Trimerotropis tessellata,* Caudell 1911. *Trimerotropis fontana,* Hebard 1928: 260. *Trimerotropis cincta,* Hebard 1935: 296. Synonymized by Rentz and Weissman 1980.

Trimerotropis ferruginea McNeill 1901: 420. Original type series from Soda Springs and Coeur d'Alene, Idaho; Hecla and Madison River, Wyoming; and American Fork Canyon, Utah, all destroyed (Rentz and Weissman 1980). Synonymized by Hebard 1929: 360.

Trimerotropis fratercula McNeill 1901: 438. Types from Pine Bluffs, Wyoming, lost (Hebard 1929: 352). Neotype ♂, here selected, Pine Bluffs, Wyoming. ANSP.

Trimerotropis gracilis (Thomas)

Oedipoda gracilis Thomas 1872: 461. Lectotype ♂, designated by Hebard 1927: 6, Colorado. USNM. *Trimerotropis,* Saussure 1884. *Conozoa,* Bruner 1893. *Trimerotropis,* McNeill 1900. *Trimerotropis gracilis gracilis,* Hebard 1928: 251.

Derotmema (*Tmetodera*) *lentignosum* Scudder 1900: 393. Lectotype ♂, designated by Rehn and Hebard 1912, South Park, Colorado, 8,000–10,000 feet, August 11–16, 1877. ANSP. *Trimerotropis gracilis,* Caudell 1909. *Trimerotropis gracilis gracilis,* Hebard 1928: 251.

Derotmema (*Tmetodera*) *lichenosum* Scudder 1900: 394. Lectotype ♂, designated by Rehn and Hebard 1912, Alkali Stage Station, Green River, Wyoming, July 27, 1877, 6,000 feet. ANSP. *Trimerotropis gracilis gracilis,* Hebard 1928: 251.

Trimerotropis sordida Walker 1902: 9. Holotype ♀, Waldeck, Assiniboia, Saskatchewan, August 30, 1897. Royal Ontario Museum. *Trimerotropis gracilis sordida,* Hebard 1928: 250.

Trimerotropis huroniana E. M. Walker. 1902: 6. Lectotype ♂, designated by Rentz and Weissman 1980, Southampton, Ontario. Royal Ontario Museum.

Trimerotropis inconspicua Bruner 1904: 59. Lectotype ♂, designated by Rehn and Hebard 1912, Palisade, Colorado. ANSP.

Trimerotropis viriditibialis Henderson 1924: 104. Lectotype ♀, designated by Rentz and Weissman, Junction, Utah. USNM. Evidently synonymized by Thompson and Buxton 1964: 28.

Trimerotropis infantilis Rentz and Weissman 1984. Holotype ♂, 1 mile east of the intersection of East Zyante and Graham Hill roads, Felton, Santa Cruz County, California July 16, 1977 (Rentz and Weissman). CAS.

Trimerotropis inyo Rentz and Weissman 1984. Holotype ♂, 12.8 km west of Lone Pine on Whitney Portal Road, 1,920 meters, August 6, 1978 (Weissman and Lightfoot). CAS.

Trimerotropis koebelei (Bruner)

Conozoa koebelei Bruner 1890: 67. Holotype ♂, Placer County, California, September (A. Koebele). USNM. *Trimerotropis,* McNeill 1900: 30.

Trimerotropis latifasciata Scudder 1881: 26. Lectotype ♀, designated by Rentz and Weissman, Lake Point, Utah, August 14. Scudder's type 1875 (a). Type no. 15873. ANSP.

Trimerotropis laticincta Saussure 1884: 169. Lectotype ♀, designated by Rentz and Weissman, "Texas, Boll, 01,05, *laticincta* Ss. *Trimer. sulfurea* Sa type *Trimer. laticincta* Saussure." GM. NEW SYNONYM.

Trimerotropis snowi Rehn 1905: 223. Holotype ♀, Congress Junction, Yavapai County, Arizona. University of Kansas Entomological Collection.

Trimerotropis lauta Scudder 1876: 26. Holotype ♂, Guadalupe Island, Baja California, Mexico. ANSP. *Circotettix,* Kirby 1910. *Trimerotropis,* Rentz and Weissman 1980.

Trimerotropis leucophaea Rentz and Weissman 1984. Holotype ♂, Westgard Pass, 7,200 feet, Inyo County, California, July 26, 1962 (Rentz and McNeill). CAS.

Trimerotropis maritima (Harris)

Locusta maritima Harris 1841: 143. Type lost. Neotype ♂, here selected, Nantucket, Massachusetts, September 10, 1913. ANSP. *Oedipoda,* Uhler 1862. *Trimerotropis,* Stål 1873. *Trimerotropis maritima maritima* Walker 1898.

Trimerotropis santabarbara Rentz and Weissman 1981: 81. Holotype ♂, Graveyard Canyon, Santa Barbara Island, California, May 9, 1970. CAS.

Trimerotropis saxatilis McNeill 1900: 34. Lectotype ♂, designated by Frison 1927, Union County, Illinois, July 23, 1884. Illinois Natural History Survey, Urbana, Illinois.

Trimerotropis schaefferi Caudell 1904: 112. Lectotype ♂, designated by Caudell and Hebard 1912: 161, Topo, near Brownsville, Texas, May 1903, USNM.

Trimerotropis sparsa (Thomas)

Oedipoda sparsa Thomas 1875: 883. Types lost (Rentz and Weissman 1980). Neotype ♂, here selected, Cedar Creek, Montrose County, Colorado, 6,500–7,000 feet, September 1, 1921 (Rehn and Hebard). ANSP. *Circotettix,* Scudder 1876.

Trimerotropis azurescens Bruner 1889: 69. Lectotype ♀, designated by Rehn and Hebard 1912, Alkali Station, Green River, Wyoming, 6,000 feet, July 27, 1877. ANSP. *Circotettix,* Caudell 1902. Synonymized by Hebard 1928: 252.

Trimerotropis perplexa Bruner 1889: 74. Lectotype ♀, designated by Rehn and Hebard 1912, badlands, 5–6 miles north of Chadron, Nebraska, July 1888. ANSP. *Circotettix,* Scudder 1900. *Trimerotropis azurescens,* McNeill 1901. *Circotettix azurescens,* Caudell 1902. *Trimerotropis perplexa,* Hebard 1925. *Trimerotropis sparsa,* Hebard 1928: 252.

Trimerotropis thalassica Bruner 1890: 72. Neotype ♂, designated by Rentz and Weissman 1980, Los Angeles County, California. USNM.

Trimerotropis santamonica Rentz and Weissman 1981: 83. Holotype ♂, 22 km east of Oxnard on Potrero Road, Ventura County, California, September 9, 1973. CAS. NEW SYNONYM.

Trimerotropis titusi Caudell 1906: 125. Lectotype ♂, designated by Caudell and Hebard 1912: 161. Spreckles, Monterey County, California, September 20, 1904. USNM.

Trimerotropis tolteca (Saussure)

Oedipoda tolteca Saussure 1861: 397. Lectotype ♂, designated by Hebard 1932: 260, Oaxaca, Mexico, GM. *Trimerotropis,* Saussure 1884.

Hippiscus australis Bruner 1905: 157. Lectotype ♂, designated by Rehn and Hebard 1912, Eslava, D. F., Mexico. ANSP. NEW SYNONYM.

Trimerotropis topanga Rentz and Weissman 1981: 86. Holotype ♂, Topanga Canyon, Los Angeles County, California, May 8, 1930. ANSP.

Trimerotropis verruculata verruculata (Kirby)

Locusta verruculatus Kirby 1837: 250. Type, lat. 57°N, probably in Hope Museum, Oxford University (Rentz and Weissman 1980). *Acridium* (*Locusta*), Haan. *Oedipoda,* Scudder 1862. *Trimerotropis,* Scudder 1874. *Circotettix,* Saussure 1884 and Scudder 1900. *Trimerotropis,* White 1951. *Circotettix,* Helwig 1955. *Trimerotropis verruculatus verruculatus,* Vickery, Johnstone, and Kevan 1974.

Trimerotropis verruculata suffusa Scudder

Trimerotropis suffusus Scudder 1876: 265. Lectotype ♀, here designated, Vancouver Island, British Columbia. ANSP. *Circotettix,* Scudder 1900. *Trimerotropis,* Hebard 1928: 261. *Trimerotropis verruculatus suffusa,* Vickery and Nagy 1974. *Trimerotropis suffusa,* Rentz and Weissman 1973.

Trimerotropis obscura Scudder 1876: 294. Types from northern New Mexico apparently lost, not at ANSP with Scudder material. *Trimerotropis suffusus,* Hebard 1929: 362.

Trimerotropis nubila McNeill 1900: 34. Holotype ♂, from Hot Springs, New Mexico, 7,000 feet, is lost (Rentz and Weissman 1980). *Trimerotropis suffusus,* Hebard 1929: 362.

Trimerotropis conspersa McNeill 1901: 443. Holotype ♂, from Mt. Shasta, California, September 1885 (J. Behrens), was destroyed; not at ANSP as claimed by Rentz and Weissman. *Trimerotropis suffusa,* Hebard 1929: 362.

Trimerotropis variegata McNeill 1901: 443. Holotype ♀, from Tighes Station, San Diego County, California, is lost from USNM (Rentz and Weissman 1980). *Trimerotropis suffusus* Hebard 1929: 362.

Trimerotropis whitei Rentz and Weissman 1984. Holotype ♂, Zion National Park, near east boundary, 1,646 meters (Weissman). CAS.

Tropidolophus formosus (Say). *Gryllus formosus* Say 1825: 138. Neotype ♂, here designated, Baxter, Pueblo County, Colorado, 4,624 feet, September 13, 1921. ANSP.

Xanthippus aquilonius Otte, NEW SPECIES. Holotype ♂, Vernon, British Columbia, May 31, 1929 (E. R. Buckell). ANSP.

Xanthippus brooksi Vickery 1967: 263. Holotype ♂, Canada, Northwest Territories, Reindeer Depot, Mackenzie River delta, July 7, 1948 (W. J. Brown). Canadian National Collection. NEW STATUS.

Xanthippus corallipes (Haldeman)

Oedipoda corallipes Haldeman 1852: 371. Type lost, valley of Great Salt Lake, Utah (Stansbury's expedition). *Hippiscus,* Scudder 1876a. *Xanthippus,* Saussure 1884. *Hippiscus* (*Xanthippus*), Scudder 1892. *Xanthippus,* Kirby 1910. *Hippiscus,* Caudell 1911. *Xanthippus corallipes corallipes,* Hebard 1925b: 78.

Oedipoda pardalina Saussure 1861: 324. Type lost? not seen in GM or VM. *Xanthippus,* Saussure 1884. *Hippiscus* (*Xanthippus*), Scudder 1892. *Xanthippus,* Kirby 1910. *Hippiscus,* Caudell 1911. *Xanthippus corallipes pardalinus,* Hebard 1932: 247.

Oedipoda paradoxa Thomas 1871: 457. Lectotype ♀, designated by Hebard 1927, between Ogden and Smithfield, Utah. USNM. *Hippiscus rugosus paradoxus,* Blatchley 1891: 79. *Hippiscus* (*Xanthippus*) *paradoxus,* Kirby 1910. *Hippiscus,* Caudell 1911. *Xanthippus,* Hebard 1927. *Xanthippus corallipes corallipes,* Hebard 1935: 288.

Xanthippus leprosus Saussure 1884: 92. Lectotype ♀, designated by Hebard 1929: 340, Taos Valley, New Mexico. GM. *Hippiscus* (*Xan-*

thippus), Scudder 1892. *Xanthippus,* Kirby 1910. *Hippiscus,* Caudell 1911. *Xanthippus corallipes leprosus,* Hebard 1928: 245.

Xanthippus toltecus Saussure 1884: 91. Types 2♀, Mexico City. GM. *Hippiscus* (*Xanthippus*), Scudder 1892. *Xanthippus,* Kirby 1910. *Hippiscus,* Caudell 1911. *Xanthippus corallipes zapotecus,* Hebard 1932: 248.

Xanthippus zapotecus Saussure 1884: 91. Type ♀, Puebla, Mexico. GM. *Hippiscus* (*Xanthippus*), Scudder 1892. *Xanthippus,* Caudell 1911. *Xanthippus corallipes zapotecus,* Hebard 1932: 248.

Hippiscus (*Xanthippus*) *affrictus* Scudder 1892: 336. Lectotype ♀, designated by Hebard 1929: 337, Yuma, Arizona. ANSP. *Xanthippus,* Kirby 1910. *Hippiscus,* Caudell 1911. *Xanthippus,* Hebard 1929. NEW SYNONYM.

Hippiscus (*Xanthippus*) *albulus* Scudder 1892: 351. Types ♂, ♀, Prescott Mountain District, central Arizona. ANSP. *Xanthippus,* Kirby 1910. *Xanthippus corallipes albulus,* Hebard 1925. *Xanthippus corallipes leprosus,* Hebard 1935: 289. *Xanthippus corallipes,* Strohecker 1952: 684.

Hippiscus (*Xanthippus*) *altivolus* Scudder 1892: 347. Lectotype ♀, designated by Hebard 1929: 341, Mt. Lincoln, Colorado, 11,000–13,000 feet. ANSP. *Xanthippus,* Kirby 1910. *Hippiscus,* Caudell 1911. *Xanthippus corallipes altivolus,* Hebard 1928. *Xanthippus corallipes,* Strohecker 1942: 684.

Hippiscus (*Xanthippus*) *conspicuus* Scudder 1892: 319. Lectotype ♀, designated by Hebard 1929: 338, Silver City, New Mexico. ANSP. *Xanthippus,* Kirby 1910. *Hippiscus,* Caudell 1911. *Xanthippus corallipes corallipes* Hebard 1929: 338.

Hippiscus (*Xanthippus*) *cupidus* Scudder 1892: 348. Holotype ♀, Pinal Mountains, Arizona. ANSP. *Xanthippus,* Kirby 1910. *Hippiscus,* Caudell 1911. *Xanthippus corallipes cupidus,* Hebard 1935: 290. *Xanthippus corallipes,* Strohecker 1952: 684.

Hippiscus (*Xanthippus*) *eremitus* Scudder 1892: 320. Holotype ♀, Arizona. ANSP. *Xanthippus,* Kirby 1910. *Hippiscus,* Caudell 1911. *Xanthippus corallipes,* Hebard 1935: 288.

Hippiscus (*Xanthippus*) *latefasciatus* Scudder 1892: 359. Lectotype ♂, here designated, Calgary, Alberta, June 11, 1890. ANSP. *Xanthippus,* Kirby 1910. *Hippiscus,* Caudell 1911. *Xanthippus corallipes latefasciatus,* Hebard 1925: 78.

Hippiscus (*Xanthippus*) *maculatus* Scudder 1892: 333. Lectotype ♂, designated by Hebard 1929: 337, Pueblo, Colorado. ANSP. *Xanthippus,* Kirby 1910. *Hippiscus,* Caudell 1911. *Xanthippus corallipes pantherinus, Hebard 1929: 337.*

Hippiscus (*Hippiscus*) *pantherinus* Scudder 1892: 285. Holotype ♂, Pecos River, Texas, June 18. ANSP. *Pardalophora,* Kirby 1910. *Hippiscus,* Caudell 1911. *Xanthippus corallipes pantherinus* Hebard 1929: 338. *Xanthippus corallipes,* Strohecker 1952: 686.

Hippiscus (*Xanthippus*) *pumilus* Scudder 1892: 349. Lectotype ♀, designated by Hebard 1929: 341, South Park, Colorado, 8,000–9,000 feet, June 16–23. ANSP. *Xanthippus,* Kirby 1910. *Hippiscus,* Caudell 1911. *Xanthippus corallipes altivolus,* Hebard 1929: 341. *Xanthippus corallipes,* Strohecker 1952: 686.

Xanthippus corallipes miniatus Strohecker 1952: 686. Holotype ♂, June Lake, California, June 20, 1948. Location of type not known.

Xanthippus corallipes buckelli Hebard 1928: 241. Holotype ♀, Chilcotin District, British Columbia, May 29, 1921. ANSP.

Xanthippus montanus (Thomas)

Oedipoda montana Thomas 1871: 462. Lectotype ♀, designated by Hebard 1927: 4, southern Montana. USNM. *Hippiscus,* Bruner 1885a. *Hippiscus* (*Xanthippus*) Scudder 1892. *Cratypedes?,* Kirby 1910. *Hippiscus,* Caudell 1911. *Xanthippus,* Hebard 1925: 78.

Xanthippus olancha (Caudell)

Hippiscus olancha Caudell 1921: 149. Holotype ♂, Olancha, Inyo County, California, April 24, 1917. USNM. *Xanthippus,* Strohecker, Middlekauff, and Rentz 1968: 131.

Agymnastus haemopterus Strohecker 1952: 687. Holotype ♂, Hesperia, California, April 14, 1948. University of California at Davis. *Xanthippus olancha,* Strohecker, Middlekauff, and Rentz 1968.

APPENDIX 7

Pronunciation of Names

In the names of grasshoppers, vowels have the following sounds: *a* is sounded either as in *back* (ă), as in *what* (a), as in *alms* (ah), or as in *may* (ay); *e* is either as in *end* (e) or as in *peek* (ee); *i* is either as in *pin* (i), as in *peek* (ee), or as in *pie* (ie); *o* is as in *on* (o), as in *oar* (ō), or as in *toe* (oh); *u* is sounded as in *push* (u), as in *moon* (oo), or as in *cue* (ew); *y* is sounded as in *tin* (i) or as in *peek* (ee). Italics indicate which syllables are stressed.

Agymnastus Ăgim*năs*tùs
Anconia Ăn*coh*neeah
Arphia *Ahr*feeah
Camnula Căm*noo*lah
Chimarocephala *Kim*mahroh*sef*fahlah
Chortophaga Kōr*to*fahgah
Circotettix Sirkoh*tet*tix
Conozoa Kono*zoh*wah
Cratypedes Krătti*pee*deez
Derotmema Derott*mee*mah
Dissosteira Dissos*tie*rah
Encoptolophus *En*copto*loh*fus
Eximacris Exim*ă*criss
Hadrotettix *Hăd*rohtettix
Heliastus Heelee*ăs*tus
Hippiscus Hip*pis*cus
Hippopedon *Hip*poh*pee*don
Lactista Lăc*tiss*tah
Leprus *Lep*prus
Leuronotina *Loo*rohnoh*tie*nah
Mestobregma *Mess*to*breg*mah
Metator *Meetay*tōr
Microtes Mie*croh*teez
Pardalophora *Par*dah*lof*forah
Psinidia Sin*id*eeah
Shotwellia Shot*wel*leeah

Spharagemon Sfăr*ă*gemon
Sphingonotus *Sfin*gohn*oh*tus
Sticthippus Stict*hip*pus
Tomonotus *Toh*mohn*oh*tus
Trachyrhachys *Tră*ckeer*ă*ckiss
Trepidulus Trep*pi*dewlus
Trimerotropus T*rie*merohtr*oh*pus
Tropidolophus *Tro*pidohl*oh*fus
Xanthippus Zăn*tip*pus

References

Alexander, G. 1941. Keys for the identification of Colorado Orthoptera. *University of Colorado Studies. Series D (Physical and Biological Sciences)* 1: 129–164.

Alexander, G. 1967. Cold hardiness in overwintering juvenile grasshoppers. *Entomological News* 78: 147–154.

Alexander, G., and J. R. Hilliard. 1969. Altitudinal and seasonal distribution of Orthoptera in the Rocky Mountains of northern Colorado. *Ecological Monographs* 39: 385–431.

Anderson, N. L., and J. C. Wright. 1952. Grasshopper investigations on Montana range lands. *Montana State College Agricultural Experimental Station Bulletin* 486: 1–46.

Ball, E. D., E. R. Tinkham, R. Flock, and C. T. Vorhies. 1942. The grasshoppers and other Orthoptera of Arizona. *Technical Bulletin Arizona College of Agriculture* 93: 275–373.

Banfil, J. C., and M. A. Brusven. 1973. Food habits and ecology of grasshoppers in the Seven Devils Mountains and Salmon River breaks of Idaho. *Melanderia* 12: 1–21.

Beamer, R. 1917. Grasshoppers of Kansas. II. The Oedipodinae of Kansas. *Bulletin of the University of Kansas Biological Series* 18: 51–126.

Bei-Bienko, G. Y. 1930. A monograph of thegenus Bryodema Fieber (Orthoptera, Acrididae) and its nearest allies. *Annuaire du Musée Zoologique de l'Académie des Sciences de L'URSS* 1930: 71–127.

Bei-Bienko, G. Y., and L. L. Mishchenko. 1964. *Locusts and grasshoppers of the U.S.S.R. and adjacent countries.* II. *Keys to the fauna of the U.S.S.R.* Zoological Institute of the U.S.S.R. Academy of Sciences, no. 40.

Blatchley, W. S. 1891. Some Indiana Acrididae. *Canadian Entomologist* 23: 74–81.

——— 1903. The Orthoptera of Indian. *27th Annual Report of the Department of Geology and Natural Resources of Indiana,* 1902: 123–471.

——— *1920. Orthoptera of northeastern America.* Indianapolis.

Brooks, A. R. 1958. Acridoidea of southern Alberta, Saskatchewan and Manitoba (Orthoptera). *Canadian Entomologist* 90: suppl. 9: 92.

Bruner, L. 1876. New species of Nebraska Acrididae. *Canadian Entomologist* 8: 123–125.

——— 1883. *On reference to the Rocky Mountain locust.* Chap. 4. Notes on other locusts and on the western cricket. U.S. Entomological Commission, report no. 3, pt. 1.

——— 1885a. Contributions of the North Trans-Continental Survey. Orthoptera. *Canadian Entomologist* 17: 9–19.

——— 1885b. First contribution to a knowledge of the Orthoptera of Kansas. *Bulletin Washburn College Laboratory National History* 1: 125–139.

——— 1890 (1889 vol.). New North American Acrididae, found north of the Mexican boundary. *Proceedings of the United States National Museum* 12: 47–82.

——— 1893a. Descriptions of new species of Orthoptera from the Death Valley expedition. *North American Fauna* 7: 266–268.

——— 1893b. A list of Nebraska Orthoptera. *Publications of the Nebraska Academy of Sciences* 3: 19–33.

——— 1893c. Notes on the wing color of North American locusts belonging to the subfamily Oedipodinae and its seeming relation to climatic conditions. *Science* 21: 133.

——— 1897. The grasshoppers that occur in Nebraska. *Annual Report Entomology, Nebraska State Building of Agriculture* 1897: 105–135.

——— 1900–1909. *Biologia Centrali-Americana,* vol. 2: 1–14 (1900); 14–25 (1901); 25–32 (1902); 33–105 (1904); 105–176 (1905); 177–208 (1906); 208–248 (1907); 249–342 (1908); 343–412 (1909).

——— 1905. Concerning the locust genus *Mestobregma,* Scudder. *Entomological News* 16: 259–260.

Brunner, C. de Wattenwyl. 1897. Tomonotus theresiae sp.n. *Berliner entomologischer Zeitschrift* 40 (Heft 2): 277.

Buckell, E. R. 1920. Life-history notes on some species of Acrididae (Orthoptera) found in British Columbia. *Report of the Entomological Society of Ontario* 1919: 53–61.

——— 1922a. Notes on the ecological distribution of some Orthoptera from the Chilcotin district of British Columbia. *Proceedings of the Entomological Society of British Columbia* 18: 32–38.

——— 1922b. A list of the Orthoptera and Dermaptera recorded from British Columbia prior to the year 1922, with annotations. *Proceedings of the Entomological Society of British Columbia* 20: 941.

——— 1930. The Dermaptera and Orthoptera of Vancouver Island, British Columbia. *Proceedings of the Entomological Society of British Columbia* 27: 17–51.

Burmeister, H. 1838. *Handbuch der Entomologie, Orthoptera,* vol. 2, pt. 2. Berlin. 397–756.

Campbell, J. B., W. H. Arnett, J. D. Lambley, O. K. Jantz, and H. Knutson. 1974. *Grasshoppers (Acrididae) of the Flint Hills native tallgrass prairie in Kansas.* Kansas State University Agricultural Experiment Station Research Paper no. 19: 1–146.

Cantrall, I. J. 1943. *The ecology of the Orthoptera and Dermaptera of the George Reserve, Michigan.* Miscellaneous Publications of the University of Michigan Museum of Zoology 54: 1–184.

Carothers, E. E. 1917. The segregation and recombination of homologous chromosomes as found in two genera of Acrididae (Orthoptera). *Journal of Morphology* 29: 445–520.

——— 1921. Genetical behavior of heteromorphic homologous chromosomes of *Circotettix* (Orthoptera). *Journal of Morphology* 35: 457–483.

Catesby, M. 1743. *The natural history of Carolina, Florida and the Bahama Islands* 2: 89.

Caudell, A. N. 1901. On some Arizona Acrididae. *Canadian Entomologist* 33: 102–106.

——— 1903a. Some new or unrecorded Orthoptera from Arizona. *Proceedings of the Entomological Society of Washington* 5: 162–166.

——— 1903b. Notes on Orthoptera from Colorado, New Mexico, Arizona, and Texas with descriptions of new species. *Proceedings of the United States National Museum* 26: 775–809.

——— 1904a. Orthoptera from southwestern Texas collected by the museum expeditions of 1903, 1904. *Bulletin of the Museum of the Brooklyn Institute of Arts and Sciences* 1: 105–116.

——— 1904b. Two Orthoptera hitherto unrecorded from the United States. *Proceedings of the United States National Museum* 27: 949–952.

——— 1905. On a collection of Orthoptera from southern Arizona, with descriptions of new species. *Proceedings of the United States National Museum* 28: 461–477.

——— 1906. The species of the genus *Chimarocephala* and description of two new species of California Orthoptera. *Proceedings of the Entomological Society of Washington* 7: 123–126.

——— 1911. Some remarks on Kirby's synonymic catalogue of Orthoptera, vol. III, with additional notes on vol. 5. I and II. *Entomological News* 22: 158–167.

Caudell, A. N., and M. Hebard. 1912. Fixation of the single type (lectotypic) specimens of species of American Orthoptera. *Proceedings of the Academy of Natural Sciences of Philadelphia* 1912: 157–168.

Cockerell, T. D. A. 1902. Notes on the Orthopteran genus *Leprus. Entomological News* 13: 305–306.

Coleman, L. C. 1948. The cytology of some western species of *Trimerotropis* (Acrididae). *Genetics* 33: 519–528.

Coppock, S. 1962. The grasshoppers of Oklahoma (Orthoptera: Acrididae). Ph. D. thesis, Oklahoma State University.

Dakin, M. E., Jr., and K. L. Hays. 1970. A synopsis of Orthoptera (sensu latu) of Alabama. *Auburn University Agricultural Experiment Station Bulletin* no. 404: 1–118.

Davis, W. T. 1889. List of the Orthoptera found on Staten Island. *Entomologica Americana* 5: 78–81.

DeGeer, C. 1773. *Memoires pour servir à l'Histoire des Insectes,* vol. 3. Stockholm.

DeHaan, W. 1842. *Bijdragen tot de kennis de orthoptera* (Zool., pt. 2): 45–238.

Dirsh, V. M. 1956. The public complex in Acridoidea (Orthoptera) in relation to taxonomy. *Transactions of the Royal Entomological Society of London* 108: 223–356.

——— 1961. A preliminary revision of the families and subfamilies of Acridoidea (Orthoptera, Insecta). *Bulletin of the British Museum (Natural History), Entomology* 10: 351–419.

——— 1975. *Classification of the acridomorphoid insects.* Oxford.

Erichson, W. F. 1848. Versuch einer Fauna und Flora von Britisch-Guiana. Insekten. In R. Schomburgk, *Reisen in den Jahren 1840–1849, British-Guiana* 3: 553–617.

Fabricius, J. C. 1775. *Systema entomologiae.* Leipzig.

——— 1781. *Species insectorum exhibentes eorum differentias specificas, synonyma auctorum, loca natalia, metamorphosin adjectis observationibus, adumbrationibus.* Tome I.

Fieber, F. X. 1843. Synopsis der europäischen Orthoptera. *Lotos* 3: 115–131.

Fox, H. 1914. Data on the Orthoptera faunistics of eastern Pennsylvania and southern New Jersey. *Proceedings of the Academy of Natural Sciences of Philadelphia* 1914: 441–534.

——— 1917. Field notes on Virginia Orthoptera. *Proceedings of the United States National Museum* 52: 199–234.

——— 1928. A revised annotated list of the Dermaptera and Orthoptera of New Jersey. *State of New Jersey Department of Agriculture circular* 138: 3–58.

Frison, T. H. A list of the insect types in the collection of the Illinois State Natural History Survey and the University of Illinois. *Illinois Natural History Survey Bulletin* 16: 131–309.

Froeschner, R. C. 1954. The grasshoppers and other Orthoptera of Iowa. *Iowa State College Journal of Science* 29 (2): 163–354.

Gurney, A. B. 1940. Notes on certain genera of North American grasshoppers of the subfamily Oedipodinae, with the description of a new genus and species (Orthoptera: Acrididae). *Proceedings of the Entomological Society of Washington* 42: 1–12.

Hahn, C. W. 1835. *Icones orthoptorum. Albildungen der haut flügligen Insekten.* Nuremberg.

Harris, T. W. 1835. A catalogue of the animals and plants in Massachusetts. VIII. Insects, 553–602. In E. Hitchcock, *Report on the geology, mineralogy, botany and zoology of Massachusetts.* Amherst, Mass.

——— 1841. *A report on the insects of Massachusetts, injurious to vegetation.* Cambridge, Mass.

Hart, C. A. 1906. Notes of a winter trip in Texas, with an annotated list of the Orthoptera. *Entomological News* 87: 154–166.

Hauke, H. A. 1953. An annotated list of the Orthoptera of Nebraska. II. The Tettigidae and Acrididae. *Bulletin of the University of Nebraska State Museum* 3 (9): 1–79.

Hebard, M. 1924 (1923 vol.). Studies in the Dermaptera and Orthoptera of Colombia. *Transactions of the American Entomological Society* 49: 165–313.

——— 1925a. Dermaptera and Orthoptera from the state of Sinaloa, Mexico. II. *Transactions of the American Entomological Society* 51: 265–310.

——— 1925b. The Orthoptera of South Dakota. *Proceedings of the Academy of Natural Sciences of Philadelphia* 77: 33–155.

——— 1927. Fixation of the single types of species of Orthoptera described by Cyrus Thomas. *Proceedings of the Academy of Natural Sciences of Philadelphia* 79: 1–11.

——— 1928. The Orthoptera of Montana. *Proceedings of the Academy of Natural Sciences* 80: 211–306.

——— 1929. The Orthoptera of Colorado. *Proceedings of the Academy of Natural Sciences of Philadelphia* 81: 303–425.

——— 1931a. The Orthoptera of Kansas. *Proceedings of the Academy of Natural Sciences of Philadelphia* 83: 119–227.

——— 1931b. Studies in lower California Orthoptera. *Transactions of the American Entomological Society* 57: 113–127.

——— 1931c. The Orthoptera of Alberta. *Proceedings of the Academy of Natural Sciences of Philadelphia* 82: 377–403.

——— 1932. New species and records of Mexican Orthoptera. *Transactions of the American Entomological Society* 58: 201–371.

——— 1934a. The Dermaptera and Orthoptera of Illinois. *Bulletin of the Illinois State Natural History Survey* 20: 125–279.

——— 1934b. Notes on Orthoptera from northwestern Minnesota. *Entomological News* 45: 103–106.

——— 1935a. Studies in the Orthoptera of Arizona. I. New genera, species and geographical races. *Transactions of the American Entomological Society* 61: 111–153.

——— 1935b. Studies in the Orthoptera of Arizona. II. *Transactions of the American Entomological Society* 61: 269–316.

——— 1935c. Orthoptera of the upper Rio Grande valley and the adjacent mountains in northern New Mexico. *Proceedings of the Academy of Natural Sciences of Philadelphia* 57: 45–82.

——— 1937. Studies in the Orthoptera which occur in North America north of the Mexican boundary. IX. *Transactions of the American Entomological Society* 63: 347–379.

——— 1938. An ecological survey of the Orthoptera of Oklahoma. *Oklahoma Agricultural and Mechanical College Agricultural Experiment Station Technical Bulletin* no. 5: 5–31.

Helfer, J. R. 1963. *How to know the grasshoppers, cockroaches and their allies.* Dubuque, Iowa.

Helwig, E. R. 1929. Chromosomal variations correlated with geographical distribution in *Circotettix verruculatus* (Orthoptera). *Journal of Morphology* 47: 1–36.

——— 1955. Spermatogenesis in hybrids between *Circotettix verruculatus* and *Trimerotropis suffusa* (Orthoptera: Oedipodinae). *University of Colorado Studies, Series Biology* 3: 47–64.

Henderson, W. W. 1924. A taxonomic and ecological study of the species of the subfamily Oedipodinae (Orthoptera-Acrididae) found in Utah. *Utah Agricultural Experiment Station Technical Bulletin* 191: 7–150.

——— 1931. Crickets and grasshoppers in Utah. *Utah Agricultural Experiment Station Circular* 96: 5–38.

Hewitt, G. B., and W. F. Barr. 1967. The banded-wing grasshoppers of Idaho (Orthoptera: Oedipodinae). *University of Idaho Agricultural Experiment Station, Research Bulletin* no. 72: 1–64.

Hubbell, T. H. 1922a. Note on the Orthoptera of North Dakota. *Occasional Papers of the University of Michigan Museum of Zoology* 113: 1–56.

——— 1922b. Dermaptera and Orthoptera of Berrien County, Michigan. *Occasional Papers of the University of Michigan Museum of Zoology* 116: 1–77.

——— 1929. The distribution of the beach-grasshoppers *Trimerotropis huroniana* and *Trimerotropis maritima interior* in the Great Lakes region. *Journal of the New York Entomological Society* 37: 31–38.

Isley, F. B. 1935. Acridian researches within northeastern Texas. *Entomological News* 46: 37–43, 69–75.

——— 1937. Seasonal succession, soil relations, numbers, and regional distribution of northeastern Texas acridians. *Ecological Monographs* 3: 317–344.

——— 1938. The relations of Texas Acrididae to plants and soils. *Ecological Monographs* 8: 551–604.

Joern, A. 1977. Guild and community structure in primary consumers: resource utilization in arid grassland grasshopper communities (Orthoptera: Acrididae). Doctoral dissertation, University of Texas at Austin.

——— 1979. Feeding patterns in grasshoppers (Orthoptera: Acrididae): Factors influencing diet specialization. *Oecologia* (Berlin) 38: 325–347.

——— 1982a. Vegetation structure and microhabitat selection in grasshoppers (Orthoptera, Acrididae). *Southwestern Naturalist* 27: 197–209.

——— 1982b. Distributions, densities, and relative abundances of grasshoppers (Orthoptera: Acrididae) in a Nebraska sandhills prairie. *Prairie Naturalist* 14: 37–45.

——— Host plant utilization by grasshoppers (Orthoptera: Acrididae) from a sandhills prairie. *Journal of Range Management* (in press).

Joern, A., and L. R. Lawlor. 1980. Food and microhabitat utilization by grasshoppers from arid grasslands. Comparisons with neutral models. *Ecology* 61: 591–599.

——— 1981. Guild structure in grasshopper assemblages based on food and microhabitat resources. *Oikos* 37: 93–104.

Johnston, H. B. 1956. *Annotated catalogue of African grasshoppers.* Cambridge.

——— 1968. *Annotated catalogue of African grasshoppers. Supplement.* Cambridge.

Kelly, G. D., and W. W. Middlekauff. 1961. Biological studies of *Dissosteira spurcata* Saussure with distributional notes on related California species. *Hilgardia* 30: 395–424.

King, R. L. 1924. Heteromorphic homologous chromosomes in three species of *Pseudotrimerotropis* (Orthoptera: Acrididae). *Journal of Morphology* 38: 19–63.

Kirby, W. 1837. The insects. In Richardson, *Fauna Boreale Americana* 4.

Kirby, W. F. 1910. *A synonymic catalogue of Orthoptera,* vol. III. British Museum of Natural History Publication.

LaRivers, Ira. 1948. A synopsis of Nevada Orthoptera. *American Midland Naturalist* 39: 652–720.

Linnaeus, C. 1758. *Systema natura per regna tria nature, secundum classes, ordines, genera, species.* Tome 1 (10th ed.).

McNeill, J. 1900. The Orthopteran genus *Trimerotropis. Psyche* 9: 27–36.

——— 1901. Review of the Orthopteran genus *Trimerotropis. Proceedings of the United States National Museum* 23: 393–449.

Middlekauff, W. W. 1958. Biology and ecology of several species of California rangeland grasshoppers (Orthoptera: Acrididae). *Pan-Pacific Entomologist* 34: 1–11.

Morse, A. P. 1894. *Spharagemon:* A study of the New England species. *Proceedings of the Boston Society of Natural History* 26: 221–240.

——— 1895. Revision of the species of *Spharagemon. Psyche* 7: 287–299.

——— 1897. Notes on the New England Acrididae. III. Oedipodinae. *Psyche* 8: 6–8, 35–37, 50–51, 64–66, 80–82, 87–89, 91–114.

——— 1904a. *Researches on North American Acrididae.* Carnegie Institution of Washington Publication no. 18: 1–55.

——— 1904b. New Acrididae from the southeastern states. *Psyche* 11: 7–13.

——— 1907. *Further researches on North American Acrididae.* Carnegie Institution of Washington Publication no. 68: 1–54.

——— 1919. A list of Orthoptera of New England. *Psyche* 26: 21–39.

——— 1920. Manual of the Orthoptera of New England. *Proceedings of the Boston Society of Natural History* 35: 197–556.

——— 1921. Grasshoppers and related insects. *Bulletin of the Maine Agricultural Experiment Station, Orono* 296: 1–36.

Morse, A. P., and M. Hebard. 1915. Fixation of single type (Lectotypic) specimens of species of American Orthoptera (described by Morse). *Proceedings of the Academy of Natural Sciences of Philadelphia* 1915: 96–106.

Mulkern, G. B. 1967. Food selection by grasshoppers. *Annual Review of Entomology* 12: 59–78.

——— 1980. Population fluctuations and competitive relationships of grasshopper species (Orthoptera: Acrididae). *Transactions of the American Entomological Society* 106: 1–41.

Mulkern, G. B., K. P. Pruess, H. Knutson, H. F. Hagen, J. B. Campbell, and J. D. Lambley. 1969. Food habits and preferences of grassland grasshoppers of the north central Great Plains. *Agricultural Experimental Station North Dakota State University Bulletin* no. 481: 1–31.

Mulkern, G. B., D. R. Toczek, and M. A. Brusven. 1964. Biology and ecology of North Dakota grasshoppers. II. Food habits and preference of grasshoppers associated with the sand hills prairie. *North Dakota Agricultural Experimental Station Report* no. 11: 59.

Newton, R. C., and A. B. Gurney. 1956–1957. Distribution maps of range grasshoppers in the United States. *Cooperative Economic Insect Report of the U.S. Department of Agriculture* 6 (1956): 597–600, 710–712, 743–744, 766, 838–840, 883–884, 920, 938–940, 972, 987–988, 1002–1004, 1019–1020, 1036, 1050–1052, 1090–1092, 1122–1124, 1149–1150; 7 (1957): 50–52, 71–72, 89–90, 109–110, 129–130, 150, 188, 208, 225–226, 247–248, 263–264, 315–316, 368, 388, 409, 432, 455–456, 479–480.

Otte, D. 1968. A comparative study of communication in grasshoppers. Ph. D. dissertation, University of Michigan.

——— 1970. *A comparative study of communicative behavior in grasshoppers.* Miscellaneous Publications of the University of Michigan Museum of Zoology no. 141: 1–168.

——— 1972. Environmentally induced color dimorphisms in grasshoppers. *Annals of the Entomological Society of America* 65: 1154–1161.

——— 1976. Species richness patterns of New World desert grasshoppers in relation to plant diversity. *Journal of Biogeography* 3: 197–209.

——— 1977. Communication in Orthoptera. Chap. 16. In T. A. Sebeok, ed., *How animals communicate.* Bloomington, Ind.

——— 1978. The primary types of Orthoptera (Saltatoria, Blattodea, Phasmatodea, and Mantodea) at the Academy of Natural Sciences of Philadelphia. *Proceedings of the Academy of Natural Sciences of Philadelphia* 30: 26–87.

——— 1981. *The North American Grasshoppers.* I. *Acrididae: Gomphocerinae and Acridinae.* Cambridge, Mass.

Otte, D., and A. Joern. 1977. On feeding patterns in desert grasshoppers and the evolution of specialized diets. *Proceedings of the Academy of Natural Sciences of Philadelphia* 128: 89–126.

Pickford, R. 1953. A two-year life-cycle in grasshoppers (Orthoptera: Acrididae) overwintering as eggs and nymphs. *Canadian Entomologist* 85: 9–14.

Rehn, J. A. G. 1902. Notes on some southern Californian Orthoptera. *Canadian Entomologist* 34: 141–146.

——— 1903 (1902 vol.). Notes on the Orthoptera of New Mexico and western Texas. *Proceedings of the Academy of Natural Sciences of Philadelphia* 54: 717–727.

——— 1904a. Notes on Orthoptera from northern and central Mexico. *Proceedings of the Academy of Natural Sciences of Philadelphia* 56: 513–549.

——— 1904b. Notes on Orthoptera from Arizona, New Mexico and Colorado. *Proceedings of the Adademy of Natural Sciences of Philadelphia* 56: 562–575.

——— 1905. A contribution to the knowledge of the Acrididae (Orthoptera) of Costa Rica. *Proceedings of the Academy of Natural Sciences of Philadelphia* 57: 400–453.

——— 1907a. Notes on Orthoptera from southern Arizona, with descriptions of new species. *Proceedings of the Academy of Natural Sciences of Philadelphia* 59: 24–81.

——— 1907b. Three new species of Acrididae (Orthoptera) from California. *Entomological News* 18: 350–355.

——— 1910. Some notes on Idaho Orthoptera, with the description of a new species of *Trimerotropis*. *Proceedings of the Academy of Natural Sciences of Philadelphia* 62: 12–17.

——— 1911. Orthoptera from the Santa Rita Mountains, Arizona, collected by the University of Kansas Expedition. *Kansas University Science Bulletin* 5 (17): 299–307.

——— 1919. Descriptions of new and critical notes upon previously known forms of North American Oedipodinae (Orthoptera: Acrididae). *Transactions of the American Entomological Society* 45: 229–255.

——— 1921. Descriptions of new and critical notes upon previously known forms of North American Oedipodinae (Orthoptera: Acrididae). *Transactions of the American Entomological Society* 47: 171–197.

——— 1940. (1939 vol.). The South American species of the oedipodine genus *Trimerotropis* (Orthoptera: Acrididae). *Transactions of the American Entomological Society* 65: 395–414.

——— 1945. The status of Stål's *Oedipoda venusta,* 1861. *Entomological News* 56: 132.

Rehn, J. A. G., and H. J. Grant, Jr. 1959. Critical remarks on a recent contribution to the taxonomy of the Acridoidea (Orthoptera) by V. M. Dirsh. *Entomological News* 70: 245–249.

——— 1961. *A monograph of the Orthoptera of North America* (*north of Mexico*), vol. 1. Monographs of the Academy of Natural Sciences of Philadelphia 12: 1–255.

Rehn, J. A. G., and M. Hebard. 1905a (1904 vol.). The Orthoptera of Thomas County, Georgia, and Leon County, Florida. *Proceedings of the Academy of Natural Sciences of Philadelphia* 56: 774–802.

——— 1905b. A contribution to the knowledge of Orthoptera of south and central Florida. *Proceedings of the Academy of Natural Sciences of Philadelphia* 57: 29–55.

——— 1906. A contribution to the knowledge of the Orthoptera of Montana, Yellowstone Park, Utah and Colorado. *Proceedings of the Academy of Natural Sciences of Philadelphia* 58: 360–418.

——— 1907. Orthoptera from Northern Florida. *Proceedings of the Academy of Natural Sciences of Philadelphia* 59: 279–319.

——— 1908. An orthopterological reconnaissance of the southwestern United States. I: Arizona. *Proceedings of the Academy of Natural Sciences of Philadelphia* 60: 365–402.

——— 1909. An orthopterological reconnaissance of the southwestern United States. II: New Mexico and western Texas. *Proceedings of the Academy of Natural Sciences of Philadelphia* 61: 111–175.

——— 1910 (1909 vol.). An orthopterological reconnaissance of the southwestern United States. III: California and Nevada. *Proceedings of the Academy of Natural Sciences of Philadelphia* 61: 409–483.

——— 1911a (1910 vol.). Records of Georgia and Florida Orthoptera, with the descriptions of one new species and one new subspecies. *Proceedings of the Academy of Natural Sciences of Philadelphia* 62: 585–589.

——— 1911b (1910 vol.). Preliminary studies of North Carolina Orthoptera.

Proceedings of the Academy of Natural Sciences of Philadelphia 62: 615–650.

——— 1912a. Fixation of single type (lectotype) specimens of species of American Orthoptera. I. *Proceedings of the Academy of Natural Sciences of Philadelphia* 64: 60–128.

——— 1912b. On the Orthoptera found on the Florida Keys and in extreme southern Florida. I. *Proceedings of the Academy of Natural Sciences of Philadelphia* 64: 235–276.

——— 1914. On the Orthoptera found on the Florida Keys and in extreme southern Florida. II. *Proceedings of the Academy of Natural Sciences of Philadelphia* 66: 373–412.

——— 1916. Studies in the Dermaptera and Orthoptera of the coastal plain and piedmont region of the southeastern United States. *Proceedings of the Academy of Natural Sciences of Philadelphia* 68: 87–314.

——— 1938. New genera and species of West Indian Acrididae, with notes on previously known species. *Transactions of the American Entomological Society* 64: 201–226.

Rehn, J. W. H. 1939. Notes on the Orthoptera of Nova Scotia and Newfoundland. *Canadian Entomologist* 71: 175–178.

Rentz, D. C. 1977. Two new grasshoppers from California (Orthoptera, Acrididae, Oedipodinae). *Acrida* 6: 141–151.

——— 1978. New species and records of western Orthoptera. *Pan-Pacific Entomologist* 54: 81–97.

Rentz, D. C., and D. B. Weissman. 1980. An annotated checklist of the grasshopper species of *Aerochoreutes, Circotettix,* and *Trimerotropis* (Orthoptera: Acrididae: Oedipodinae). *Transactions of the American Entomological Society* 106: 223–252.

——— 1981. *Faunal affinities, systematics and bionomics of the Orthoptera of the California channel islands.* University of California Publications Entomology 94: 1–240.

——— 1984. *Pan-Pacific Entomologist* (in press).

Ritchie, J. M. 1981. A taxonomic revision of the genus *Oedaleus* Fieber (Orthoptera: Acrididae). *Bulletin of the British Museum (Natural History)* 42: 83–183.

——— 1982. A taxonomic revision of the genus *Gastrimargus* Saussure (Orthoptera: Acrididae). *Bulletin of the British Museum (Natural History)* 44: 239–329.

Roberts, H. R. 1941. A comparative study of the subfamilies of the Acrididae (Orthoptera) primarily on the basis of their phallic structures. *Proceedings of the Academy of Natural Sciences of Philadelphia* 93: 201–246.

Rockwood, L. P. 1925. On night flying and attraction to light in Acridiidae and the relation of meteorological conditions thereto. *Pan-Pacific Entomologist* 2: 36–38.

Rodell, C. F. 1977. A grasshopper model for a grassland ecosystem. *Ecology* 58: 227–245.

Saussure, H. de. 1859. Orthoptera Nova Americana (Diagnoses Praeliminares). I. *Revue et Magazin de Zoologie* (2) 11: 59–63, 201–212, 315–317, 390–394.

——— 1861a. Orthoptera Nova Americana (Diagnoses Praeliminares). II. *Revue et Magazin de Zoologie* (2) 13: 126–130, 156–164, 313–324, 397–402.

——— 1861b. Orthoptera Nova Americana (Diagnoses Praeliminares) (Series IIa). *Revue et Magazin de Zoologie, pure et applique,* 13: 313–329.

——— 1884. *Prodromus Oedipodiorum Insectorum ex ordine Orthopterorum.* Mémoires de la Société de Physique et d'Histoire Naturelle de Genève 28: 1–254.

——— 1888. *Additamenta ad prodromum oedipodiorum.* Geneva.

Say, T. 1825a. Descriptions of new hemipterous insects collected in the expedition to the Rocky Mountains, performed by order of Mr. Calhoun, secretary of war, under command of Major Long. *Journal of the Academy of Natural Sciences of Philadelphia* 4: 307–345.

——— 1825b. *American entomology or descriptions of the insects of North America.* Philadelphia Museum Publications, 2.

——— 1828. *American entomology, or descriptions of the insects of North America. Illustrated by coloured figures from original drawings executed from nature,* 3 vols. (1824–1828). Philadelphia.

Schennum, W. E., and R. B. Willey. 1979. A geographical analysis of quantitative morphological variation in the grasshopper *Arphia conspersa. Evolution* 33: 64–84.

Scoggan, A. C., and M. A. Brusven. 1972. Differentiation and ecology of common immature Gomphocerinae and Oedipodinae (Orthoptera: Acrididae) of Idaho and adjacent areas. *Melanderia* 8: 1–76.

Scudder, S. H. 1862. Materials for a monograph of the North American Orthoptera. Including a catalogue of the known New England species. *Journal of the Boston Society of Natural History* 7: 409–480.

——— 1869. Descriptions of new species of Orthoptera in the collection of the American Entomological Society. *Transactions of the American Entomological Society* 2: 305–307.

——— 1872. Notes on the Orthoptera collected by Dr. F. V. Hayden in Nebraska. In F. V. Hayden, *Final Report of the U.S. Geological Survey of Nebraska and portions of adjacent territories made under the direction of the commissioner of the General Land Office:* 249–261.

——— 1874. The distribution of insects in New Hampshire. In *Hitchcock's Report on the Geology of New Hampshire* I: 331–380.

——— 1875a. A century of Orthoptera. Decade III. Acrydii. *Proceedings of the Boston Society of Natural History* 17: 472–485.

——— 1875b. A century of Orthoptera. Decade IV. Acrydii. *Proceedings of the Boston Society of Natural History* 17: 510–517.

——— 1875c. Notice on the butterflies and Orthoptera collected by Mr. George M. Dawson, as naturalist of the B. N. A. Boundary Commission. Appendix D, in G. M. Dawson, *British North American boundary commission geological report of progress for the year 1873 (in part). Report of the tertiary lignite formation in the vicinity of the 49th parallel.* Montreal.

——— 1875d. *Spharagemon.* A genus of Oedipodinae, with a revision of the

species. *Proceedings of the Boston Society of Natural History* 17: 467–471
——— 1875e. Revision of two American genera of Oedipodinae. *Proceedings of the Boston Society of Natural History* 17: 478–485.
——— 1876a. List of the Orthoptera collected by Dr. A. S. Packard in Colorado and the neighboring territories, during the summer of 1875. *Bulletin of the Geological and Geographical Survey of the Territories,* 2 (3): 261–267.
——— 1876b. Report on the Orthoptera collected by the U. S. Geological surveys west of the 100th Meridian under the direction of Lieutenant George M. Wheeler, during the season of 1875. Appendix H9 in G. M. Wheeler, *Annual report upon the geological surveys west of the 100th meridian, in California, Nevada, Utah, Colorado, Wyoming, New Mexico, Arizona and Montana.* Appendix JJ of the *Annual Report of the Chief of Engineers for 1876.*
——— 1876c. Orthoptera from the island of Guadalupe. *Proceedings of the Boston Society of Natural History* 18: 268–271.
——— 1877a. A century of Orthoptera. Decade VII: Acrydii. *Proceedings of the Boston Society of Natural History* 19: 27–35.
——— 1877b. New forms of saltatorial Orthoptera from the southern United States. *Proceedings of the Boston Society of Natural History* 19: 35–41.
——— 1892. The Orthopteran genus *Hippiscus. Psyche* 6: 265–274, 285–288, 301–304, 317–320, 333–336, 347–350, 359–363.
——— 1897. Synonymical and descriptive notes on North American Orthoptera. *Canadian Entomologist* 29: 73–76.
——— 1900. Catalogues of the described Orthoptera of the United States and Canada. *Proceedings of the Davenport Academy of Natural Sciences* 8: 1–101.
——— 1901. Four new species of *Hippiscus. Canadian Entomologist* 33: 88–92.

Scudder, S. H., and T. D. A. Cockerell. 1902. A first list of Orthoptera of New Mexico. *Proceedings of the Davenport Academy of Natural Sciences* 9: 1–60.

Serville, J. G. 1839. *Histoire Naturelle des Insects. Orthopteres.* Paris.

Shotwell, R. L. 1941. Life histories and habits of some grasshoppers of economic importance on the Great Plains. *United States Department of Agriculture Technical Bulletin* 774: 48.

Smith, C. C. 1940. Ecological distribution of Acrididae in central Oklahoma. *Proceedings of the Oklahoma Academy of Science* 20: 67–69.

Smith, S. 1981. Variation in morphology and coloration among grasshoppers related to geographical distribution, seasonal occurrence, and plant communities in the flint hills region of Kansas. Ph.D. dissertation, Kansas State University.

Somes, M. P. 1914. The Acridiidae of Minnesota. *The University of Minnesota Agricultural Experiment Station Bulletin* 141: 1–98.

Stål, C. 1873. *Recensio Orthopterorum. Revue critique des Orthoptères décrits par Linné, De Geer et Thunburg.* I: 1–154.

Strohecker, H. F. 1945. Notes on the descriptions of Mexican Orthoptera. *Annals of the Entomological Society of America* 38: 207–215.

——— 1952. Descriptions of new species and notes on North American Orthoptera. *American Midland Naturalist* 48: 683–688.

——— 1960. Several new species of North American Orthoptera. *Pan-Pacific Entomologist* 36: 31–35.

——— 1963. New Acrididae from western North America. *Pan-Pacific Entomologist* 39: 157–174.

Strohecker, H. F., and G. M. Buxton. 1963. A new Oedipodine grasshopper from California. *Pan-Pacific Entomologist* 39: 260–263.

Strohecker, H. F., W. W. Middlekauff, and D. C. Rentz. 1968. *The grasshoppers of California.* Bulletin of the California Insect Survey 10: 1–177.

Thomas C. 1870. Descriptions of grasshoppers from Colorado. *Proceedings of the Academy of Natural Sciences of Philadelphia* 22: 74–84.

——— 1871a. A list and descriptions of new species of Orthoptera. *Preliminary Report, United States Geological Survey of Wyoming and Adjacent Territories* 2, 1870: 265–284.

——— 1871b. Contribution to orthopterology. *Proceedings of the Academy of Natural Sciences of Philadelphia* 23: 149–153.

——— 1872. Notes on the Saltatorial Orthoptera of the Rocky Mountain Region. In F. V. Hayden, *Preliminary Report of the U. S. Geological Survey of Montana and portions of Adjacent Territories. 5th Annual Report of Progress, Part IV. Zoology and Botany:* 423–466.

——— 1873a. Notes on Orthoptera. *Sixth Annual Report of the United States Geological Survey of Montana, Idaho, Wyoming and Utah:* 719–725.

——— 1873b. Synopsis of the Acrididae of North America. *Report of the United States Geological Survey of the Territories* 5: 1–258.

——— 1875. Report upon the collections of Orthoptera made in portions of Nevada, Utah, California, Colorado, New Mexico and Arizona during the years 1871–1874. *Wheeler's Report on Geological and Geographical Exploration and Surveys West of the 100th Meridian* 5: 843–908.

——— 1876a. A list of Orthoptera collected by J. Duncan Putnam, of Davenport, Iowa during the summers of 1872–1875, chiefly in Colorado, Utah, and Wyoming Territories. *Proceedings of the Davenport Academy of Natural Sciences* 1: 249–264.

——— 1876b. A list of the Orthoptera of Illinois. *Illinois Museum of Natural History Bulletin* no. 1: 59–69.

——— 1880. Notes on Orthoptera. *Canadian Entomologist* 12: 221–224.

Thunberg, C. P. 1815, Hempterorum maxillosoum genera illustrata. *Mémoires de l'Academie Imperiale des Sciences de St. Petersburg* 5: 211–301.

Tinkham, E. R. 1947. New species, records and faunistic notes concerning Orthoptera in Arizona. *American Midland Naturalist* 38: 127–149.

——— 1948. Faunistic and ecological studies on the Orthoptera of the Big Bend region of Trans-Pecos, Texas. *American Midland Naturalist* 40: 521–663.

——— 1959. Notes on the self-burial habits of two Nearctic sand dune acridids (Orthoptera). *Entomologist, London* 92: 185–188.

——— 1960. Studies in Nearctic desert sand dune Orthoptera. II. Two new grasshoppers of the genus *Trimerotropis* from the Utah desert. *Great Basin Naturalist* 20: 49–58.

——— 1964. Studies in Nearctic desert dune Orthoptera. IX. A new *Trimerotropis* from southern Idaho dunes. *Great Basin Naturalist* 24: 31–35.

Treherne, R. C., and E. R. Buckell. 1924. *Grasshoppers of British Columbia, Dominion of Canada.* Department of Agriculture Bulletin 39 (Extension Bulletin 26): 1–40.

Uhler, P. R. 1862. Notes on Orthoptera. In T. W. Harris, *Insects injurious to vegetation,* 3rd ed. (ed. C. L. Flint). Boston.

Uvarov, B. 1966. *Grasshoppers and locusts. A handbook of general acridology,* vol. 1. Cambridge.

——— 1977. *Grasshoppers and locusts. A handbook of general acridology,* vol. 2. Cambridge.

Vestal. A. G. 1913. Local distribution of grasshoppers in relation to plant associations. *Biological Bulletin* 25: 141–180.

——— 1914. Notes on the habits of grasshoppers at Douglas Lake, Michigan. *Entomological News* 25: 105–109.

Vickery, V. R. 1979. Notes on some Canadian Acrididae (Orthoptera). *Canadian Entomologist* 111: 699–702.

Vickery, V. R., and D. K. McE. Kevan, 1983. *A monograph of the orthopteroid insects of Canada and adjacent regions.* Lyman Entomological Museum and Research Laboratory Memoir no. 13: 1–1462.

Vickery, V. R., and B. Nagy. 1973. Ecological notes on Orthoptera in British Columbia. *Journal of the Entomological Society of British Columbia* 70: 27–33.

Vickery, V. R., D. E. Johnstone, and D. K. McE. Kevan. 1974. *The orthopteroid insects of Quebec and the Atlantic Provinces of Canada.* Lyman Entomological Museum Research Laboratory Memoir 1: 1–204.

Walker, E. M. 1898. Notes on some Ontario Acrididae. I. *Canadian Entomologist* 30: 122–126. II, 30: 258–263.

——— 1902. The Canadian species of *Trimerotropis. Canadian Entomologist* 34: 1–11.

——— 1906. Records in Orthoptera from the Canadian Northwest. *Canadian Entomologist* 38: 55–59.

——— 1909. On the Orthoptera of northern Ontario. *Canadian Entomologist* 41: 137–144, 173–178, 205–212.

——— 1910. The Orthoptera of Western Canada. *Canadian Entomologist* 42: 293–300.

——— 1915. Notes on a collection of Orthoptera from Prince Edward Island and the Magdalen Islands, Quebec. *Canadian Entomologist* 47: 339–347.

——— 1902. The Canadian species of *Trimerotropis. Canadian Entomologist* 41: 173–212.

Walker, F. 1870–1871. *Catalogue of the specimens of Dermaptera and Saltatoria in the collection of the British Museum.* Pt. 3: 425–604; pt. 4: 606–801; pt. 5: 810–850; suppl., 1–116.

Weissman, D. B., and E. French. 1980. Autecology and population structure of *Trimerotropis occidentalis* (Bruner) (Orthoptera: Acrididae: Oedipodinae), a grasshopper with a reproductive dormancy. *Acrida* 9: 145–157.

Weissman, D. B., and D. C. F. Rentz. 1978. The Orthoptera of Stanford University's Jasper Ridge and neighboring Palo Alto, California. *Wasmann Journal of Biology* 35: 87–114.

——— 1980. Cytological, morphological and crepitational characteristics of the Trimerotropine (*Aerochoreutes, Circotettix,* and *Trimerotropis*) grasshoppers (Orthoptera: Oedipodinae). *Transactions of the American Entomological Society* 106: 253–272.

White, M. J. D. 1949. A cytological survey of wild populations of *Trimerotropis* and *Circotettix.* (Orthoptera, Acrididae). I. The chromosomes of twelve species. *Genetics* 34: 537–563.

——— A cytological survey of wild populations of *Trimerotropis* and *Circotettix* (Orthoptera, Acrididae). II. Racial differentiation in *T. sparsa. Genetics* 36: 31–53.

——— 1954. An extreme form of chiasma localization in a species of *Bryodema* (Orthoptera, Acrididae). *Evolution* 8: 350–358.

——— 1973. *Animal cytology and evolution.* New York.

Willey, R. B., and R. L. Willey. 1967a. Barriers to gene flow in natural populations of grasshoppers. I. The Black Canyon of the Gunnison River and *Arphia conspersa. Psyche* 74: 42–57.

——— 1967b. Barriers to gene flow in natural populations of grasshoppers. II. Maintenance of narrow hybrid-zones between morphs of *Arphia conspersa* on Black Mesa, Colorado. *Psyche* 78: 330–340.

——— 1969. Visual and acoustical displays by the grasshopper *Arphia conspersa* (Orthoptera: Acrididae). *Psyche* 76: 280–305.

——— 1970. The behavioral ecology of desert grasshoppers. I. Presumed sex-role reversal in flight displays of *Trimerotropis agrestis. Animal Behaviour* 18: 473–477.

Taxonomic Index

Names in bold type are valid genera and species; names in italics are junior synonyms or subspecies. Following each species and genus is the author. In parentheses after the species name is the sequence of genera to which it has been assigned; the last name is the present assignment.

GENERA

SPECIES

www.ingramcontent.com/pod-product-compliance
Lightning Source LLC
LaVergne TN
LVHW091518200926
848483LV00005B/446/J